T0281697

Beginning Microsoft Power BI

A Practical Guide to Self-Service Data Analytics

Third Edition

Dan Clark

Apress®

Beginning Microsoft Power BI: A Practical Guide to Self-Service Data Analytics

Dan Clark
Camp Hill, PA, USA

ISBN-13 (pbk): 978-1-4842-5619-0 ISBN-13 (electronic): 978-1-4842-5620-6
https://doi.org/10.1007/978-1-4842-5620-6

Managing Director, Apress Media LLC: Welmoed Spahr
Acquisitions Editor: Joan Murray
Development Editor: Laura Berendson
Coordinating Editor: Jill Balzano

Cover image designed by Freepik (www.freepik.com)

Distributed to the book trade worldwide by Springer Science+Business Media New York, 233 Spring Street, 6th Floor, New York, NY 10013. Phone 1-800-SPRINGER, fax (201) 348-4505, e-mail orders-ny@springer-sbm.com, or visit www.springeronline.com. Apress Media, LLC is a California LLC and the sole member (owner) is Springer Science + Business Media Finance Inc (SSBM Finance Inc). SSBM Finance Inc is a **Delaware** corporation.

For information on translations, please e-mail rights@apress.com, or visit http://www.apress.com/rights-permissions.

Apress titles may be purchased in bulk for academic, corporate, or promotional use. eBook versions and licenses are also available for most titles. For more information, reference our Print and eBook Bulk Sales web page at http://www.apress.com/bulk-sales.

Any source code or other supplementary material referenced by the author in this book is available to readers on GitHub via the book's product page, located at www.apress.com/9781484256190. For more detailed information, please visit http://www.apress.com/source-code.

Printed on acid-free paper

Table of Contents

About the Author

Dan Clark is a senior business intelligence (BI) and programming consultant specializing in Microsoft technologies. He is focused on learning new BI and data technologies and training others on how to best implement the technology. Dan has published several books and numerous articles on .NET programming and BI development. He is a regular speaker at various developer and database conferences and user group meetings and enjoys interacting with the Microsoft communities. In a previous life, Dan was a physics teacher. He is still inspired by the wonder and awe of studying the universe and figuring out why things behave the way they do. Dan can be reached at Clark.drc@gmail.com.

About the Technical Reviewer

Al MacKinnon is a principal customer success manager with CDW, focusing on Microsoft Cloud solutions with customers in the Mid-Atlantic region. He has a long background in IT, including 10 years with Microsoft. That experience includes technical training, consulting, project management, solutions architecture, and course development. He's held numerous certifications from Novell, Microsoft, and CompTIA and has been a CISSP since 2003. He gets the most satisfaction from helping organizations map technology to business goals and challenges. Al is supported by his wife of over 30 years, and, when not working, he often discusses our place in the universe with the author. Finally, he's gratified that his BA in English is finally starting to pay off.

Acknowledgments

Once again, thanks to the team at Apress for making the writing of this book an enjoyable experience. A special thanks goes to my good friend and technical reviewer, Al—thank you for your attention to detail and excellent suggestions while reviewing this book.

Introduction

Self-service business intelligence (BI)—you have heard the hype, seen the sales demos, and are ready to give it a try. Now what? You have probably checked out a few web sites for examples, given them a try, and learned a thing or two, but you are probably still left wondering how all the pieces fit together and how you go about creating a complete solution. If so, this book is for you. It takes you step by step through the process of analyzing data using the tools that are at the core of Microsoft's self-service BI offering: Power Query and Power BI Desktop.

Quite often, you need to take your raw data and transform it in some way before you load it into the data model. You may need to filter, aggregate, or clean the raw data. I show you how Power Query allows you to easily transform and refine data before incorporating it into your data model. Next, I show you how to create robust, scalable data models using Power BI Desktop. Because creating robust Power BI models is essential to creating self-service BI solutions, I cover it extensively in this book. Next up, I show you how to use Power BI Desktop to easily build interactive visualizations that allow you to explore your data to discover trends and gain insight. Finally, I show you how to deploy your solution to the Power BI Service for your colleagues to use.

I strongly believe that one of the most important aspects of learning is doing. You can't learn how to ride a bike without jumping on a bike, and you can't learn to use the BI tools without actually interacting with them. Any successful training program includes both theory and hands-on activities. For this reason, I have included a hands-on activity at the end of every chapter designed to solidify the concepts covered in the chapter. I encourage you to work through these activities diligently. It is well worth the effort.

As you start working with Power BI, you will soon realize it is constantly evolving (for the better). Every month there is an update to Power BI Desktop that introduces new features and changes to the interface. For this reason, some of the screenshots may not look exactly like the current version of Power BI. They will be, however, close enough to figure out what has changed.

Good luck on your journey, and don't hesitate to provide feedback and suggestions for improving the experience!

CHAPTER 1

Introducing Power BI

The core of Microsoft's self-service business intelligence (BI) toolset is the Power BI data engine (also known as Power Pivot). It is integrated into both Power BI Desktop and Excel (2010 and later) and forms the foundation on top of which you will build your analytical reports and dashboards. This chapter provides you with some background information on why Power BI is such an important tool and what makes it perform so well. The chapter also provides you with an overview of the Power BI Desktop interface and gives you some experience using the different areas of the interface.

After reading this chapter, you will be familiar with the following:

- Why you should use Power Pivot
- The xVelocity in-memory analytics engine
- Exploring the Power BI Desktop interface
- Analyzing data in a Power BI report

Why Use Power BI?

You may have been involved in a traditional BI project consisting of a centralized data warehouse where the various data stores of the organization are loaded, scrubbed, and then moved to an online analytical processing (OLAP) database for reporting and analysis. Some goals of this approach are to create a data repository for historical data, create one version of the truth, reduce silos of data, clean the company data and make sure it conforms to standards, and provide insight into data trends through dashboards. Although these are admirable goals and are great reasons to provide a centralized data warehouse, there are some downsides to this approach. The most notable is the complexity of building the system and implementing change. Ask anyone who has tried to get new fields or measures added to an enterprise-wide data warehouse. Typically, this is a long, drawn-out process requiring IT involvement along with data steward committee reviews, development, and testing cycles.

D. Clark, *Beginning Microsoft Power BI*, https://doi.org/10.1007/978-1-4842-5620-6_1

What is needed is a solution that allows for agile data analysis without so much reliance on IT and formalized processes. To solve these problems, many business analysts have used Excel to create pivot tables and perform ad hoc analysis on sets of data gleaned from various data sources. Some problems with using isolated Excel workbooks for analysis are conflicting versions of the truth, silos of data, and data security.

So how can you solve this dilemma of the centralized data warehouse being too rigid while the Excel solution is too loose? This is where Microsoft's self-service BI toolset comes in. These tools do not replace your centralized data warehouse solution but rather augment it to promote agile data analysis. Using Power BI, you can pull data from the data warehouse, extend it with other sources of data such as text files or web data feeds, build custom measures, and analyze the data using powerful visuals to gain insight into the data. You can create quick proofs of concepts that can be easily promoted to become part of the enterprise-wide solution. Power BI also promotes one-off data analysis projects without the overhead of a drawn-out development cycle. When combined with the Power BI Service portal, reports and dashboards can be shared, secured, and managed. This goes a long way to satisfying IT's need for governance without impeding the business user's need for agility.

Here are some of the benefits of Power BI:

- Power BI Desktop is a free tool for creating reports.
- Easily integrates data from a variety of sources.
- Handles large amounts of data, upward of tens to hundreds of millions of rows.
- Includes a powerful Data Analysis Expressions (DAX) language.
- Has data in the model that is read-only, which increases security and integrity.

When Power BI reports are hosted in the Power BI Service portal, some added benefits are

- Enables sharing and collaboration
- Scheduling and automation of data refresh
- Can audit changes through version management
- Can secure users for read-only and updateable access

Now that you know some of the benefits of Power BI, let's see what makes it tick.

The xVelocity In-Memory Analytics Engine

The special sauce behind Power BI is the xVelocity in-memory analytics engine (yes, that is really the name). xVelocity allows Power BI to provide fast performance on large amounts of data. One of the keys to this is it uses a columnar database to store the data. Traditional row-based data storage stores all the data in the row together and is efficient at retrieving and updating data based on the row key, for example, updating or retrieving an order based on an order ID. This is great for the order-entry system but not so great when you want to perform analysis on historical orders (say you want to look at trends for the past year to determine how products are selling, for example). Row-based storage also takes up more space by repeating values for each row; if you have a large number of customers, common names like John or Smith are repeated many times. A columnar database stores only the distinct values for each column and then stores the row as a set of pointers back to the column values. This built-in indexing saves a lot of space and allows for significant optimization when coupled with data-compression techniques that are built into the xVelocity engine. It also means that data aggregations (like those used in typical data analysis) of the column values are extremely fast.

Another benefit provided by the xVelocity engine is the in-memory analytics. Most processing bottlenecks associated with querying data occur when data is read from or written to a disk. With in-memory analytics, the data is loaded into the RAM memory of the computer and then queried. This results in much faster processing times and limits the need to store preaggregated values on disk. This advantage is especially apparent when you move from 32-bit to 64-bit operating systems and applications, which are the norm these days.

Another benefit worth mentioning is the tabular structure of the Power BI data model. The model consists of tables and table relationships. This tabular model is familiar to most business analysts and database developers. Traditional OLAP databases such as SQL Server Analysis Server (SSAS) present the data model as a three-dimensional cube structure that is difficult to work with and requires a complex query language called Multidimensional Expressions (MDX). I find that in most cases (but not all), it is easier to work with tabular models and DAX than OLAP cubes and MDX.

Setting Up the Power BI Environment

Power BI Desktop is a free tool used to create visual analytic reports that can be hosted in the Power BI portal. You can download it from the Power BI web site at `https://powerbi.microsoft.com/en-us/desktop/`. If you sign up for the Power BI portal or have an Office 365 subscription, you can log into the portal (`https://powerbi.microsoft.com`) and download the tool (see Figure 1-1).

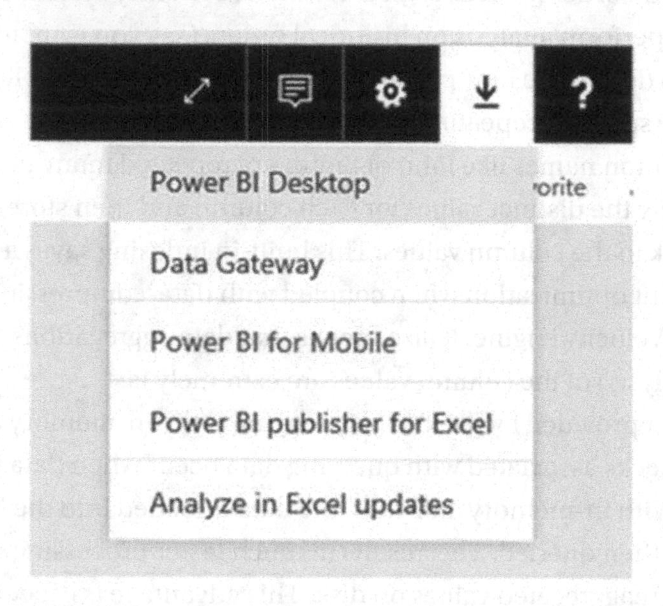

Figure 1-1. *Downloading Power BI Desktop*

Once you download Power BI Desktop, click Run to begin the install. Follow the installation wizard, which is straightforward. After the install, launch Power BI Desktop. Click the File tab ➤ Options and Settings ➤ Options to set up the various options for your development environment (see Figure 1-2).

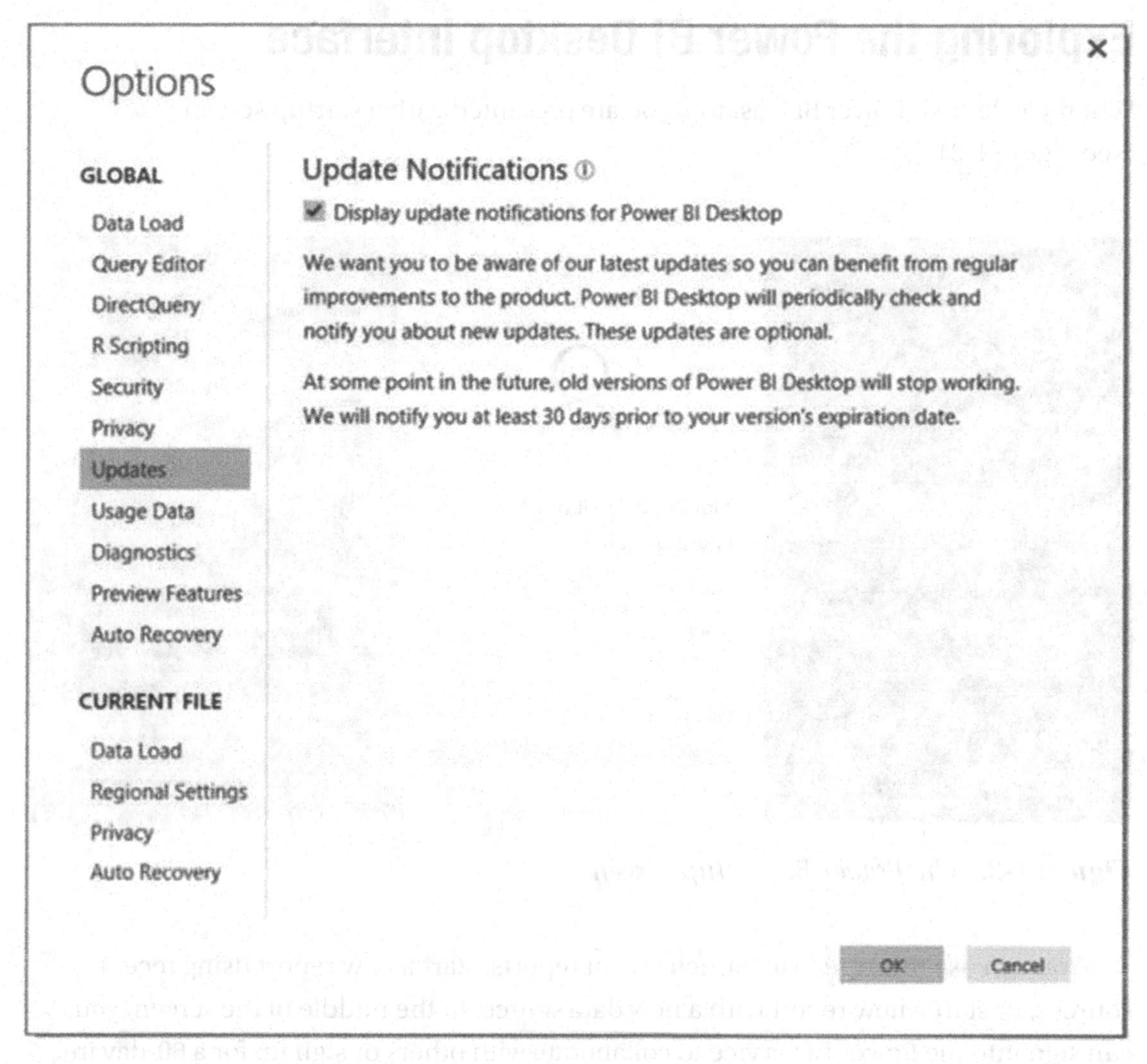

Figure 1-2. *Setting options for Power BI Desktop*

One thing to be aware of is that Microsoft has been releasing monthly updates that include new features. Make sure you get notified and install these updates when they are released. You can check which version you have installed by selecting the Diagnostics tab in the Options window. If you are like me and want to play with upcoming features that are still in development, you can turn them on in the Preview Features tab in the Options window.

Now that you have installed and set up the Power BI Desktop development environment, you are ready to explore the interface.

Exploring the Power BI Desktop Interface

When you launch Power BI Desktop, you are presented with a startup screen (see Figure 1-3).

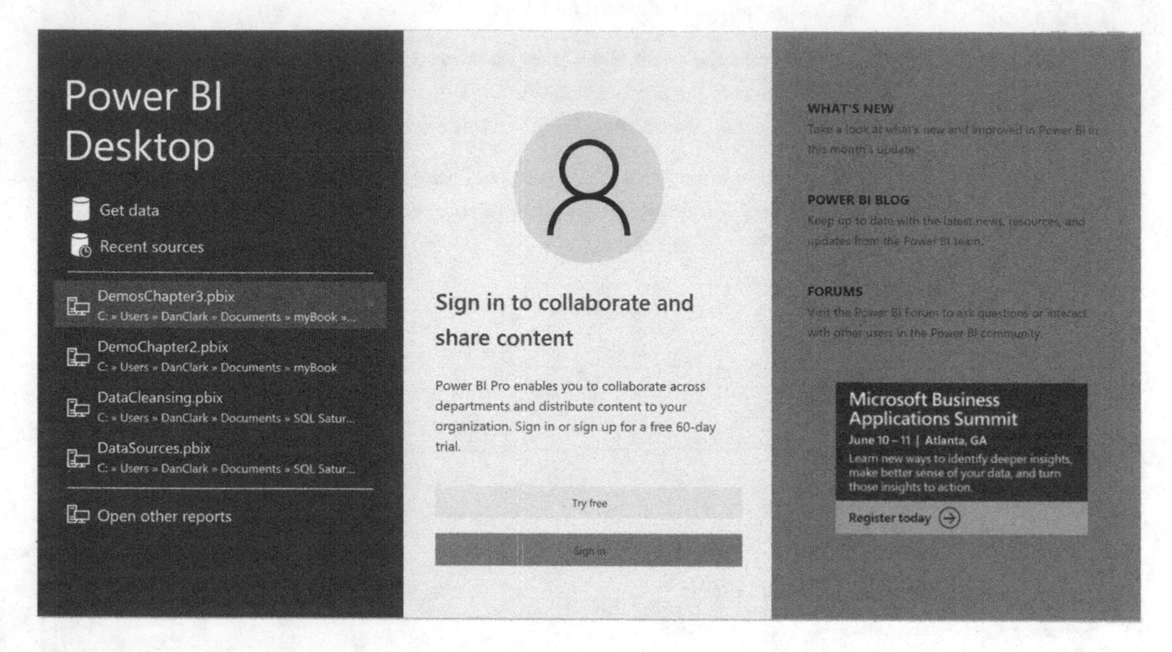

Figure 1-3. *The Power BI startup screen*

You can use this screen to launch recent reports, start a new report using recent sources, or start a new report with a new data source. In the middle of the screen, you can sign into the Power BI Service to collaborate with others or sign up for a 60-day free trial of Power BI Pro. On the left side of the screen, you can link to the Power BI blog and the Power BI forums.

Note There is a limited Power BI Free version; however this book is based on the Power BI Pro version.

Figure 1-4 shows a resent Power BI report opened in the Power BI Desktop.

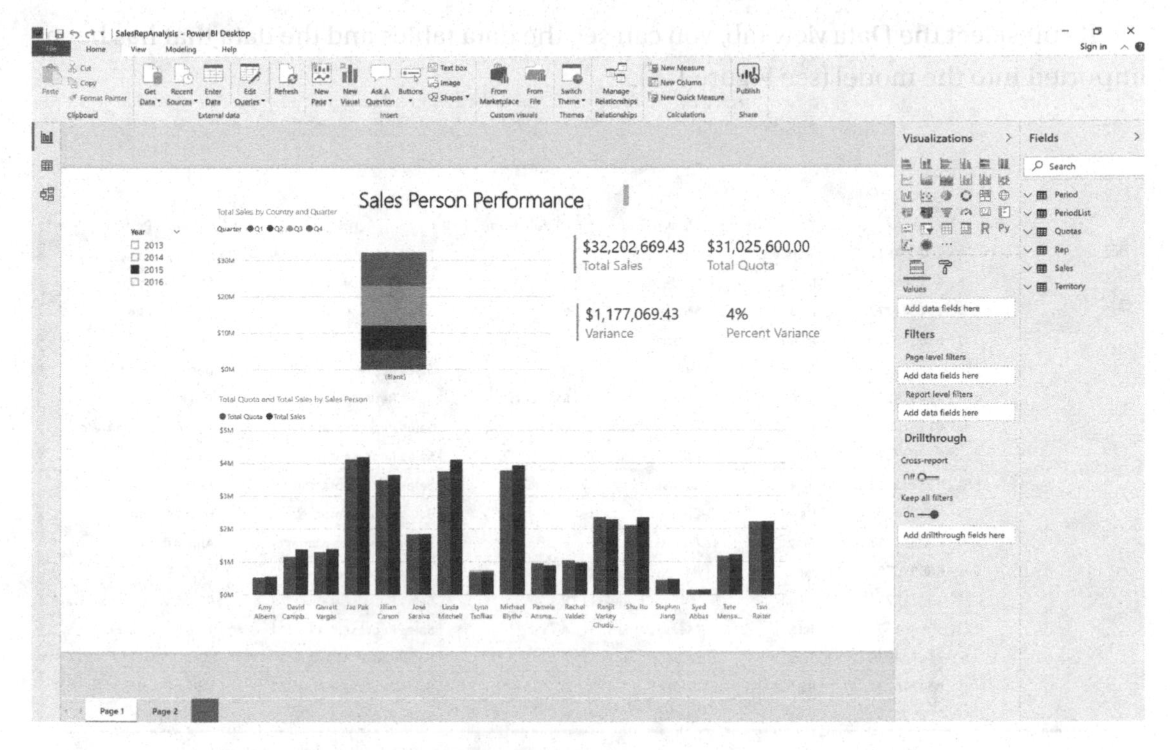

Figure 1-4. Creating a report in Power BI Desktop

When you first open a report in Power BI Desktop, you see the Report view. There are two other views you can select the Data view and the Model view. To switch views, you use the tabs on the left side of the designer (see Figure 1-5).

Figure 1-5. Switching views in the Power BI Desktop

If you select the Data view tab, you can see the data tables and the data that has been imported into the model (see Figure 1-6).

EmployeeKey	Employee ID	First Name	Last Name	Middle Name	Title	Sales Person
272	502097814	Stephen	Jiang	Y	North American Sales Manager	Stephen Jiang
281	841560125	Michael	Blythe	G	Sales Representative	Michael Blythe
282	191644724	Linda	Mitchell	C	Sales Representative	Linda Mitchell
283	615389812	Jillian	Carson		Sales Representative	Jillian Carson
284	234474252	Garrett	Vargas	R	Sales Representative	Garrett Vargas
285	716374314	Tsvi	Reiter	Michael	Sales Representative	Tsvi Reiter
286	61161660	Pamela	Ansman-Wolfe	O	Sales Representative	Pamela Ansman-Wolfe
287	139397894	Shu	Ito	K	Sales Representative	Shu Ito
288	399771412	José	Saraiva	Edvaldo	Sales Representative	José Saraiva
289	987554265	David	Campbell	R	Sales Representative	David Campbell
290	982310417	Amy	Alberts	E	European Sales Manager	Amy Alberts
291	668991357	Jae	Pak	B	Sales Representative	Jae Pak
292	134219713	Ranjit	Varkey Chudukatil	R	Sales Representative	Ranjit Varkey Chudukatil
293	90836195	Tete	Mensa-Annan	A	Sales Representative	Tete Mensa-Annan
294	481044938	Syed	Abbas	E	Pacific Sales Manager	Syed Abbas
295	954276278	Rachel	Valdez	B	Sales Representative	Rachel Valdez
296	758596752	Lynn	Tsoflias	N	Sales Representative	Lynn Tsoflias

***Figure 1-6.** Viewing data in the Data view tab*

The Model view tab shows the relationships and the filter direction between the tables in the model (see Figure 1-7).

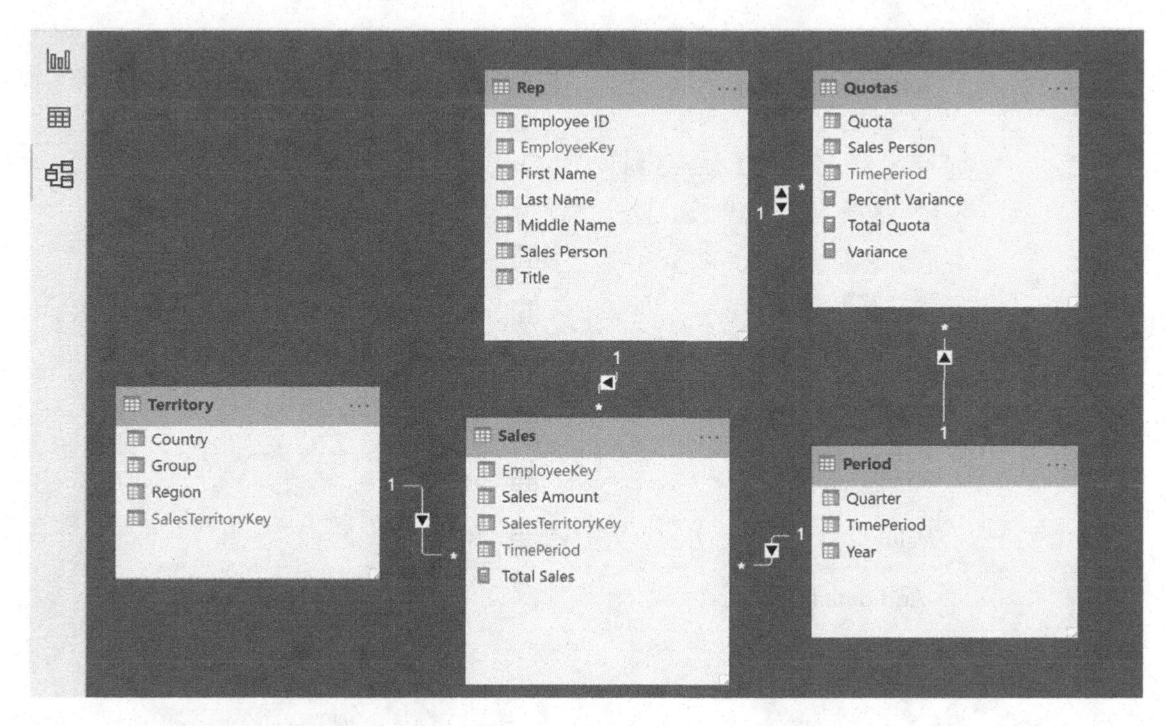

Figure 1-7. *The Model view tab*

The menus at the top of the designer will change depending on what view you have selected. Figure 1-8 shows the menus available when you are in the Report view tab. You will become intimately familiar with the menus in the designer as you progress through this book. For now, suffice it to say that this is where you initiate various actions such as connecting to data sources, creating data queries, formatting data, setting default properties, and editing visual interactions.

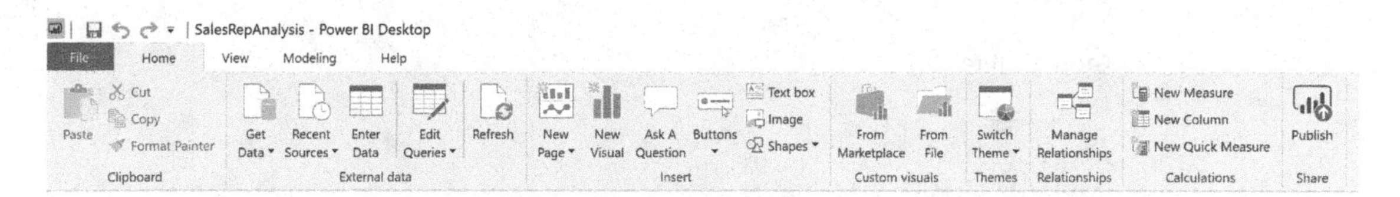

Figure 1-8. *The Home menu in the Report view*

On the right side of the report designer are the Visualizations and the Fields windows (see Figure 1-9). This is where you select the visualizations you want on the report, add fields to the visualizations, and set the properties of the visualizations.

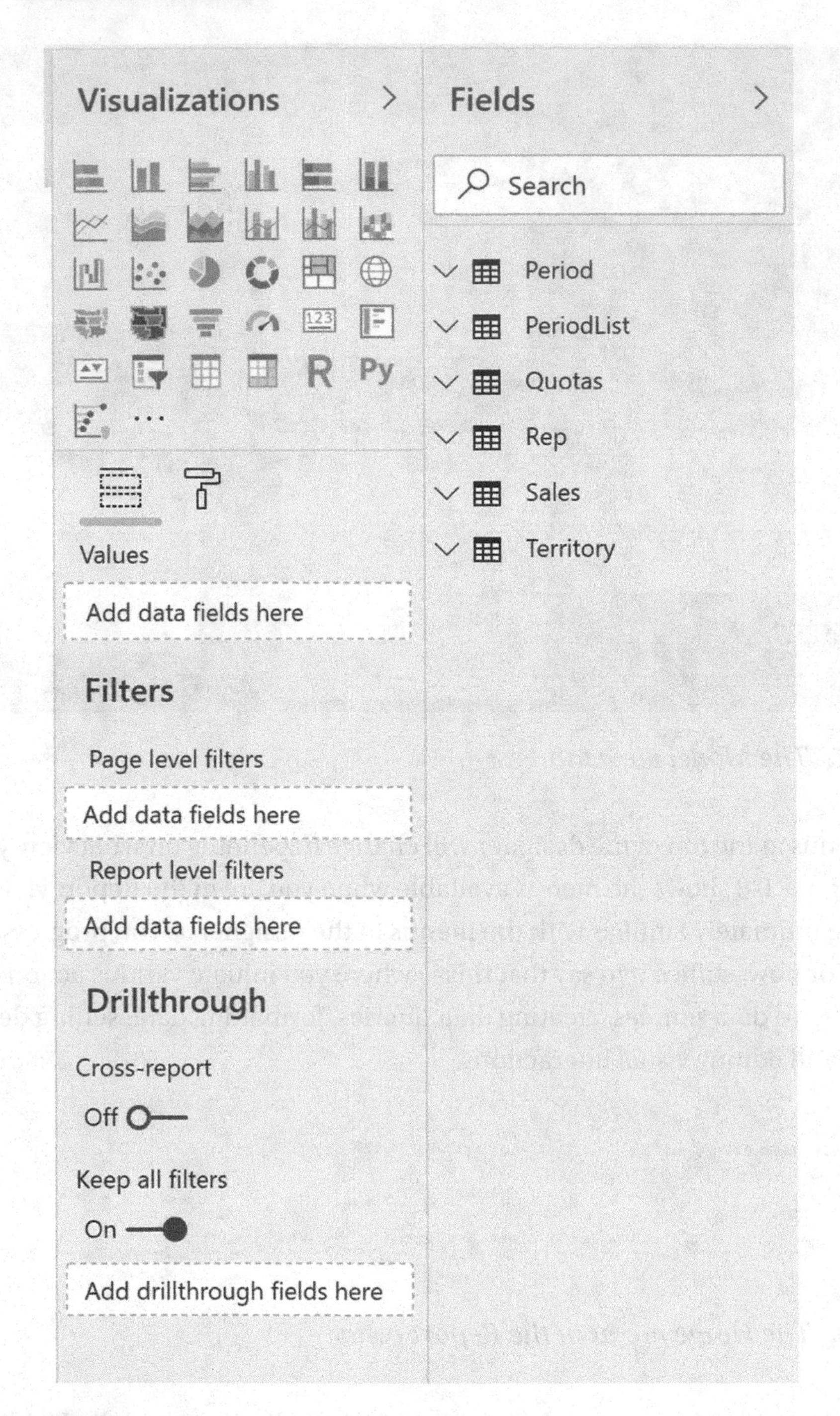

***Figure 1-9.** The Visualizations and Fields windows*

Now that you are familiar with the various parts of the Power BI Desktop report designer, it's time to get your hands dirty and complete the following hands-on lab. This lab will help you become familiar with working in Power BI Desktop.

Note To complete the labs in this book, make sure you download the starter files from `https://github.com/Apress/beginning-power-bi-3ed`.

HANDS-ON LAB: EXPLORING POWER PIVOT

In this following lab, you will

- Install Power BI Desktop
- View the various tabs of Power BI Desktop
- Explore the data using a matrix

1. Go to `https://powerbi.microsoft.com/en-us/desktop/` and download and install Power BI Desktop.
2. Launch Power BI Desktop and dismiss the startup screen.
3. On the File menu, select Options and Settings and then Options. You should see the Options window (see Figure 1-10).

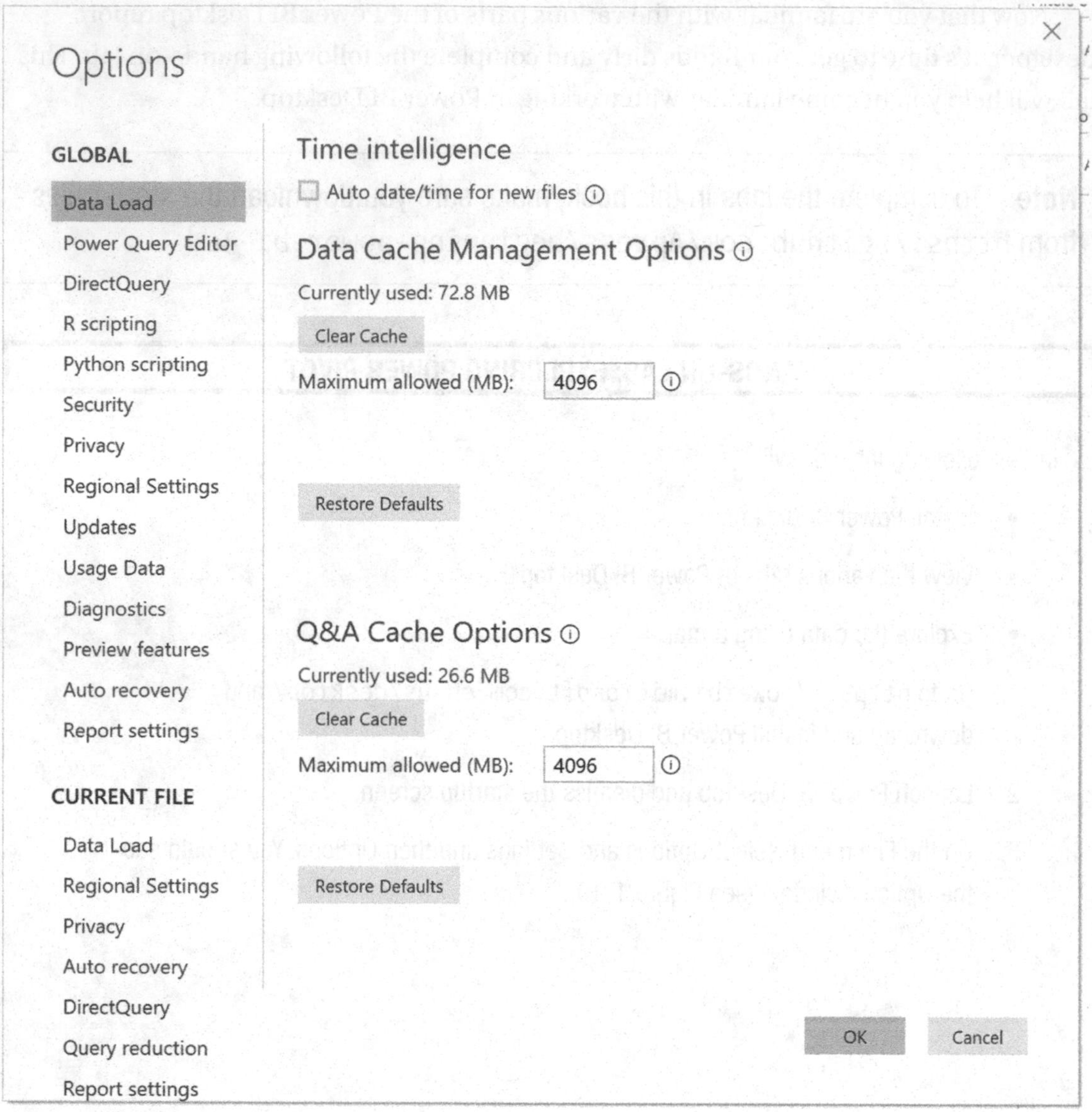

Options

GLOBAL
Data Load
Power Query Editor
DirectQuery
R scripting
Python scripting
Security
Privacy
Regional Settings
Updates
Usage Data
Diagnostics
Preview features
Auto recovery
Report settings

CURRENT FILE
Data Load
Regional Settings
Privacy
Auto recovery
DirectQuery
Query reduction
Report settings

Time intelligence
Auto date/time for new files

Data Cache Management Options
Currently used: 72.8 MB
Clear Cache
Maximum allowed (MB): 4096
Restore Defaults

Q&A Cache Options
Currently used: 26.6 MB
Clear Cache
Maximum allowed (MB): 4096
Restore Defaults

OK Cancel

Figure 1-10. *Viewing the Options settings*

4. On the Global Data Load tab, make sure the Auto date/time for new files option is unchecked.

5. View some of the other setting options available.

6. Open the Chapter1Lab1.pbix file located in the LabStarterFiles folder.

7. You should see a basic matrix showing sales by year and country, as shown in Figure 1-11.

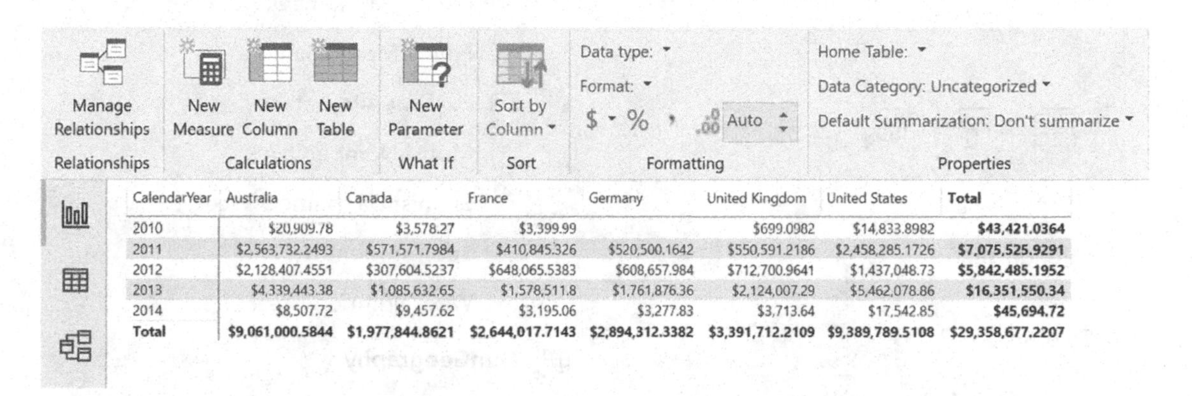

Manage Relationships
New Measure
New Column
New Table
New Parameter
Sort by Column
Data type:
Format:
$ - % , .00 Auto
Home Table:
Data Category: Uncategorized
Default Summarization: Don't summarize
Relationships
Calculations
What If
Sort
Formatting
Properties
CalendarYear Australia Canada France Germany United Kingdom United States Total
2010 $20,909.78 $3,578.27 $3,399.99 $699.0982 $14,833.8982 $43,421.0364
2011 $2,563,732.2493 $571,571.7984 $410,845.326 $520,500.1642 $550,591.2186 $2,458,285.1726 $7,075,525.9291
2012 $2,128,407.4551 $307,604.5237 $648,065.5383 $608,657.984 $712,700.9641 $1,437,048.73 $5,842,485.1952
2013 $4,339,443.38 $1,085,632.65 $1,578,511.8 $1,761,876.36 $2,124,007.29 $5,462,078.86 $16,351,550.34
2014 $8,507.72 $9,457.62 $3,195.06 $3,277.83 $3,713.64 $17,542.85 $45,694.72
Total $9,061,000.5844 $1,977,844.8621 $2,644,017.7143 $2,894,312.3382 $3,391,712.2109 $9,389,789.5108 $29,358,677.2207

***Figure 1-11.** Using a matrix table*

8. Click anywhere in the matrix. You should see the visual properties and the field list on the right side, as shown in Figure 1-12.

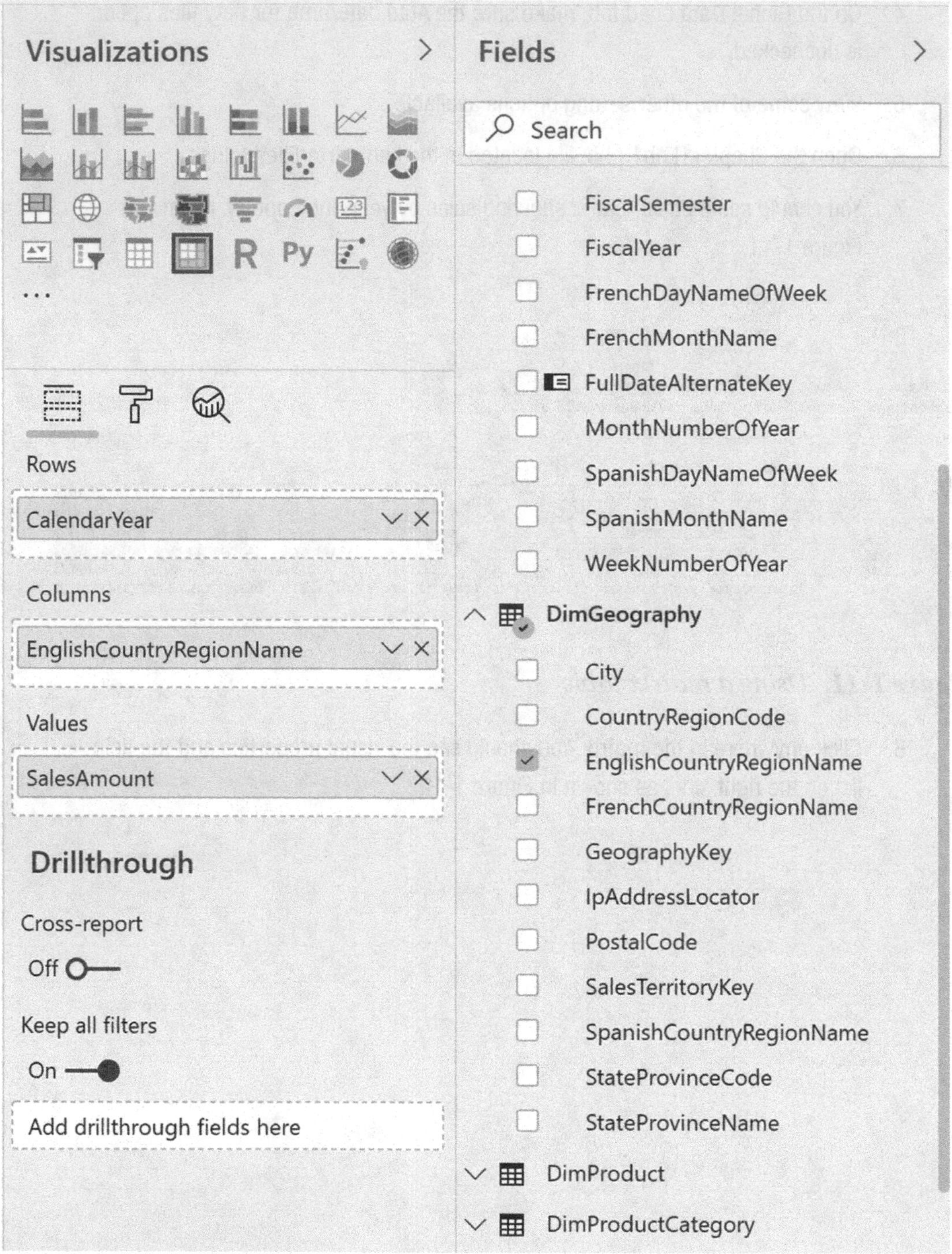

Visualizations

Fields

Search

FiscalSemester
FiscalYear
FrenchDayNameOfWeek
FrenchMonthName
FullDateAlternateKey
MonthNumberOfYear
SpanishDayNameOfWeek
SpanishMonthName
WeekNumberOfYear
DimGeography
City
CountryRegionCode
EnglishCountryRegionName
FrenchCountryRegionName
GeographyKey
IpAddressLocator
PostalCode
SalesTerritoryKey
SpanishCountryRegionName
StateProvinceCode
StateProvinceName
DimProduct
DimProductCategory

R Py

...

Rows
CalendarYear
Columns
EnglishCountryRegionName
Values
SalesAmount

Drillthrough

Cross-report
Off
Keep all filters
On
Add drillthrough fields here

Figure 1-12. *The visual properties and field list*

9. In the visual properties, there are drop areas for the rows, columns, and values referred to as wells. You drag and drop the fields into these wells to create the matrix.

10. In the Fields list, expand the DimProductCategory table. Find the EnglishProductCategoryName field and drag it to the rows well under the CalendarYear field.

11. In the Visualizations window, select the paint roller icon and expand the Row Headers tab. Turn on the +/- icons for expanding the matrix rows (see Figure 1-13).

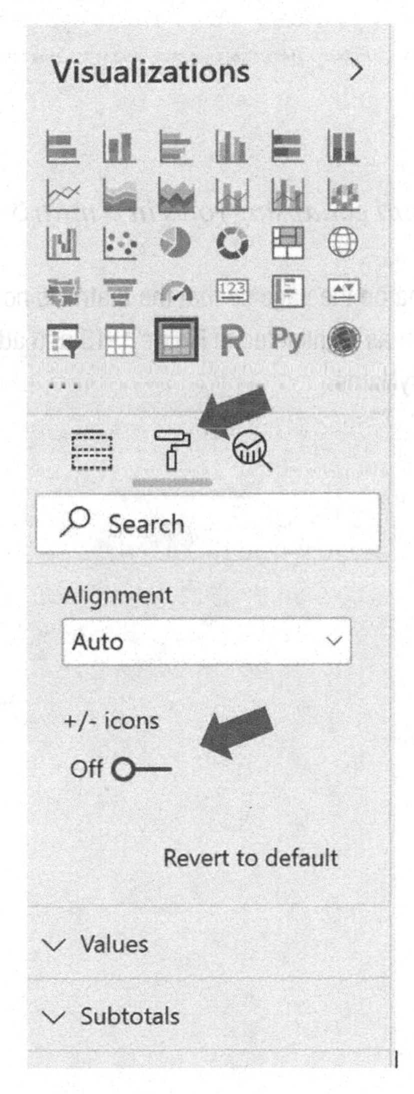

Figure 1-13. *Turning on the +/- icons*

12. Notice that you can now expand and collapse the matrix (see Figure 1-14). Investigate some of the other format settings available for the matrix.

CalendarYear	Australia	Canada	France	Germany	United Kingdom	United States	**Total**
⊞ **2010**	**$20,909.78**	**$3,578.27**	**$3,399.99**		**$699.0982**	**$14,833.8982**	**$43,421.0364**
⊞ **2011**	**$2,563,732.2493**	**$571,571.7984**	**$410,845.326**	**$520,500.1642**	**$550,591.2186**	**$2,458,285.1726**	**$7,075,525.9291**
⊟ **2012**	**$2,128,407.4551**	**$307,604.5237**	**$648,065.5383**	**$608,657.984**	**$712,700.9641**	**$1,437,048.73**	**$5,842,485.1952**
Accessories	$573.99	$56.97	$442.12	$360.17	$278.47	$435.36	**$2,147.08**
Bikes	$2,127,687.0151	$307,497.5637	$647,605.4383	$608,121.864	$712,341.5341	$1,436,441.91	**$5,839,695.3252**
Clothing	$146.45	$49.99	$17.98	$175.95	$80.96	$171.46	**$642.79**
⊟ **2013**	**$4,339,443.38**	**$1,085,632.65**	**$1,578,511.8**	**$1,761,876.36**	**$2,124,007.29**	**$5,462,078.86**	**$16,351,550.34**
Accessories	$132,763.21	$96,922.04	$60,599.81	$59,388.39	$73,967.62	$244,600.46	**$668,241.53**
Bikes	$4,139,720.96	$938,654.76	$1,491,724.96	$1,679,892.32	$2,019,210.81	$5,090,298.55	**$15,359,502.36**
Clothing	$66,959.21	$50,055.85	$26,187.03	$22,595.65	$30,828.86	$127,179.85	**$323,806.45**
⊟ **2014**	**$8,507.72**	**$9,457.62**	**$3,195.06**	**$3,277.83**	**$3,713.64**	**$17,542.85**	**$45,694.72**
Accessories	$5,353.43	$6,398.84	$2,364.85	$2,484.03	$2,383.95	$11,386.25	**$30,371.35**
Clothing	$3,154.29	$3,058.78	$830.21	$793.8	$1,329.69	$6,156.6	**$15,323.37**
Total	**$9,061,000.5844**	**$1,977,844.8621**	**$2,644,017.7143**	**$2,894,312.3382**	**$3,391,712.2109**	**$9,389,789.5108**	**$29,358,677.2207**

Figure 1-14. *Expanding and collapsing rows in a matrix*

13. Click on an empty area on the page so that the matrix is not selected. Select the Slicer visual as highlighted in Figure 1-15 and add the EnglishProdutCategoryName.

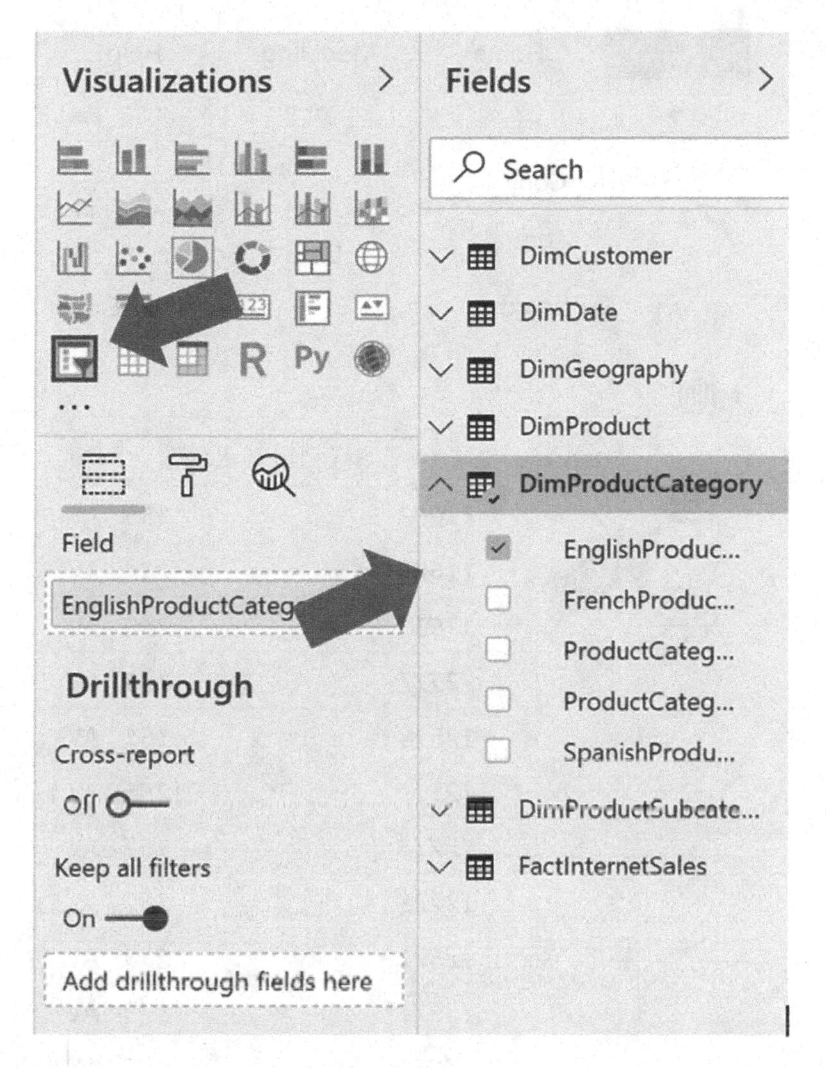

Figure 1-15. *Adding a slicer to the report*

14. Rearrange the slicer and the matrix on the page and notice selecting categories in the slicer filters the matrix.

15. On the left side of the designer, switch to the Data view (see Figure 1-16).

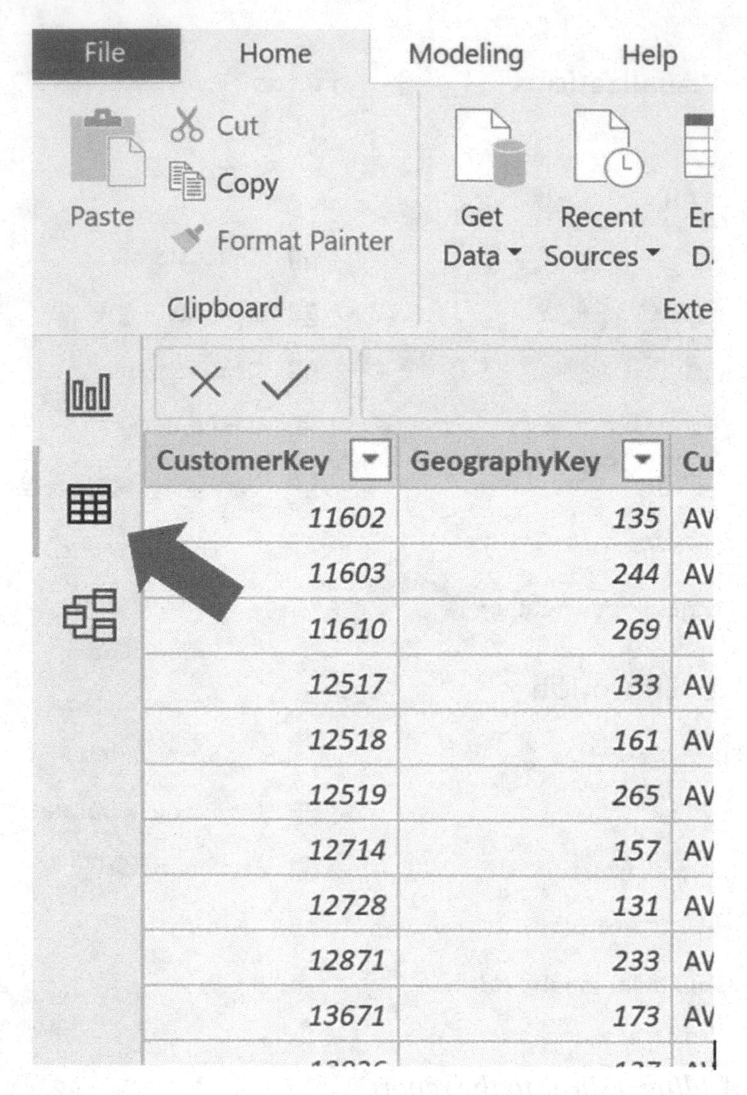

Figure 1-16. *Switching to the Data view*

16. Explore the data in the different tables using the Fields list on the right side of the designer.
17. Go to the ProductAlternateKey column in the DimProduct table. Notice that it's grayed out. This means it's hidden in the Report view. You can verify this by switching back to the Report view and verifying that you cannot see the field in the field list.

18. In the FactInternetSales table, click the Margin column. Notice this is a calculated column using the SalesAmount and the ProductStandardCost. It has also been formatted as currency.

19. There is a measure called Total Sales Amount in the FactInternetSales table. Click the measure and note that in the formula bar above the table is the DAX code used to calculate the measure.

20. On the right side of the designer, switch to the Model view. Observe the relationships between the tables which are indicated by the lines connecting the tables.

21. If you hover over the relationship with the mouse pointer, you can see the fields involved in the relationship, as shown in Figure 1-17.

Figure 1-17. *Exploring relationships*

22. Take some time to explore the Model, Data, and Report views. (Feel free to try to break things!) When you're done, save the file and close Power BI Desktop.

Summary

This chapter introduced you to Power BI Desktop. You got a little background into why Power BI can handle large amounts of data using the xVelocity engine and columnar data storage. You also got to investigate and gain some experience using the Power BI Desktop designer. Don't worry about the details of how you develop the various parts of the model and reports just yet. That will be explained in detail as you progress through the book. In the next chapter, you will learn how to get data into the model from various kinds of data sources.

CHAPTER 2

Importing Data into Power BI Desktop

One of the first steps in creating the Power BI analytics model is importing data. Traditionally, when creating a BI solution based on an OLAP cube, you need to import the data into the data warehouse and then load it into the cube. It can take quite a while to get the data incorporated into the cube and available for your consumption. This is one of the greatest strengths of the Power BI model. You can easily and quickly combine data from a variety of sources into your model. The data sources can be from relational databases, text files, web services, and OLAP cubes, just to name a few. This chapter shows you how to incorporate data from a variety of these sources into a Power BI model.

After completing this chapter, you will be able to

- Import data from relational databases
- Import data from text files
- Import data from a data feed
- Import data from an OLAP database

Importing Data from Relational Databases

One of the most common types of data sources you will run into is a relational database. Relational database management systems (RDMS) such as SQL Server, Oracle, DB2, and Access consist of tables and relationships between the tables based on keys. For example, Figure 2-1 shows a purchase order detail table and a product table. They are related by the ProductID column. This is an example of a one-to-many relationship. For each row in the product table, there are many rows in the purchase order detail table. The keys in a table are referred to as primary and foreign keys. Every table needs

D. Clark, *Beginning Microsoft Power BI*, https://doi.org/10.1007/978-1-4842-5620-6_2

a primary key that uniquely identifies a row in the table. For example, the ProductID is the primary key in the product table. The ProductID is considered a foreign key in the purchase order detail table. Foreign keys point back to a primary key in a related table. Note that a primary key can consist of a combination of columns; for example, the primary key of the purchase order detail table is the combination of the PurchaseOrderID and the PurchaseOrderDetailID.

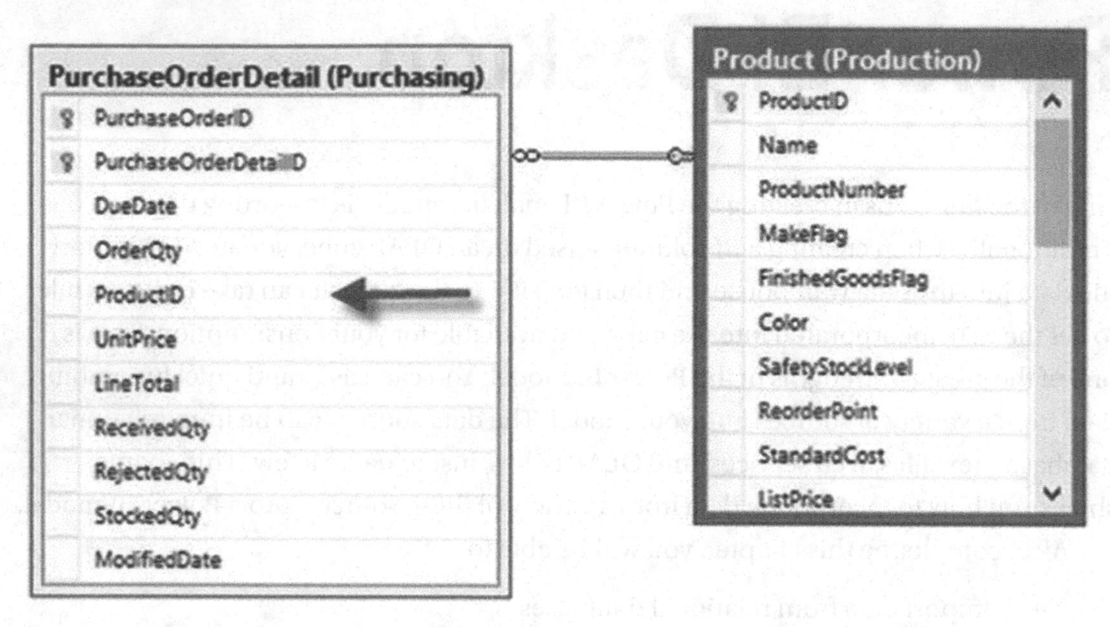

Figure 2-1. *A one-to-many relationship*

Although one-to-many relationships are the most common, you will run into another type of relationship that is fairly prevalent: the many-to-many. Figure 2-2 shows an example of a many-to-many relationship. A person may have multiple phone numbers of different types. For example, they may have a business phone and a mobile phone. You can't relate these tables directly. Instead, you need to use a *junction* table that contains the primary keys from the tables. The combination of the keys in the junction table must be unique.

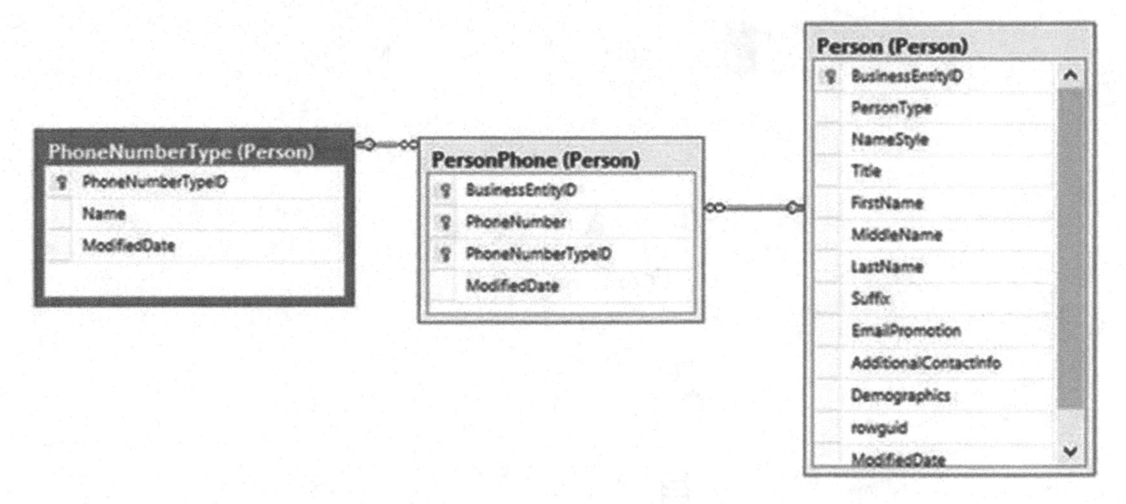

Figure 2-2. *Creating a many-to-many relationship using a junction table*

Notice that the junction table can contain information related to the association; for example, the PhoneNumber is associated with the customer and phone number type. A customer cannot have the same phone number listed as two different types.

One nice aspect of obtaining data from a relational database is that the model is very similar to a model you create in Power BI. In fact, if the relationships are defined in the database, the Power BI import wizard can detect these and set them up in the model for you.

The first step to getting data from a relational database is to create a connection. On the Home tab of Power BI Desktop, there is an External data grouping (see Figure 2-3).

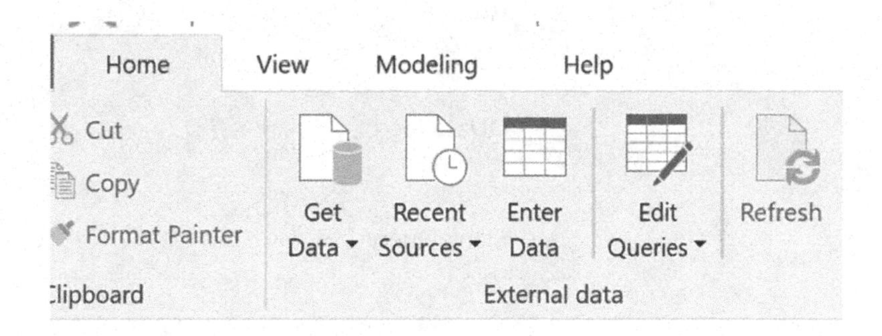

Figure 2-3. *Setting up a connection*

Selecting the Get Data drop-down allows you to connect to some of the more common data sources such as Excel, SQL Server, Analysis Services, or from another Power BI model.

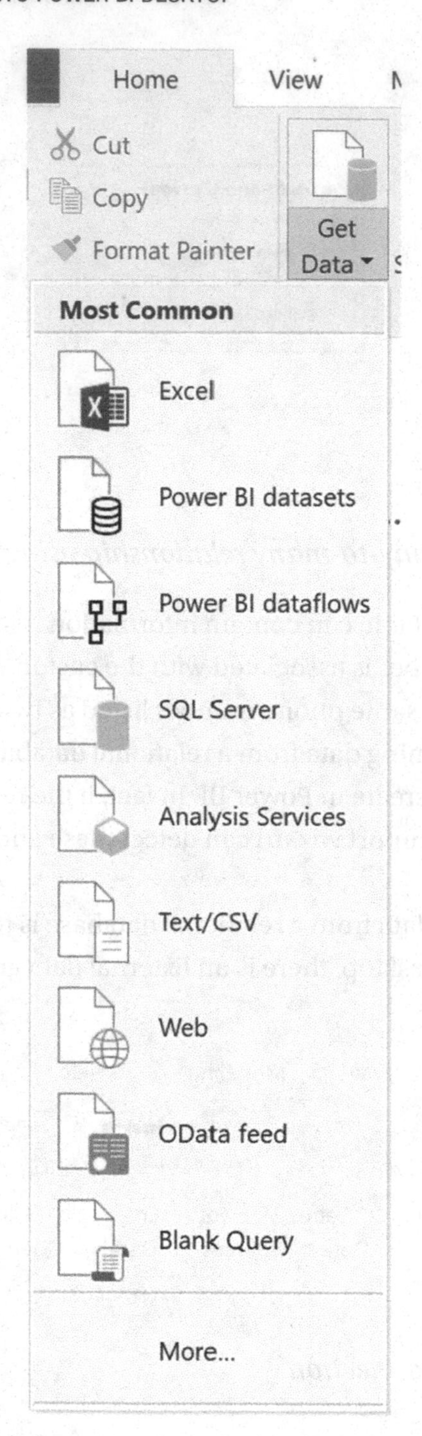

***Figure 2-4.** Selecting a data source*

If you click More in the drop-down, you can see the vast amount of data sources available to connect to (see Figure 2-5). If you don't see the one you need, you can ask the vendor if they have one or if you can use a generic driver such as the ODBC or OLEDB driver to connect to it.

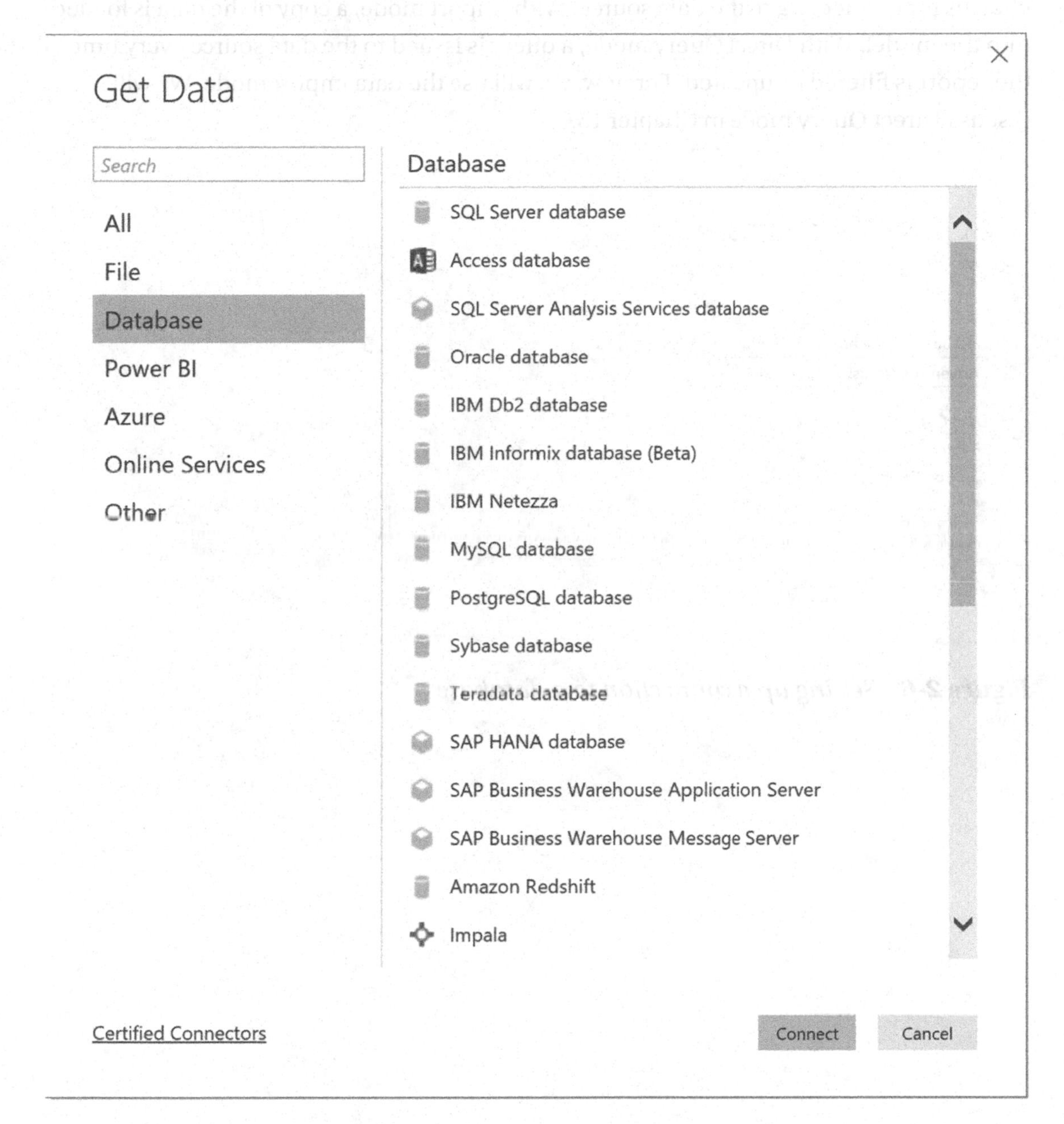

***Figure 2-5.** Selecting a data source*

After selecting a data source, you are presented with a window to enter the connection information. The connection information depends on the data source you are connecting to. For most relational databases, the information needed is very similar. Figure 2-6 shows the connection window for connecting to a SQL Server. You have two choices for connecting to the data source. With Import mode, a copy of the data is loaded into the model. With Direct Query mode, a query is issued to the data source every time the report is filtered or updated. For now, we will use the data import mode. We will discuss Direct Query mode in Chapter 15.

Figure 2-6. *Setting up a connection to a database*

After connecting to the database, you are presented with a list of tables and views (see Figure 2-7).

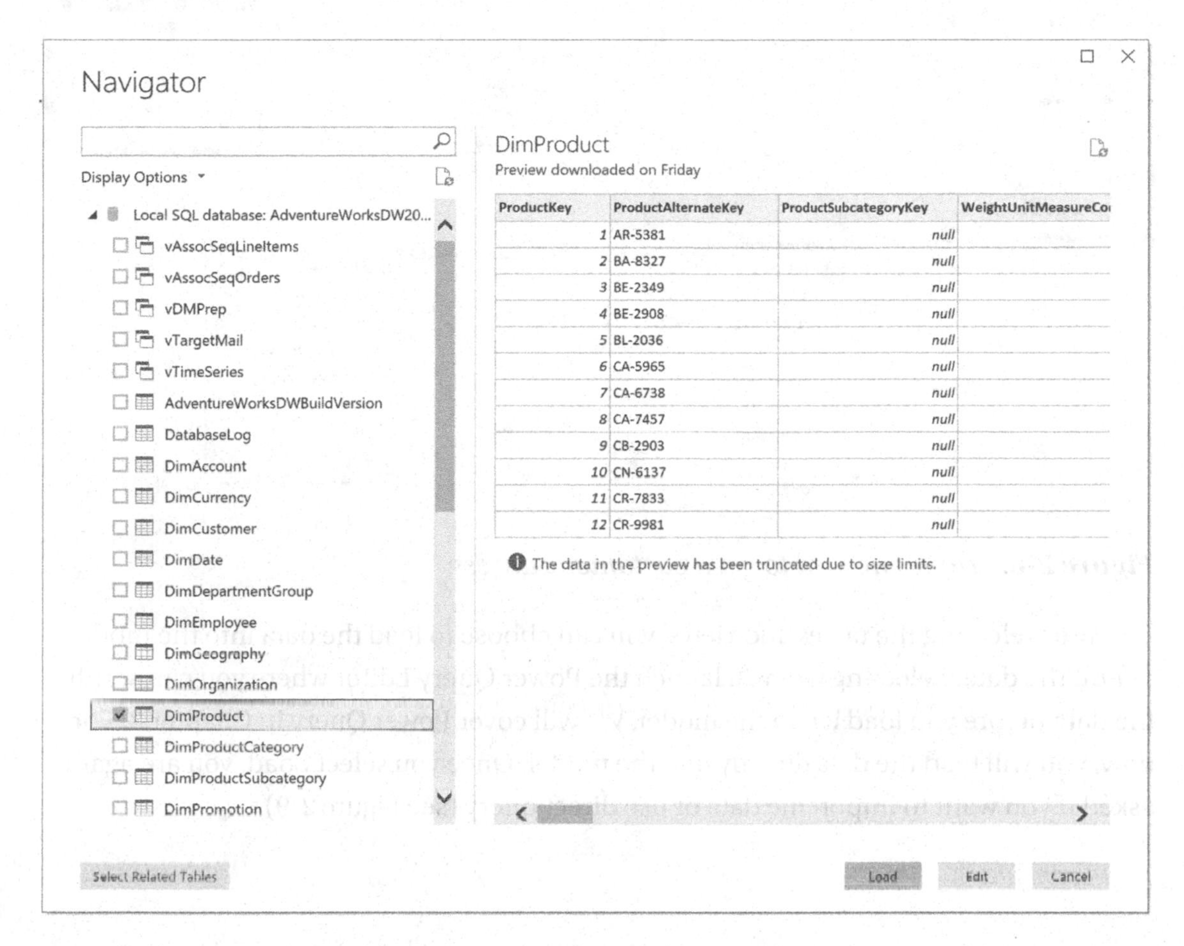

Figure 2-7. *Selecting tables and/or views*

From your perspective, a view and a table look the same. A *view* is really a stored query in the database that masks the complexity of the query from you. Views are often used to show a simpler conceptional model of the database than the actual physical model. For example, you may need a customer's address. Figure 2-8 shows the tables you need to include in a query to get the information. Instead of writing a complex query to retrieve the information, you can select from a view that combines the information in a virtual Customer Address table for you. Another common use of a view is to secure columns of the underlying table. Using a view, the database administrator can hide columns from various users.

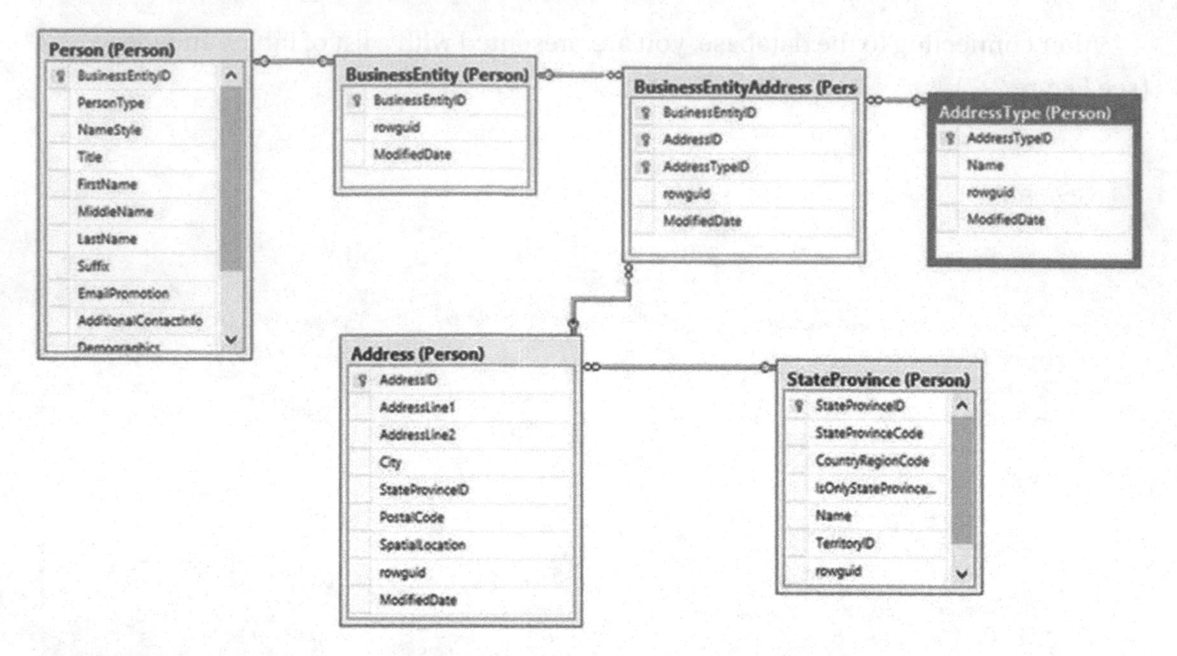

***Figure 2-8.** Tables needed to get a customer address*

After selecting the tables and views, you can choose to load the data into the model or edit the data. Selecting Edit will launch the Power Query Editor where you can scrub the data before you load it into the model. We will cover Power Query in Chapter 3. For now, you will load the data directly into the model. Once you select Load, you are again asked if you want to import the data or use direct query (see Figure 2-9).

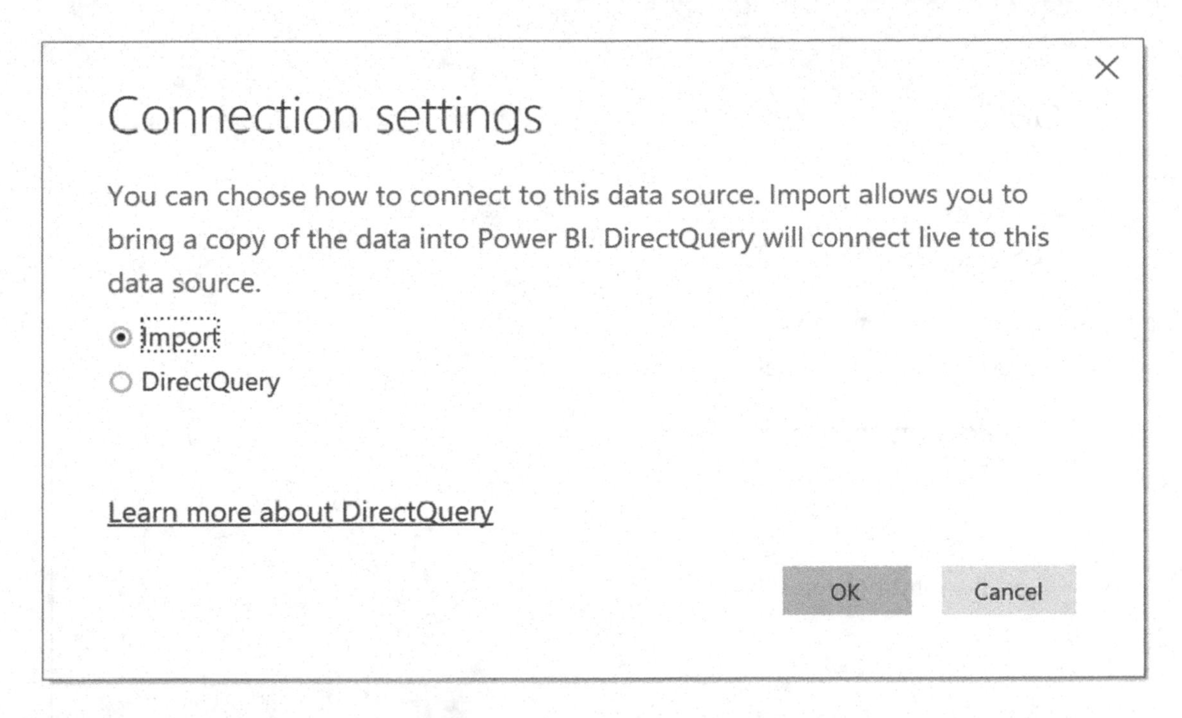

Figure 2-9. *Choosing how to connect*

After selecting Import mode, you will see a progress screen as the data is loaded into the model (see Figure 2-10).

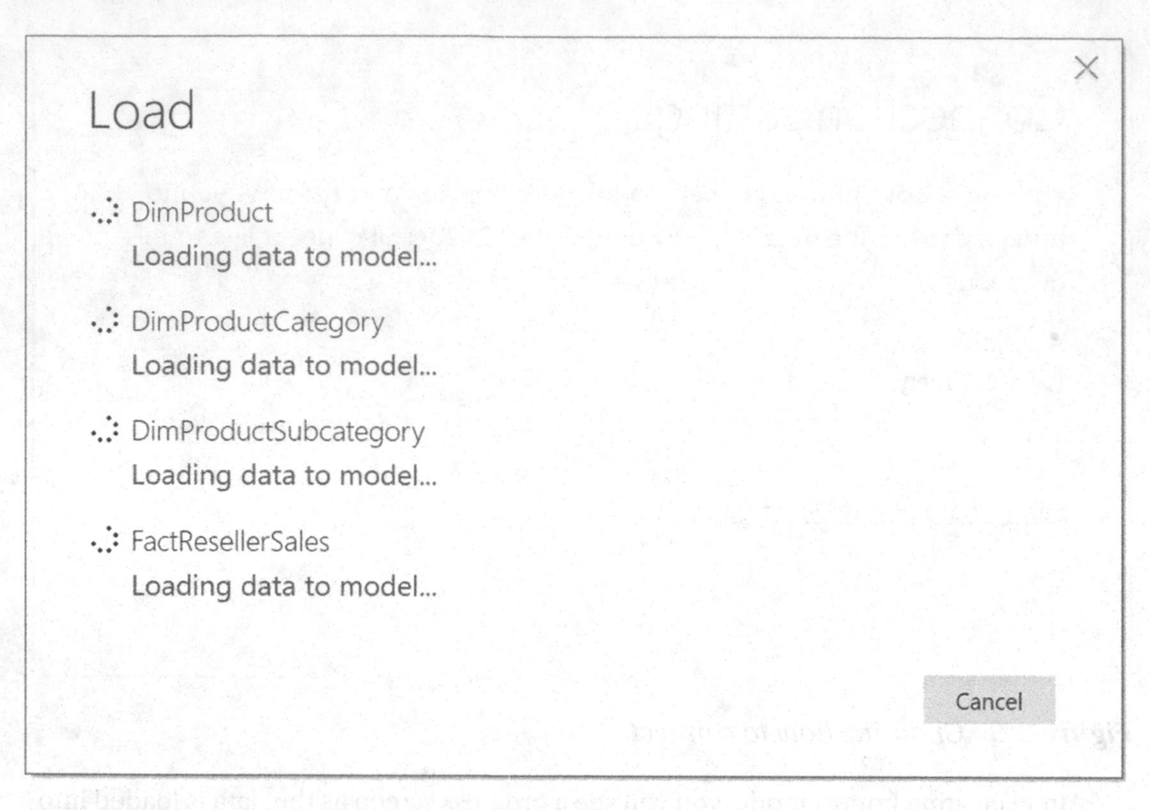

***Figure 2-10.** Importing the data into the model*

When the wizard closes, you will see the tables and columns listed in the Fields list pane in Power BI Desktop.

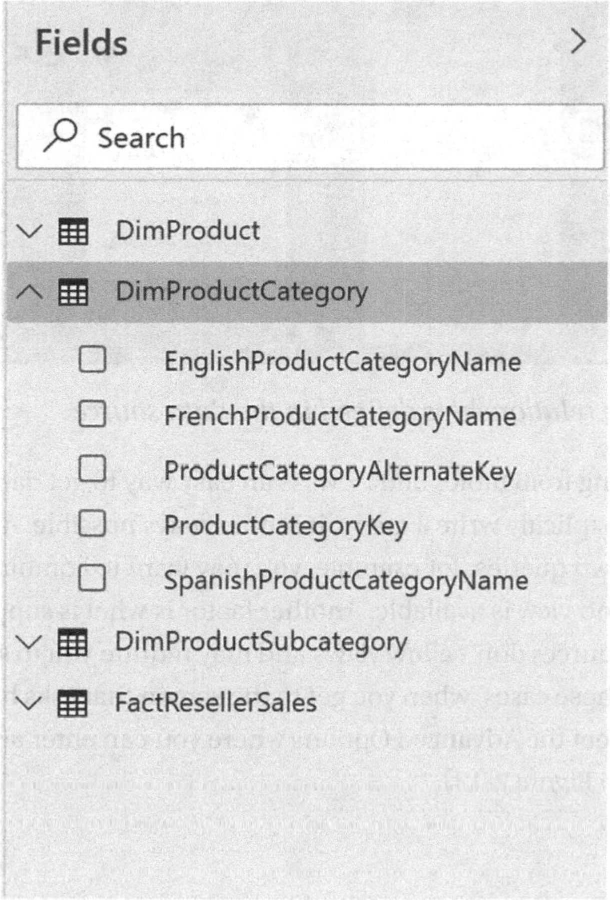

Figure 2-11. *Tables and fields imported into the model*

Note Remember that Power BI is only connected to the data source when it is retrieving the data. Once the data is retrieved, the connection is closed, and the data is part of the model.

If you switch to the Diagram View, you will see the tables, and if the table relationships were defined in the database, you will see the relationships between the tables. In Figure 2-12, you can see relationships defined between the product tables and the sales table. You can also create a relationship in the model even though one was not defined in the data source (more about this later).

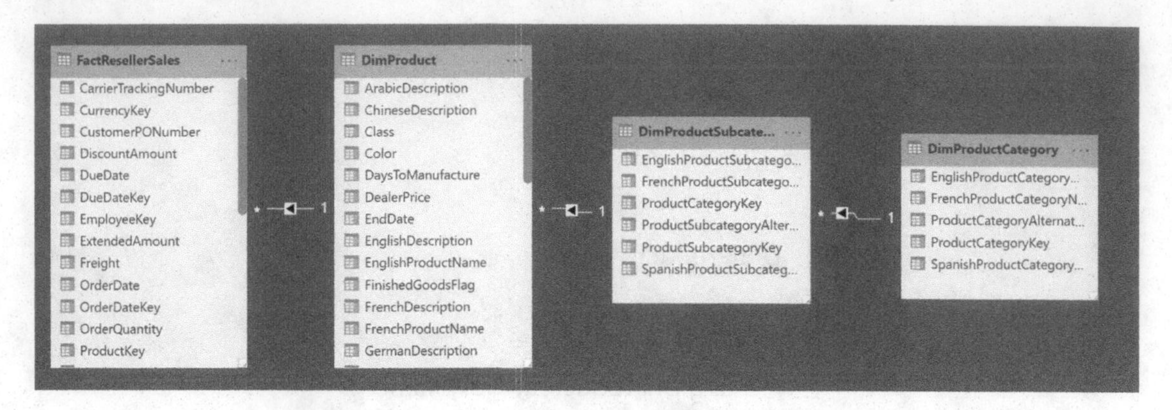

Figure 2-12. *Table relationships defined in the data source*

Although selecting from tables and views is an easy way to get data into the model without needing to explicitly write a query, it is not always possible. At times you may need to write your own queries; for example, you may want to combine data from several different tables and no view is available. Another factor is what is supported by the data source. Some data sources don't allow views and may require you to supply queries to extract the data. In these cases, when you get to the screen that asks how you want to retrieve the data, select the Advanced Options where you can enter an SQL query to retrieve the data (see Figure 2-13).

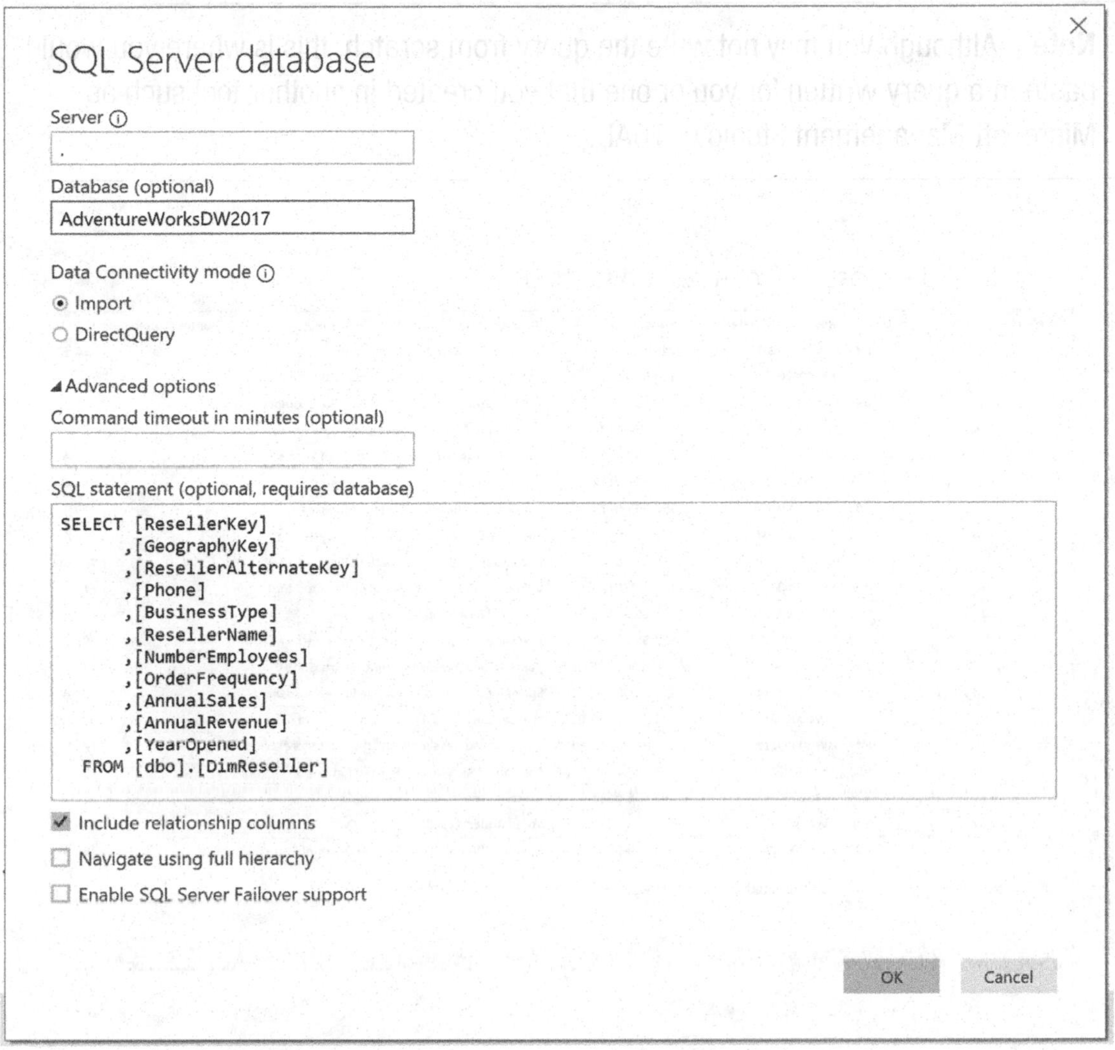

SQL Server database

Server

.

Database (optional)

AdventureWorksDW2017

Data Connectivity mode

Import

DirectQuery

Advanced options

Command timeout in minutes (optional)

SQL statement (optional, requires database)

```
SELECT [ResellerKey]
      ,[GeographyKey]
      ,[ResellerAlternateKey]
      ,[Phone]
      ,[BusinessType]
      ,[ResellerName]
      ,[NumberEmployees]
      ,[OrderFrequency]
      ,[AnnualSales]
      ,[AnnualRevenue]
      ,[YearOpened]
  FROM [dbo].[DimReseller]
```

Include relationship columns

Navigate using full hierarchy

Enable SQL Server Failover support

OK Cancel

Figure 2-13. *Entering an SQL query*

Once you enter the query and select OK, you will see a preview of the data (see Figure 2-14).

Note Although you may not write the query from scratch, this is where you would paste in a query written for you or one that you created in another tool such as Microsoft Management Studio or TOAD.

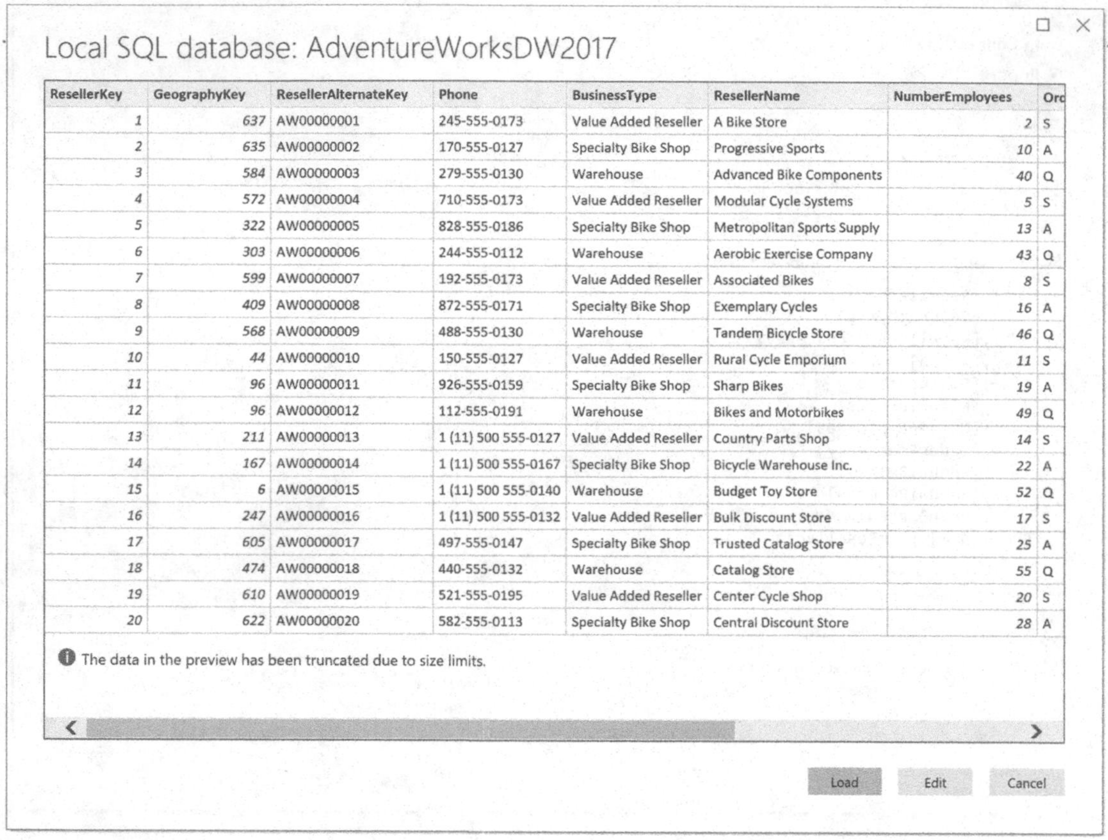

Local SQL database: AdventureWorksDW2017

ResellerKey	GeographyKey	ResellerAlternateKey	Phone	BusinessType	ResellerName	NumberEmployees	Orc
1	637	AW00000001	245-555-0173	Value Added Reseller	A Bike Store	2	S
2	635	AW00000002	170-555-0127	Specialty Bike Shop	Progressive Sports	10	A
3	584	AW00000003	279-555-0130	Warehouse	Advanced Bike Components	40	Q
4	572	AW00000004	710-555-0173	Value Added Reseller	Modular Cycle Systems	5	S
5	322	AW00000005	828-555-0186	Specialty Bike Shop	Metropolitan Sports Supply	13	A
6	303	AW00000006	244-555-0112	Warehouse	Aerobic Exercise Company	43	Q
7	599	AW00000007	192-555-0173	Value Added Reseller	Associated Bikes	8	S
8	409	AW00000008	872-555-0171	Specialty Bike Shop	Exemplary Cycles	16	A
9	568	AW00000009	488-555-0130	Warehouse	Tandem Bicycle Store	46	Q
10	44	AW00000010	150-555-0127	Value Added Reseller	Rural Cycle Emporium	11	S
11	96	AW00000011	926-555-0159	Specialty Bike Shop	Sharp Bikes	19	A
12	96	AW00000012	112-555-0191	Warehouse	Bikes and Motorbikes	49	Q
13	211	AW00000013	1 (11) 500 555-0127	Value Added Reseller	Country Parts Shop	14	S
14	167	AW00000014	1 (11) 500 555-0167	Specialty Bike Shop	Bicycle Warehouse Inc.	22	A
15	6	AW00000015	1 (11) 500 555-0140	Warehouse	Budget Toy Store	52	Q
16	247	AW00000016	1 (11) 500 555-0132	Value Added Reseller	Bulk Discount Store	17	S
17	605	AW00000017	497-555-0147	Specialty Bike Shop	Trusted Catalog Store	25	A
18	474	AW00000018	440-555-0132	Warehouse	Catalog Store	55	Q
19	610	AW00000019	521-555-0195	Value Added Reseller	Center Cycle Shop	20	S
20	622	AW00000020	582-555-0113	Specialty Bike Shop	Central Discount Store	28	A

The data in the preview has been truncated due to size limits.

Load Edit Cancel

Figure 2-14. *Previewing the data*

Clicking load will bring the data and table into the model.

Now that you know how to import data from a database, let's see how you can add data to the model from a text file.

Importing Data from Text Files

You may often need to combine data from several different sources. One of the most common sources of data is still the text file. This could be the result of receiving data as an output from another system; for example, you may need information from your company's enterprise resource planning (ERP) system, provided as a text file. You may also get data through third-party services that provide the data in a comma-separated value (CSV) format. For example, you may use a rating service to rate customers, and the results are returned in a CSV file.

Importing data into your model from a text file is similar to importing data from a relational database table. First, you select the option to get data on the Home tab. You can choose to import data from a text/csv file (see Figure 2-15).

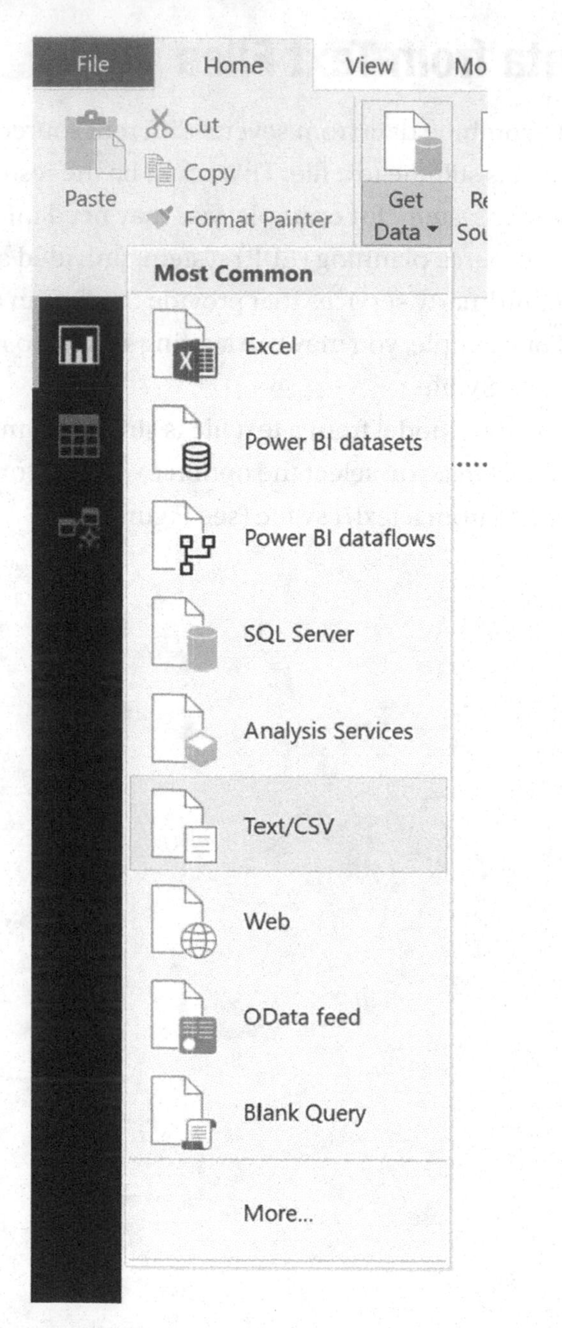

Figure 2-15. *Connecting to a text file*

Selecting the text/csv file brings up a screen where you browse to the file and load it. Once the file is loaded, you will see the sample data and the delimiter used to load the file (see Figure 2-16). Each text file is considered a table, and the file name will be the name of the table in the model.

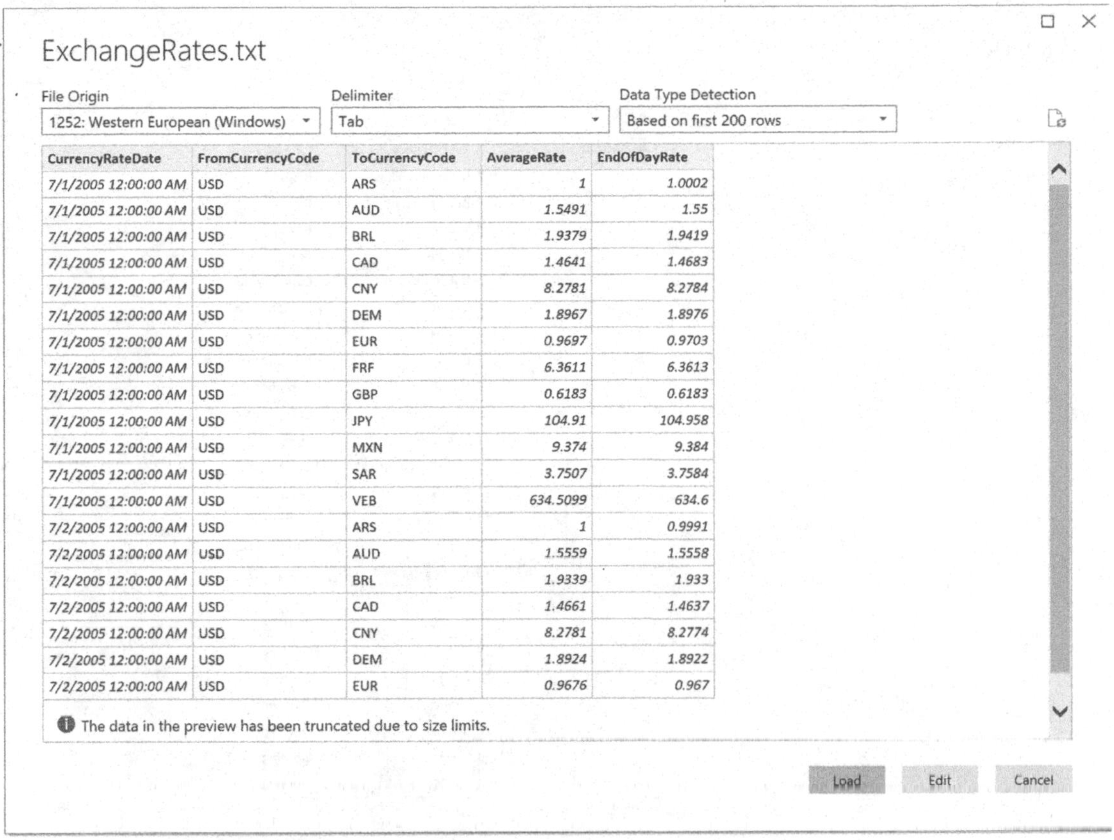

ExchangeRates.txt

File Origin: 1252: Western European (Windows)
Delimiter: Tab
Data Type Detection: Based on first 200 rows

CurrencyRateDate	FromCurrencyCode	ToCurrencyCode	AverageRate	EndOfDayRate
7/1/2005 12:00:00 AM	USD	ARS	1	1.0002
7/1/2005 12:00:00 AM	USD	AUD	1.5491	1.55
7/1/2005 12:00:00 AM	USD	BRL	1.9379	1.9419
7/1/2005 12:00:00 AM	USD	CAD	1.4641	1.4683
7/1/2005 12:00:00 AM	USD	CNY	8.2781	8.2784
7/1/2005 12:00:00 AM	USD	DEM	1.8967	1.8976
7/1/2005 12:00:00 AM	USD	EUR	0.9697	0.9703
7/1/2005 12:00:00 AM	USD	FRF	6.3611	6.3613
7/1/2005 12:00:00 AM	USD	GBP	0.6183	0.6183
7/1/2005 12:00:00 AM	USD	JPY	104.91	104.958
7/1/2005 12:00:00 AM	USD	MXN	9.374	9.384
7/1/2005 12:00:00 AM	USD	SAR	3.7507	3.7584
7/1/2005 12:00:00 AM	USD	VEB	634.5099	634.6
7/2/2005 12:00:00 AM	USD	ARS	1	0.9991
7/2/2005 12:00:00 AM	USD	AUD	1.5559	1.5558
7/2/2005 12:00:00 AM	USD	BRL	1.9339	1.933
7/2/2005 12:00:00 AM	USD	CAD	1.4661	1.4637
7/2/2005 12:00:00 AM	USD	CNY	8.2781	8.2774
7/2/2005 12:00:00 AM	USD	DEM	1.8924	1.8922
7/2/2005 12:00:00 AM	USD	EUR	0.9676	0.967

The data in the preview has been truncated due to size limits.

Load Edit Cancel

***Figure 2-16.** Previewing the data*

Another common type of file used as a data source is an Excel file. The main difference between importing data from a text file and importing data from an Excel file is that the Excel file can contain more than one table. By default, each sheet is treated as a table (see Figure 2-17). Once you select the table, you can preview the data just as you did for a text file.

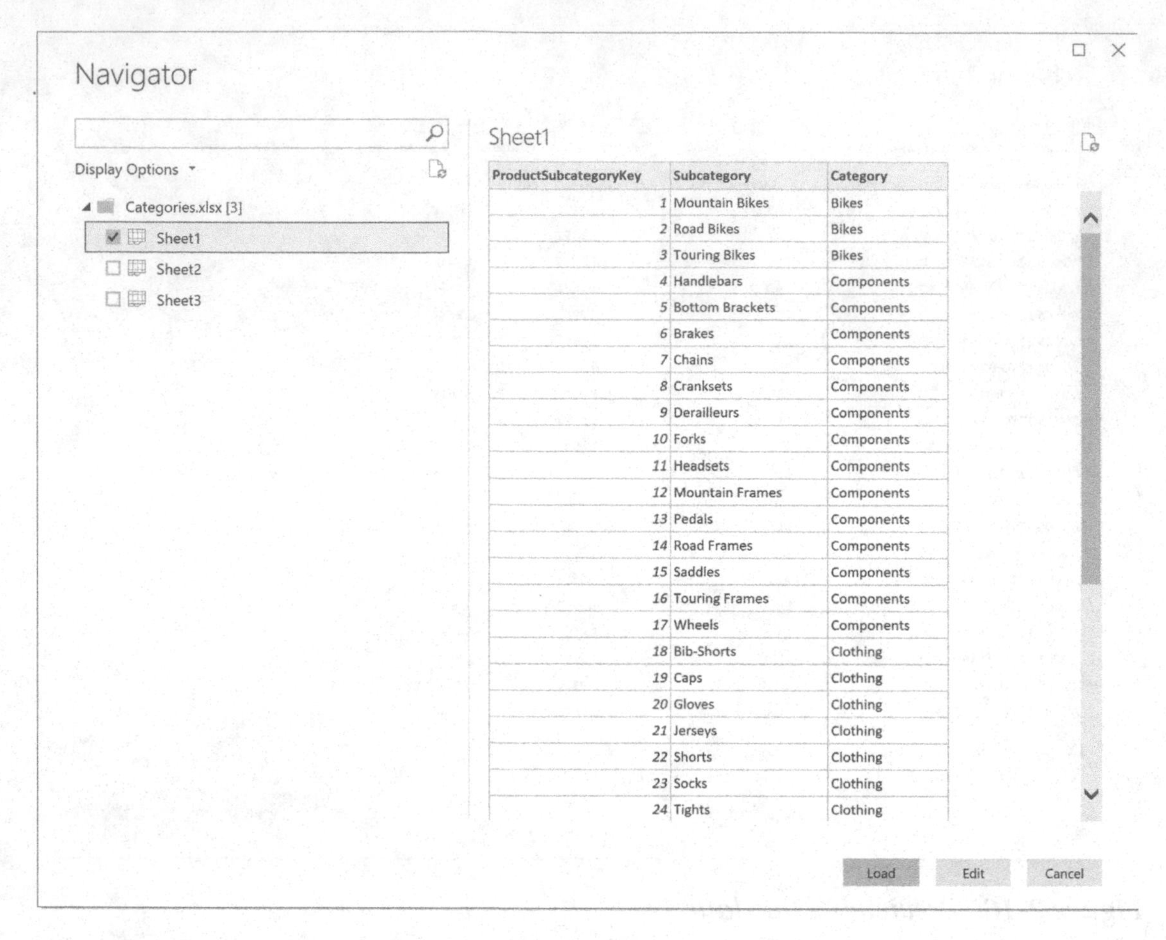

ProductSubcategoryKey	Subcategory	Category
1	Mountain Bikes	Bikes
2	Road Bikes	Bikes
3	Touring Bikes	Bikes
4	Handlebars	Components
5	Bottom Brackets	Components
6	Brakes	Components
7	Chains	Components
8	Cranksets	Components
9	Derailleurs	Components
10	Forks	Components
11	Headsets	Components
12	Mountain Frames	Components
13	Pedals	Components
14	Road Frames	Components
15	Saddles	Components
16	Touring Frames	Components
17	Wheels	Components
18	Bib-Shorts	Clothing
19	Caps	Clothing
20	Gloves	Clothing
21	Jerseys	Clothing
22	Shorts	Clothing
23	Socks	Clothing
24	Tights	Clothing

Figure 2-17. *Selecting a table in an Excel file*

In addition to importing data from a text file, you may need to supplement your data model using data imported from a data feed. This is becoming a very common way to exchange data with business partners, and you will see how to do this next.

Importing Data from a Data Feed

Although text files are one of the most popular ways of exchanging data, data feeds are becoming an increasingly prevalent way of exchanging data. *Data feeds* provide the data through web services, and to connect to the web service, you enter the web address of the web service (see Figure 2-18).

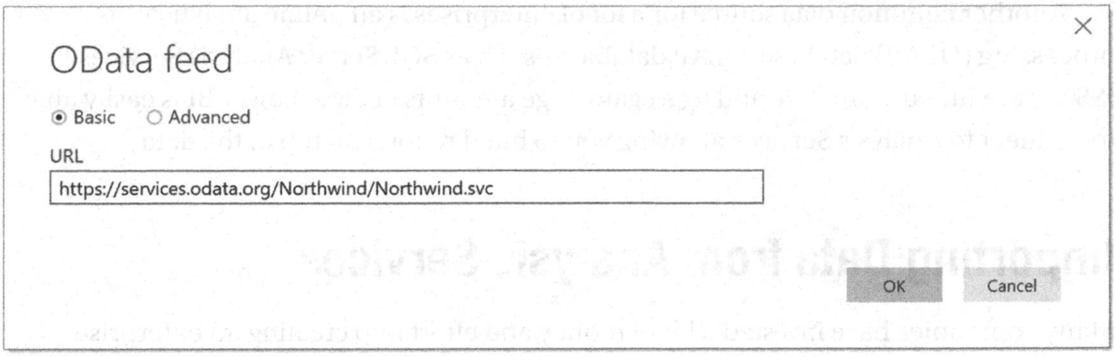

***Figure 2-18.** Connecting to a data feed*

Because the data feed contains not only the data but also the metadata (description of the data), once you make the connection, it is similar to connecting to a relational database as shown in Figure 2-19.

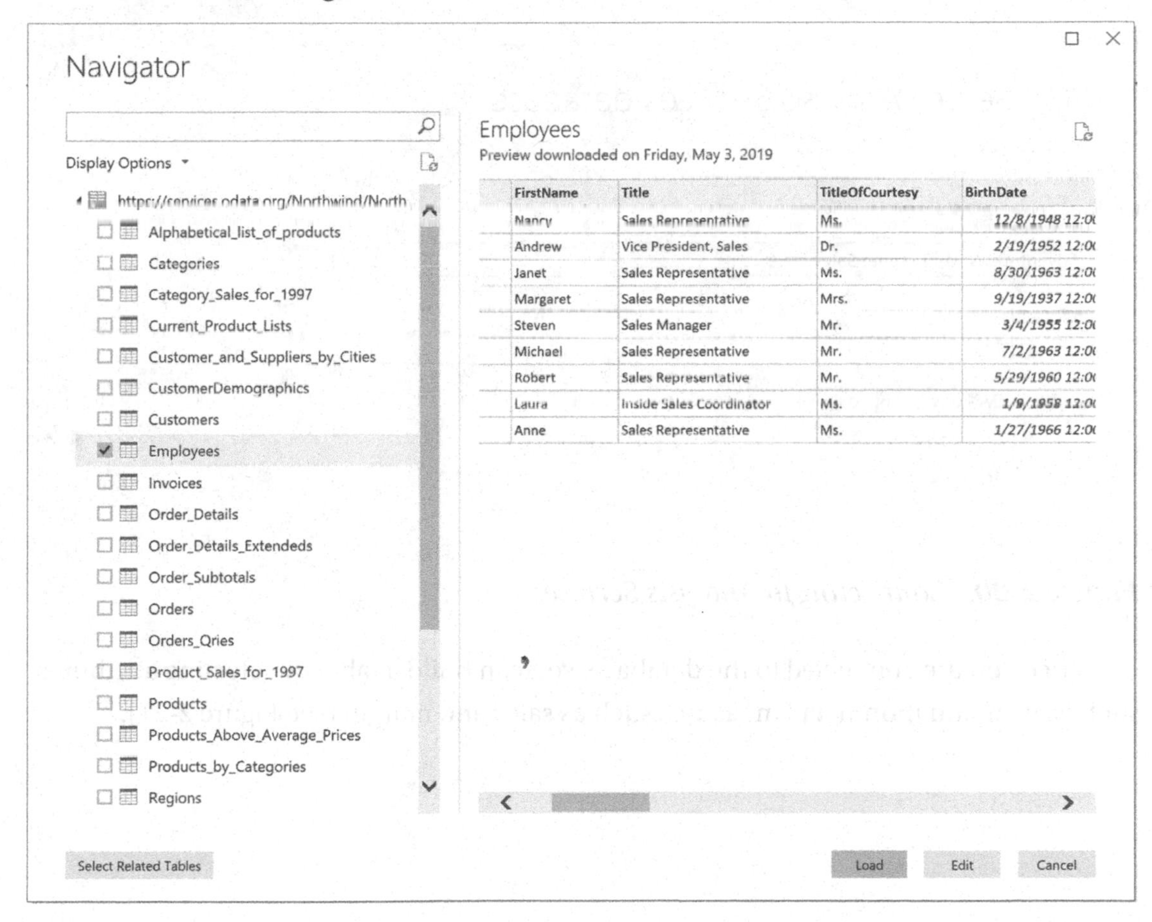

***Figure 2-19.** Previewing data from a data feed*

Another common data source for a lot of enterprises is an online analytical processing (OLAP) database. OLAP databases such as SQL Server Analysis Services (SSAS) are tuned to analyze and aggregate large amounts of data. Power BI is easily able to connect to Analysis Services allowing you to build reports on top of the data.

Importing Data from Analysis Services

Many companies have invested a lot of money and effort into creating an enterprise reporting solution consisting of an SQL Server Analysis Service (SSAS) repository that feeds various dashboards and score cards. Using Power BI, you can easily integrate data from these repositories. From the Get Data drop-down on the Home tab (Figure 2-14), choose the Analysis Services connection. This launches the connection information window as shown in Figure 2-20. Enter the server name and the database name.

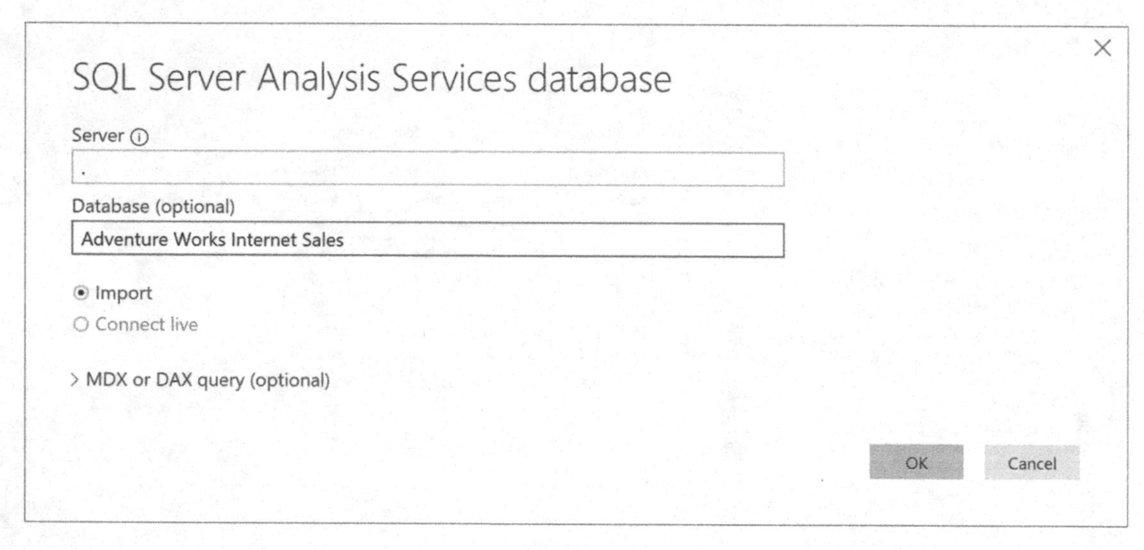

Figure 2-20. *Connecting to Analysis Services*

Once you are connected to the database, you can build a table by selecting attributes such as year and month and measures such as sales and margin (see Figure 2-21).

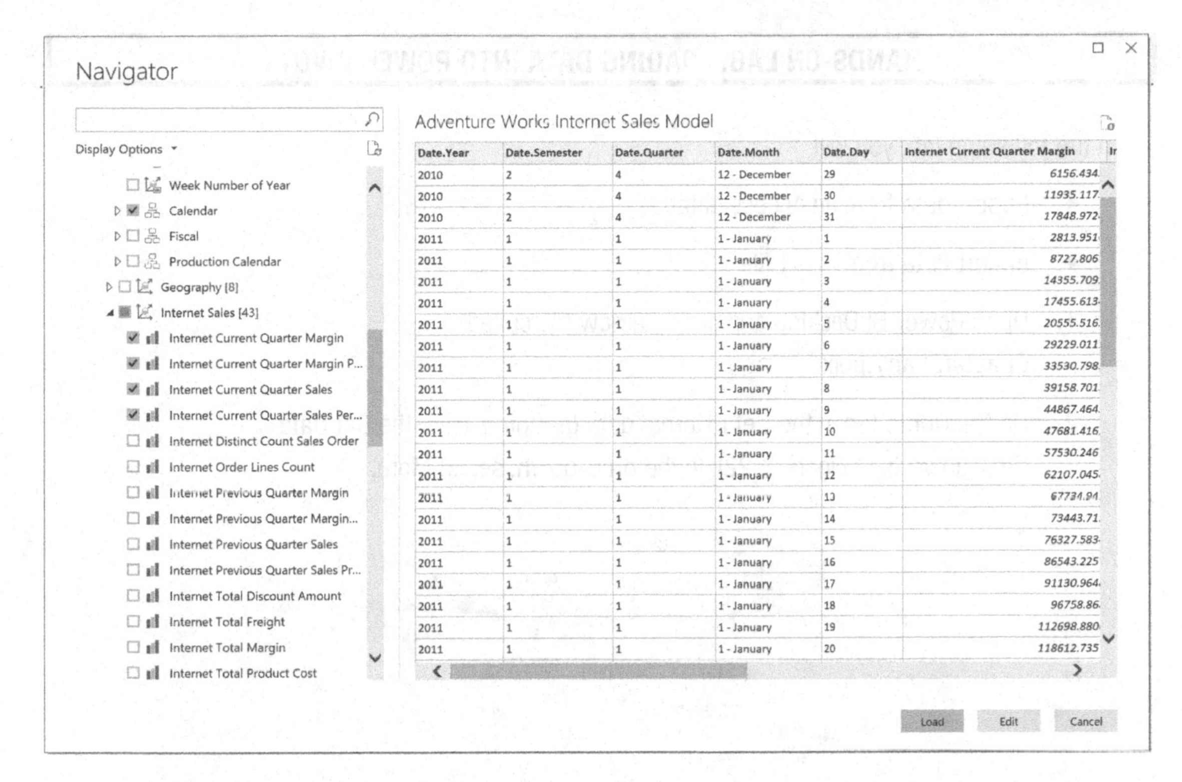

Figure 2-21. Selecting attributes and measures

Now that you have seen how to import data from various data sources into the Power BI data model, it is time to get some hands-on experience importing data.

Note The following lab uses an Access database. If you are connecting to an Access database for the first time on your computer, you may need to install Access 2013 Runtime (`https://www.microsoft.com/en-us/download/details.aspx?id=39358`).

HANDS-ON LAB: LOADING DATA INTO POWER PIVOT

In the following lab, you will

- Import data from an Access database
- Import data from a text file

1. Open Power BI Desktop and create a new file called Chapter2Lab1.pbix.
2. On the Home tab in the Get External data grouping, click the Get Data drop-down (see Figure 2-22). In the drop-down, select More.

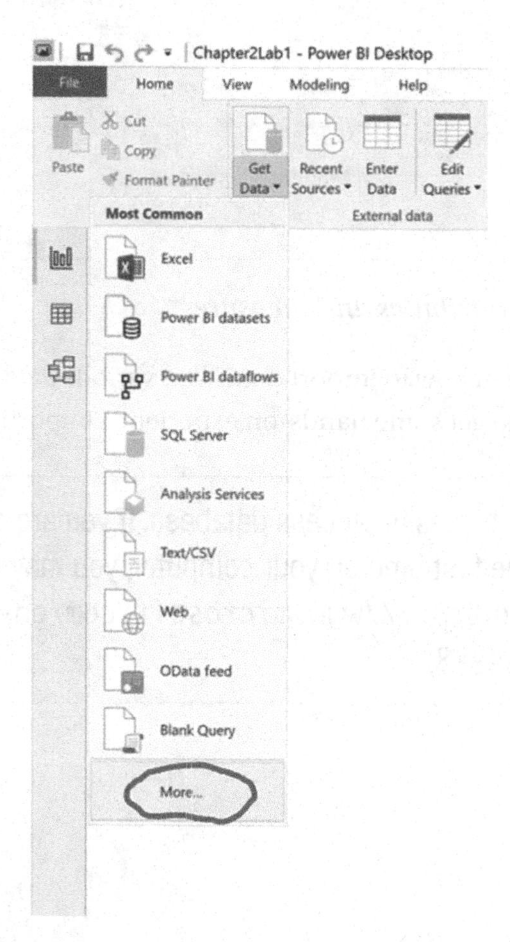

Figure 2-22. *Launching the Get Data window*

3. In the Get Data window, select the Access database and click the connect button (see Figure 2-23).

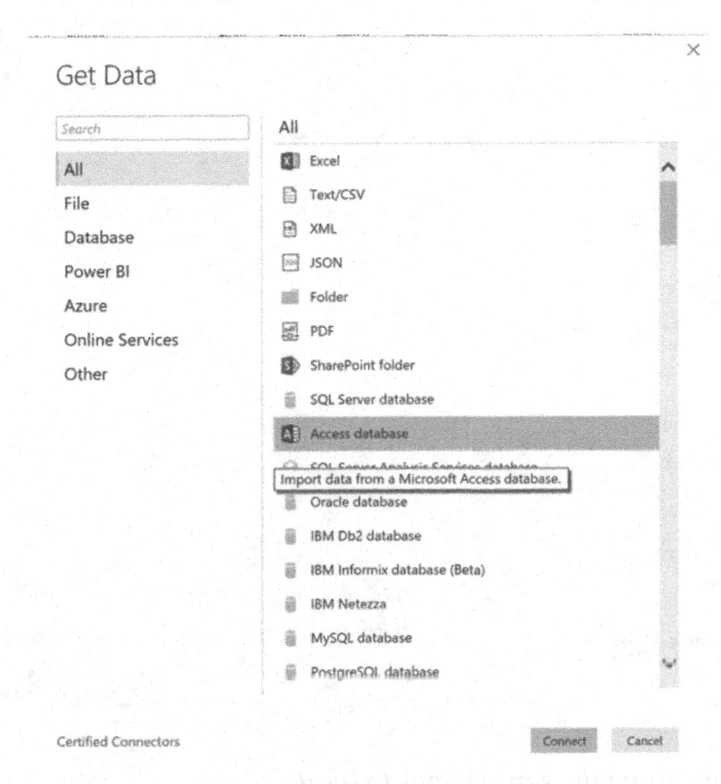

***Figure 2-23.** Connecting to an Access database*

4. Browse for the Northwind.acdb database file in the LabStarterFiles\Chapter2Lab1 and click Open.

5. In the Navigator window, you will see a list of tables and views. Select the Customers, Orders, and Employees table. At the bottom of the window, click the load button (see Figure 2-24).

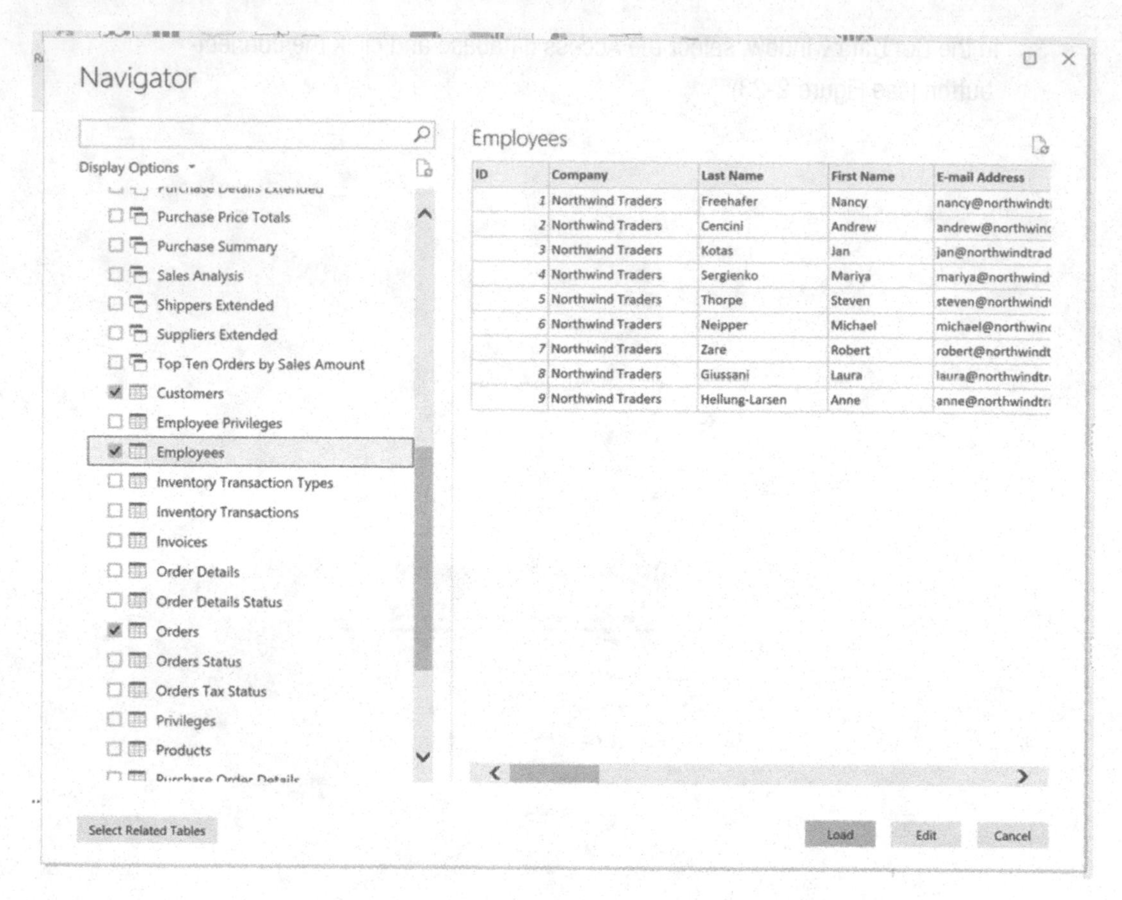

Figure 2-24. *Selecting tables and views to load*

6. The final table you are going to import is contained in a tab-delimited text file. On the Home tab in the External data section, click the Get Data drop-down. In the data sources list, select Text/CSV.
7. Browse to and open the ProductList.txt file in the LabStarterFiles\Chapter2Lab1 folder. You should see a preview of the data (see Figure 2-25). Load the data into the model.

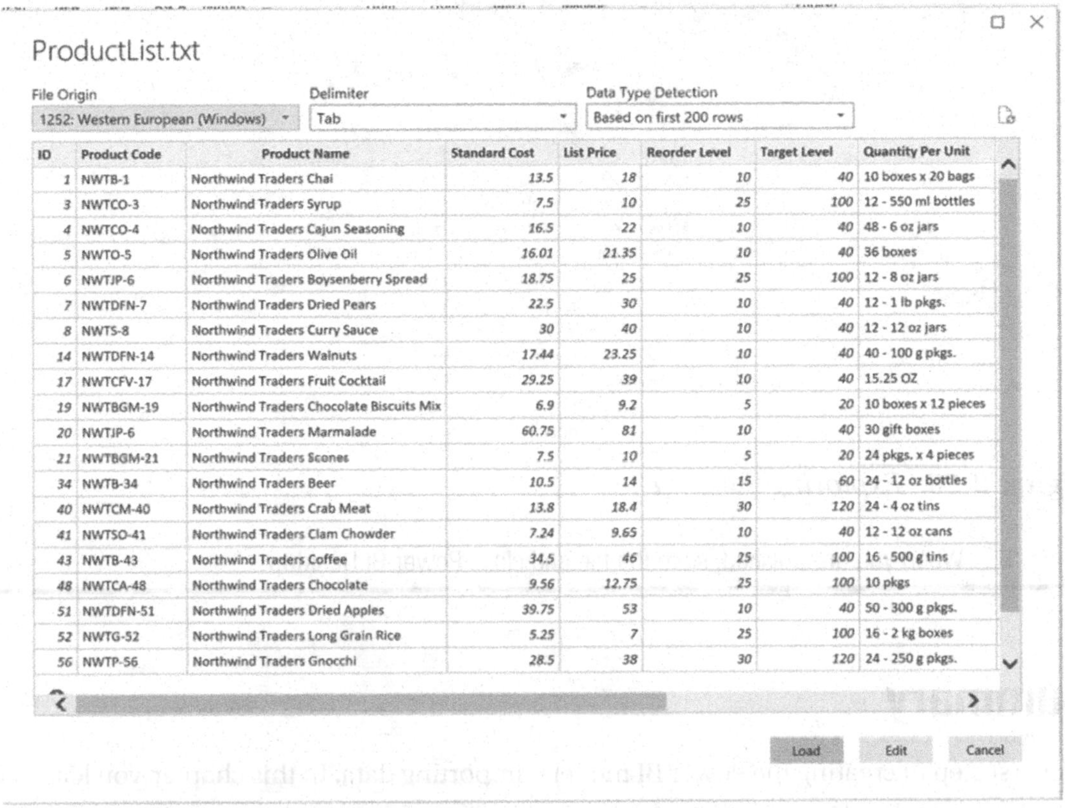

ProductList.txt

File Origin: 1252: Western European (Windows)
Delimiter: Tab
Data Type Detection: Based on first 200 rows

ID	Product Code	Product Name	Standard Cost	List Price	Reorder Level	Target Level	Quantity Per Unit
1	NWTB-1	Northwind Traders Chai	13.5	18	10	40	10 boxes x 20 bags
3	NWTCO-3	Northwind Traders Syrup	7.5	10	25	100	12 - 550 ml bottles
4	NWTCO-4	Northwind Traders Cajun Seasoning	16.5	22	10	40	48 - 6 oz jars
5	NWTO-5	Northwind Traders Olive Oil	16.01	21.35	10	40	36 boxes
6	NWTJP-6	Northwind Traders Boysenberry Spread	18.75	25	25	100	12 - 8 oz jars
7	NWTDFN-7	Northwind Traders Dried Pears	22.5	30	10	40	12 - 1 lb pkgs.
8	NWTS-8	Northwind Traders Curry Sauce	30	40	10	40	12 - 12 oz jars
14	NWTDFN-14	Northwind Traders Walnuts	17.44	23.25	10	40	40 - 100 g pkgs.
17	NWTCFV-17	Northwind Traders Fruit Cocktail	29.25	39	10	40	15.25 OZ
19	NWTBGM-19	Northwind Traders Chocolate Biscuits Mix	6.9	9.2	5	20	10 boxes x 12 pieces
20	NWTJP-6	Northwind Traders Marmalade	60.75	81	10	40	30 gift boxes
21	NWTBGM-21	Northwind Traders Scones	7.5	10	5	20	24 pkgs. x 4 pieces
34	NWTB-34	Northwind Traders Beer	10.5	14	15	60	24 - 12 oz bottles
40	NWTCM-40	Northwind Traders Crab Meat	13.8	18.4	30	120	24 - 4 oz tins
41	NWTSO-41	Northwind Traders Clam Chowder	7.24	9.65	10	40	12 - 12 oz cans
43	NWTB-43	Northwind Traders Coffee	34.5	46	25	100	16 - 500 g tins
48	NWTCA-48	Northwind Traders Chocolate	9.56	12.75	25	100	10 pkgs
51	NWTDFN-51	Northwind Traders Dried Apples	39.75	53	10	40	50 - 300 g pkgs.
52	NWTG-52	Northwind Traders Long Grain Rice	5.25	7	25	100	16 - 2 kg boxes
56	NWTP-56	Northwind Traders Gnocchi	28.5	38	30	120	24 - 250 g pkgs.

Load Edit Cancel

Figure 2-25. *Loading data from a text file*

8. After both sources have been loaded, select the Data tab in Power BI Desktop and explore the data. Verify all the tables were imported (see Figure 2-26).

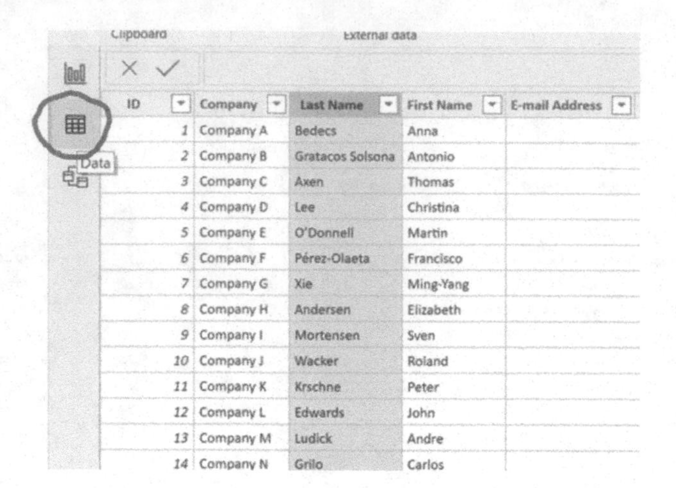

ID	Company	Last Name	First Name	E-mail Address
1	Company A	Bedecs	Anna	
2	Company B	Gratacos Solsona	Antonio	
3	Company C	Axen	Thomas	
4	Company D	Lee	Christina	
5	Company E	O'Donnell	Martin	
6	Company F	Pérez-Olaeta	Francisco	
7	Company G	Xie	Ming-Yang	
8	Company H	Andersen	Elizabeth	
9	Company I	Mortensen	Sven	
10	Company J	Wacker	Roland	
11	Company K	Krschne	Peter	
12	Company L	Edwards	John	
13	Company M	Ludick	Andre	
14	Company N	Grilo	Carlos	

***Figure 2-26.** Exploring the data*

9. When you are finished, save the file and close Power BI Desktop.

Summary

The first step in creating the Power BI model is importing data. In this chapter, you learned how to import data from a variety of data sources. One of the nice features of importing data in Power BI is that the experience is similar when you import the data from the various data sources. You create a connection, preview the data, and then import it into the model. This works well for importing data that has been cleaned and transformed into a central repository maintained by IT. However, data is coming increasingly from various sources both structured and unstructured. Data from these sources often needs to be scrubbed and transformed before it can be useful and imported into a model. In the next chapter, you will look at Power Query, a powerful tool that provides an easy-to-use interface for discovering, cleaning, and transforming data prior to importing it into your Power BI models.

CHAPTER 3

Data Munging with Power Query

Although you can import data directly into a Power BI model, quite often you need to clean and shape it (commonly called *data munging*) before loading it into the model. This is where Power Query really shines and is a very useful part of your BI arsenal. Power Query provides an easy-to-use interface for discovering and transforming data. It contains tools to clean and shape data such as removing duplicates, replacing values, and grouping data. In addition, it supports a vast array of data sources, both structured and unstructured, such as relational databases, web pages, and Hadoop, just to name a few. Once the data is extracted and transformed, you can then easily load it into a Power BI model.

After completing this chapter, you will be able to

- Discover and import data from various sources
- Cleanse data
- Merge, shape, and filter data
- Group and aggregate data
- Insert calculated columns

Discovering and Importing Data

Traditionally, if you needed to combine and transform data from various disparate data sources, you would rely on the IT department to stage the data for you using a tool such as SQL Server Integration Services (SSIS). This can often be a long, drawn-out effort of data discovery, cleansing, and conforming the data to a relational structure.

D. Clark, *Beginning Microsoft Power BI*, https://doi.org/10.1007/978-1-4842-5620-6_3

Although this type of formal effort is needed to load and conform data for the corporate operational data store, there are many times when you just want to add data to your Power BI model from a variety of sources in a quick, intuitive, and agile manner. To support this effort, you can use Power Query as your self-service BI extract, transform, and load (ETL) tool.

To launch the Power Query Editor, select the Edit Queries drop-down on the Home tab and select Edit Queries (see Figure 3-1).

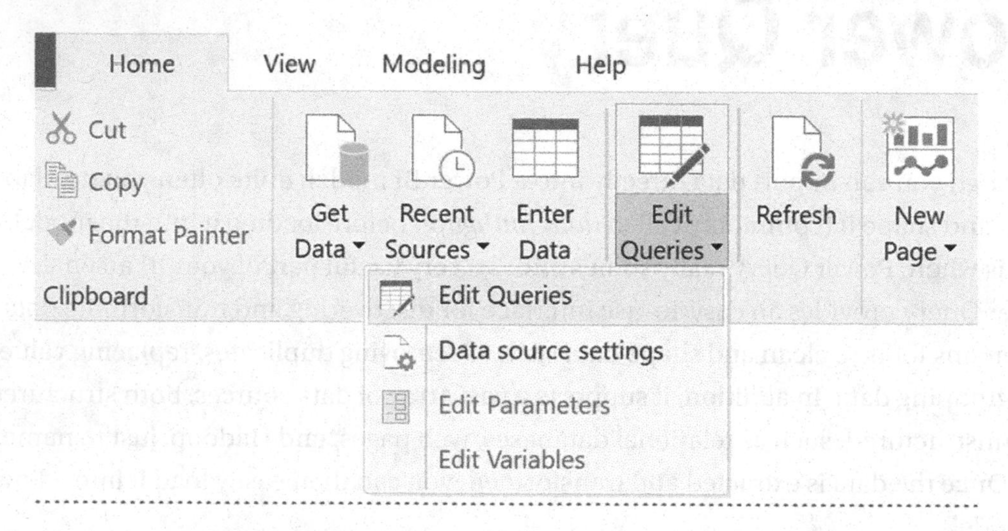

***Figure 3-1.** Launching Power Query*

After launching the Query Editor, to create a new query, select the New Source drop-down on the Home tab (see Figure 3-2).

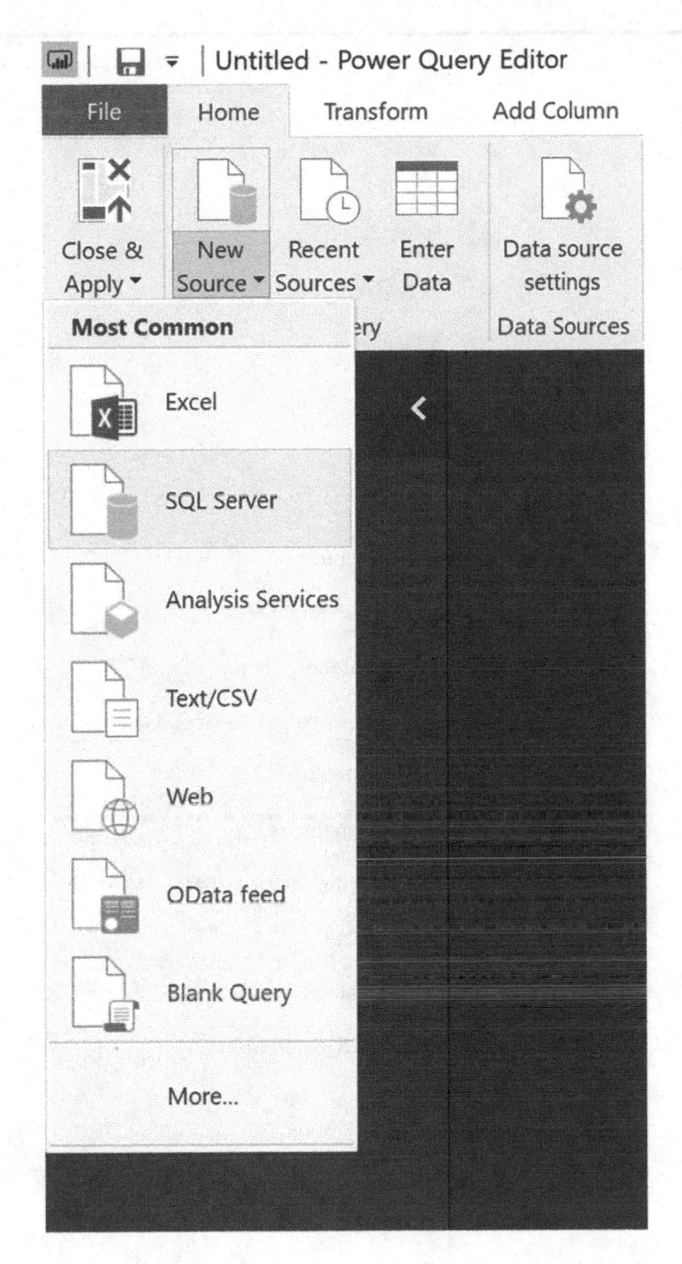

***Figure 3-2.** Selecting a new data source*

The drop-down shows some common data sources available. To see the complete list, select the More option (see Figure 3-3).

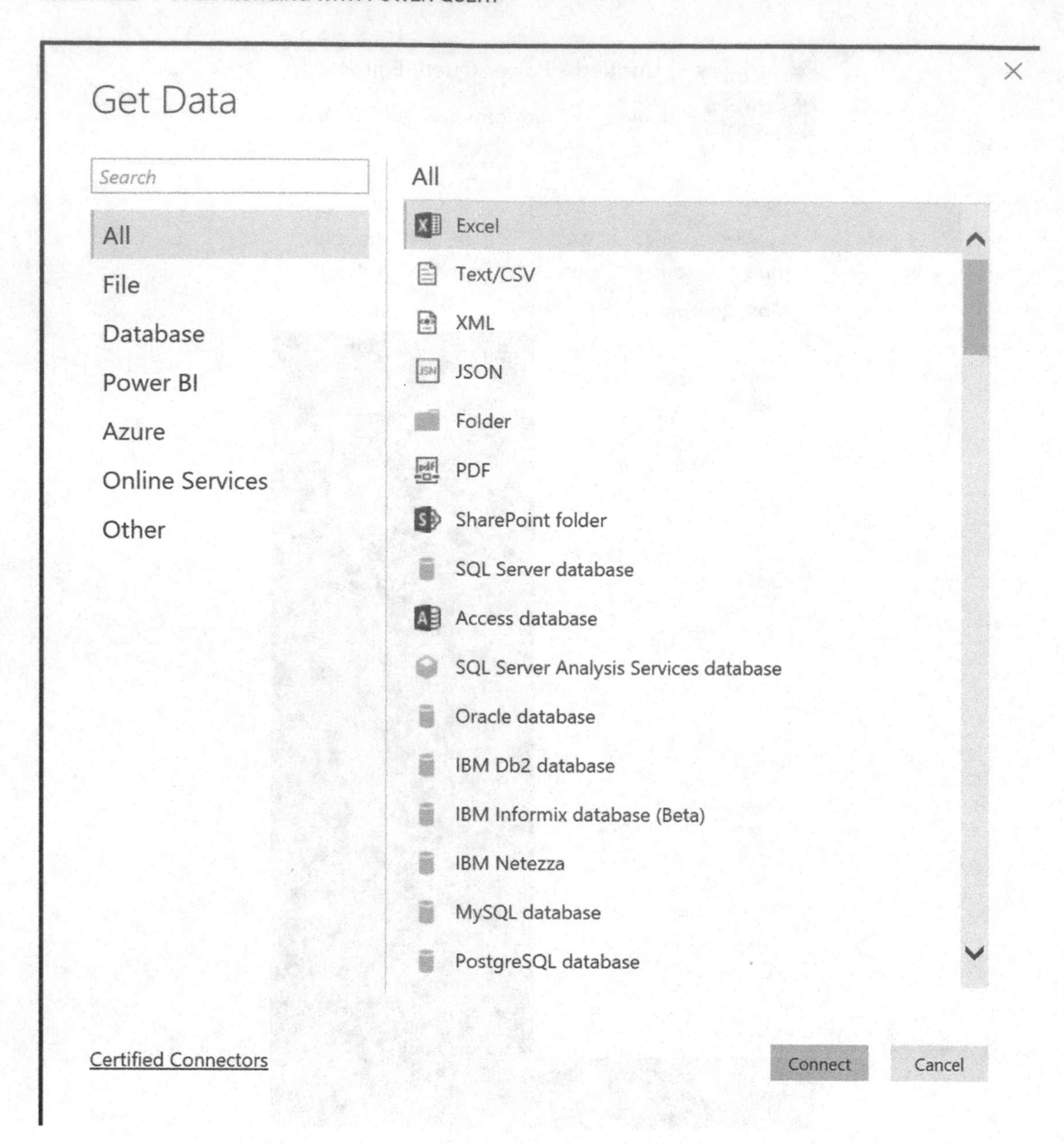

Figure 3-3. *Viewing the list of data sources available*

Once you select a data source, you need to enter the connection information. The type of data source connection will dictate what information you need to supply to gain access to the data source. For example, an SQL Database connection requires the server name and the database name (see Figure 3-4), whereas a CSV file requires the file path.

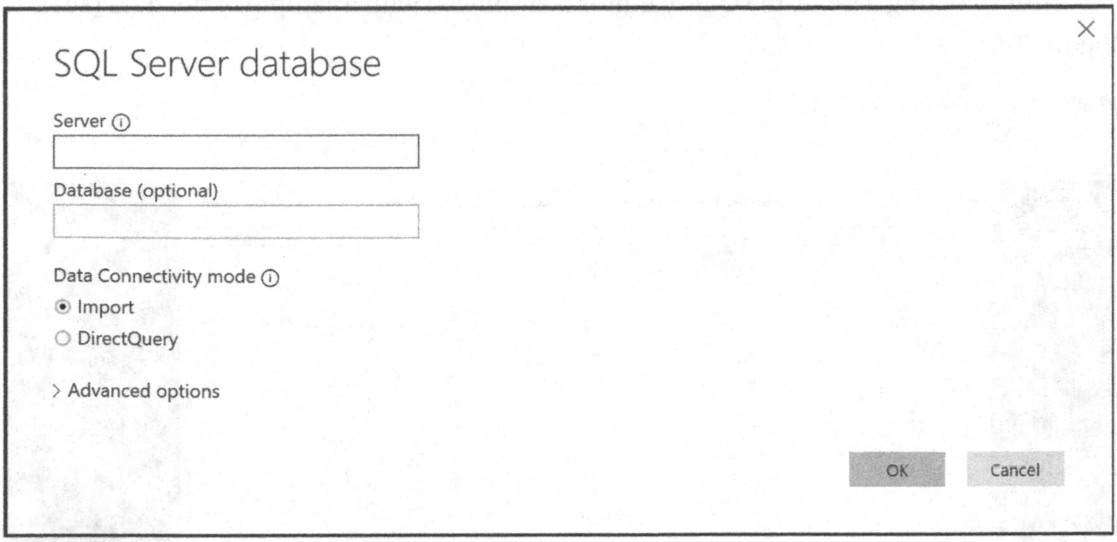

Figure 3-4. Connecting to an SQL Database

If you select a data source with multiple tables, you will see a Navigator pane displayed. Figure 3-5 shows the Navigator pane displayed when you connect to an SQL Database source.

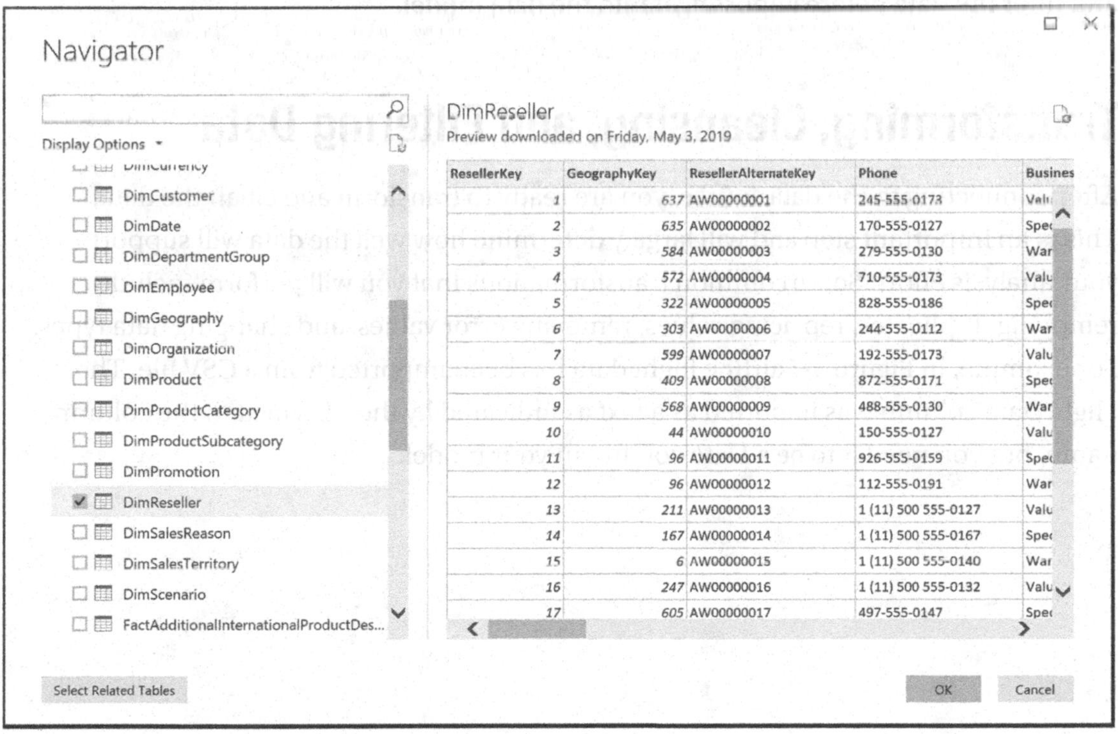

Figure 3-5. Using the Navigator pane to select a table

After selecting a table, the Query Editor is displayed with a sample of the data (see Figure 3-6).

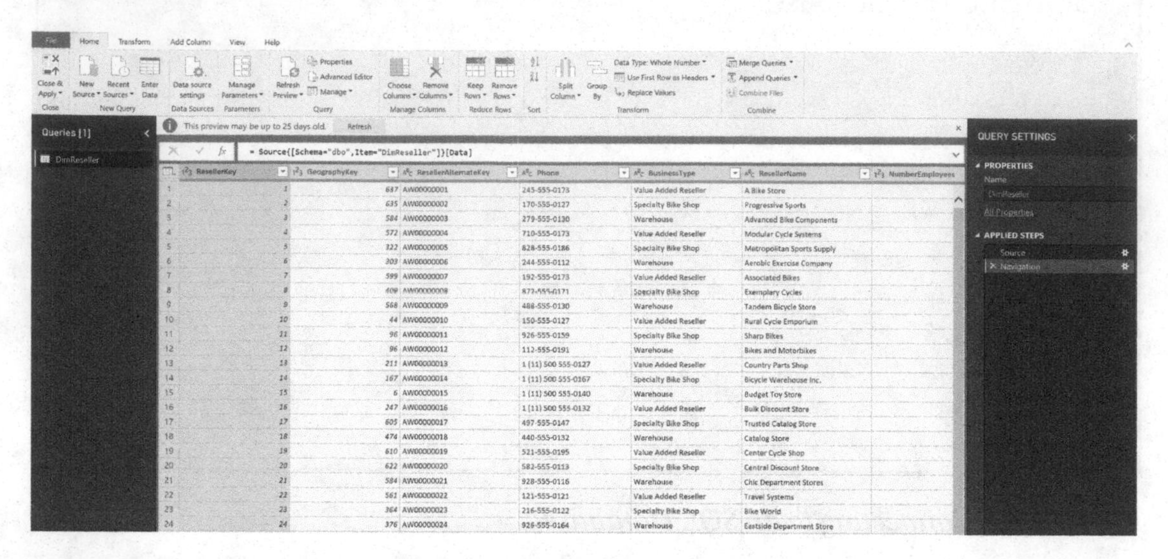

***Figure 3-6.** Viewing sample data in the Query Editor*

Once you have connected to the data source, the next step is to transform, cleanse, and filter the data before importing it into the data model.

Transforming, Cleansing, and Filtering Data

After connecting to the data source, you are ready to transform and clean the data. This is an important step and will largely determine how well the data will support your analysis effort. Some common transformations that you will perform include removing duplicates, replacing values, removing error values, and changing data types. For example, in Figure 3-7 airline flight data has been imported from a CSV file. The FlightDate column was imported as a text as indicated by the ABC next to the column name, but you need it to be a Date column in your model.

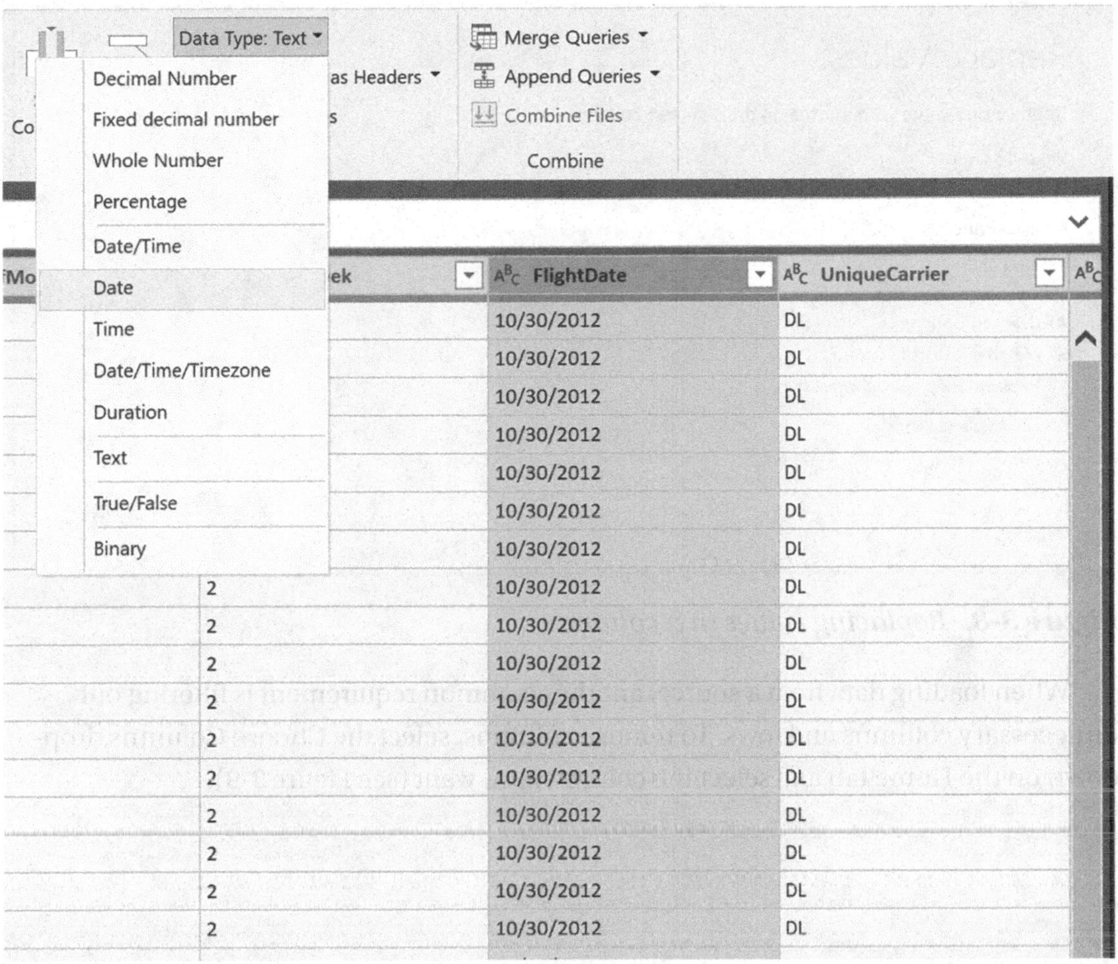

Data Type: Text
Decimal Number
Fixed decimal number
Whole Number
Percentage
Date/Time
Date
Time
Date/Time/Timezone
Duration
Text
True/False
Binary

as Headers
Merge Queries
Append Queries
Combine Files
Combine

Mo	ek	FlightDate	UniqueCarrier
		10/30/2012	DL
		10/30/2012	DL
		10/30/2012	DL
		10/30/2012	DL
		10/30/2012	DL
		10/30/2012	DL
		10/30/2012	DL
	2	10/30/2012	DL
	2	10/30/2012	DL
	2	10/30/2012	DL
	2	10/30/2012	DL
	2	10/30/2012	DL
	2	10/30/2012	DL
	2	10/30/2012	DL
	2	10/30/2012	DL
	2	10/30/2012	DL
	2	10/30/2012	DL

***Figure 3-7.** Changing the data type of a column*

Often you need to replace values from a source system so that they sync together in your model. For example, a carrier listed as VX in the CSV file has a value of VG in your existing data. You can easily replace these values as the data is imported by selecting the column and then selecting the Replace Values transformation in the menu. This launches a window in which you can enter the values to find and what to replace them with (see Figure 3-8).

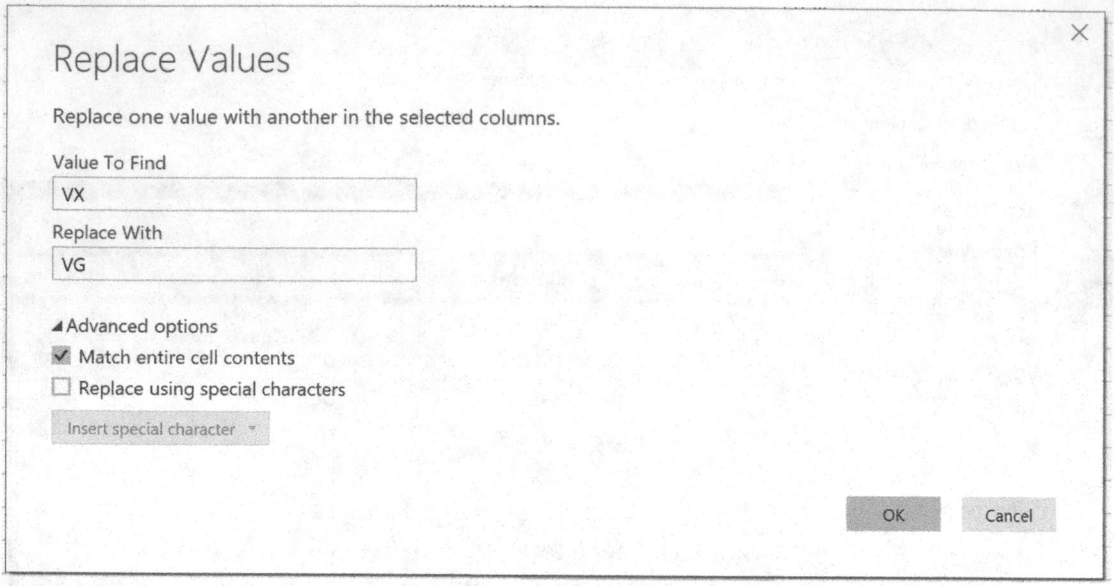

Figure 3-8. *Replacing values in a column*

When loading data from a source, another common requirement is filtering out unnecessary columns and rows. To remove columns, select the Choose Columns drop-down on the Home tab and select the columns you want (see Figure 3-9).

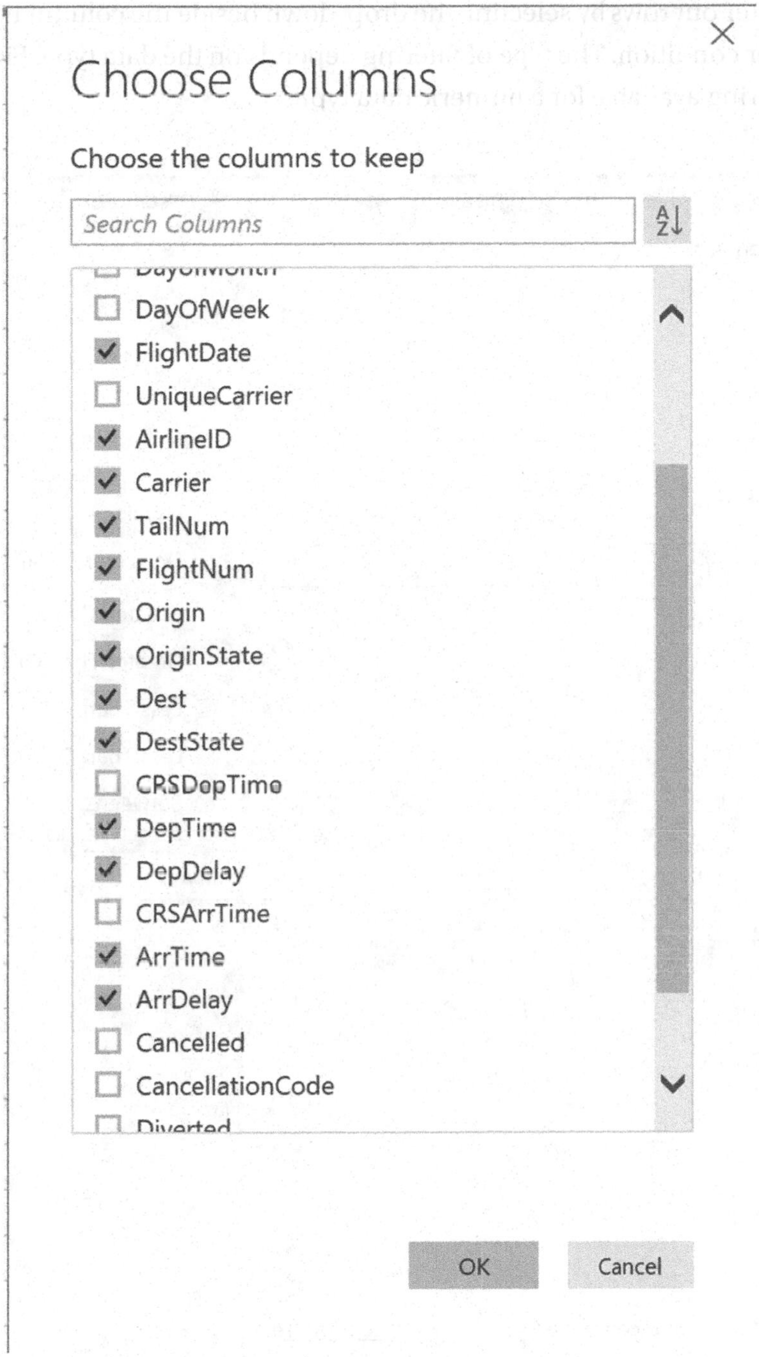

Choose Columns

Choose the columns to keep

Search Columns

DayOfWeek
FlightDate
UniqueCarrier
AirlineID
Carrier
TailNum
FlightNum
Origin
OriginState
Dest
DestState
CRSDepTime
DepTime
DepDelay
CRSArrTime
ArrTime
ArrDelay
Cancelled
CancellationCode

OK Cancel

Figure 3-9. *Filtering columns*

You can filter out rows by selecting the drop-down beside the column name and entering a filter condition. The type of filtering depends on the data type. Figure 3-10 shows the filtering available for a numeric data type.

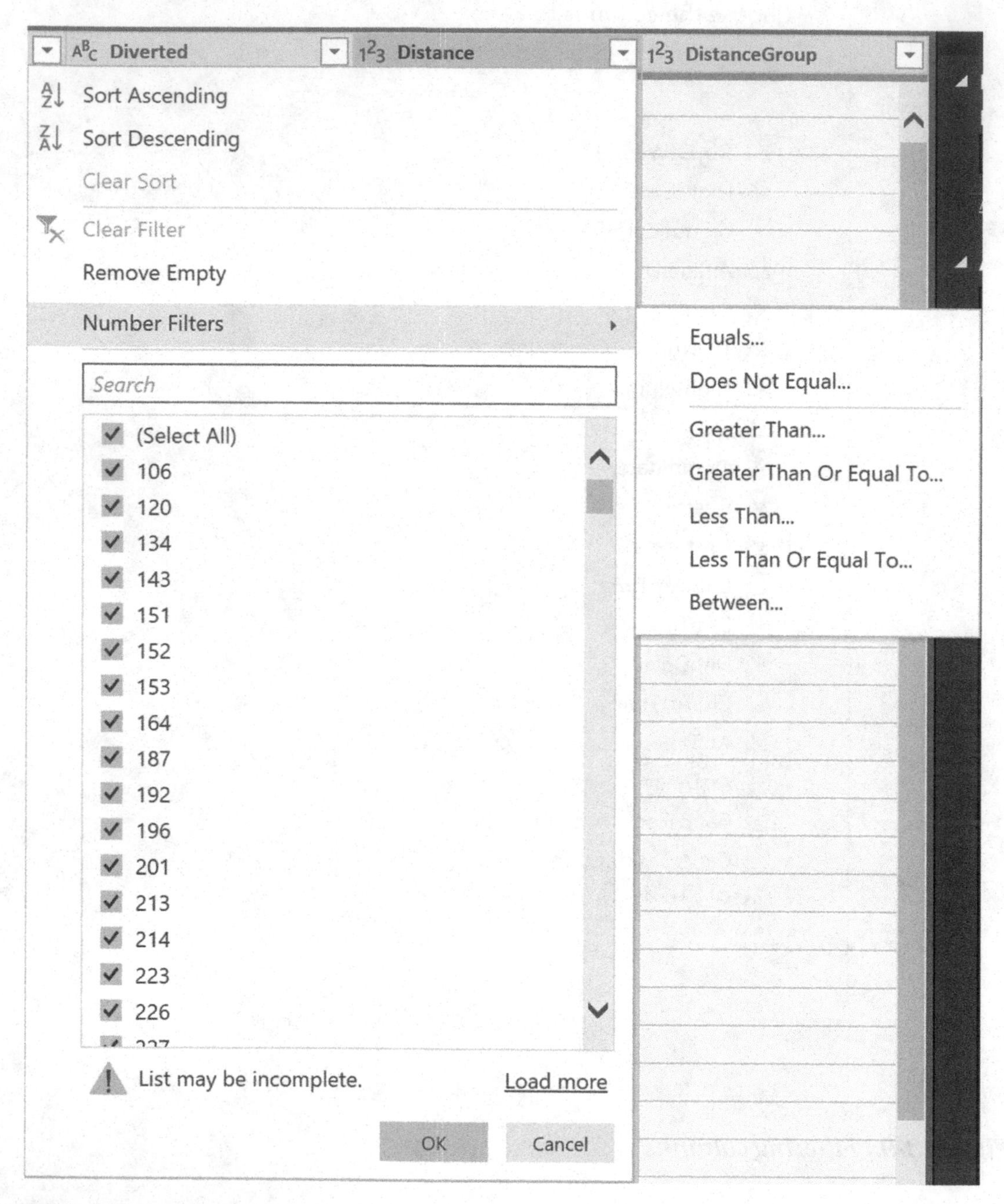

Figure 3-10. *Filtering rows*

As you apply the data transformations and filtering, the Query Editor lists the steps you have applied. This allows you to organize and track the changes you make to the data. You can rename, rearrange, and remove steps by right-clicking the step in the list (see Figure 3-11).

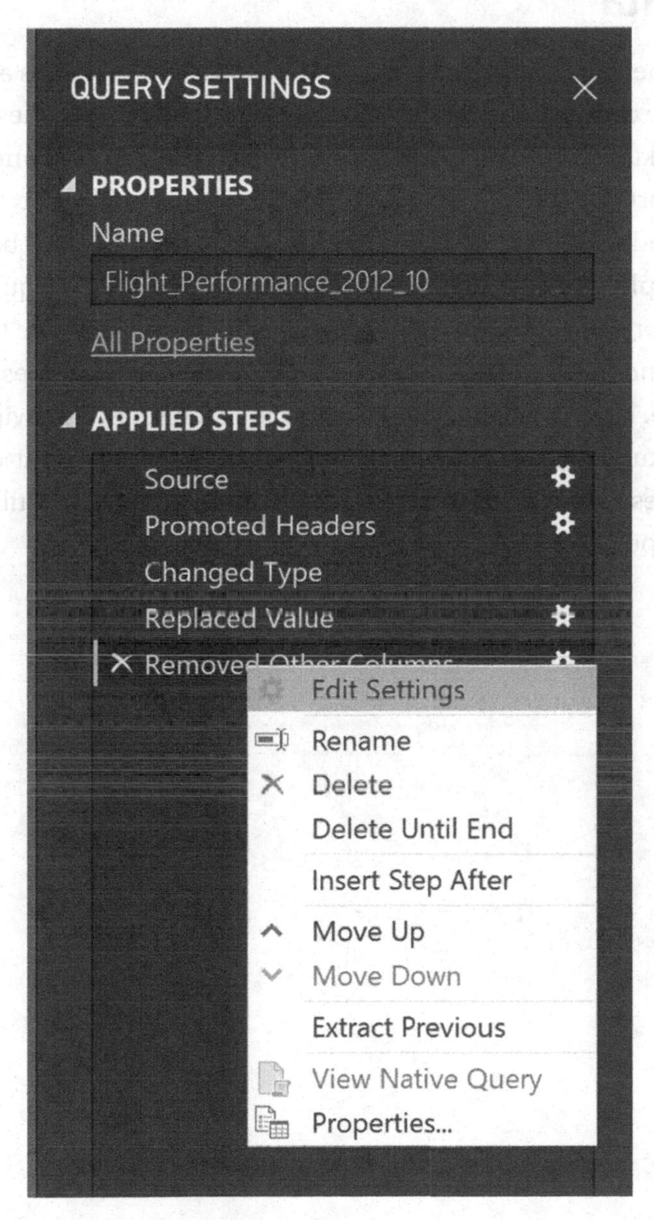

***Figure 3-11.** Managing the query steps*

After cleansing and transforming the data, you may need to combine data from several sources into one table in your data model or expand data contained in a column.

Merging Data

There are times when you may need to merge data from several tables and/or sources before you load the data into the model—for example, if you have codes in a table that link to another lookup table that contains the full value for the field. One way to deal with this is to import both tables into your model and create a link between the tables in the Power BI model. Another option is to merge the tables together before importing the data. For example, in the flight data you saw earlier, there is a UniqueCarrier column that contains carrier codes. You can merge these with another CSV file that contains the carrier codes and the carrier name, so we can use the carrier names in our reports instead of the code. First, create and save a query for each set of data with the Query Editor. For the lookup table, you can uncheck the Enable Load by right-clicking the query in the Queries list window (see Figure 3-12). You do, however, still want the query to run when the report refreshes.

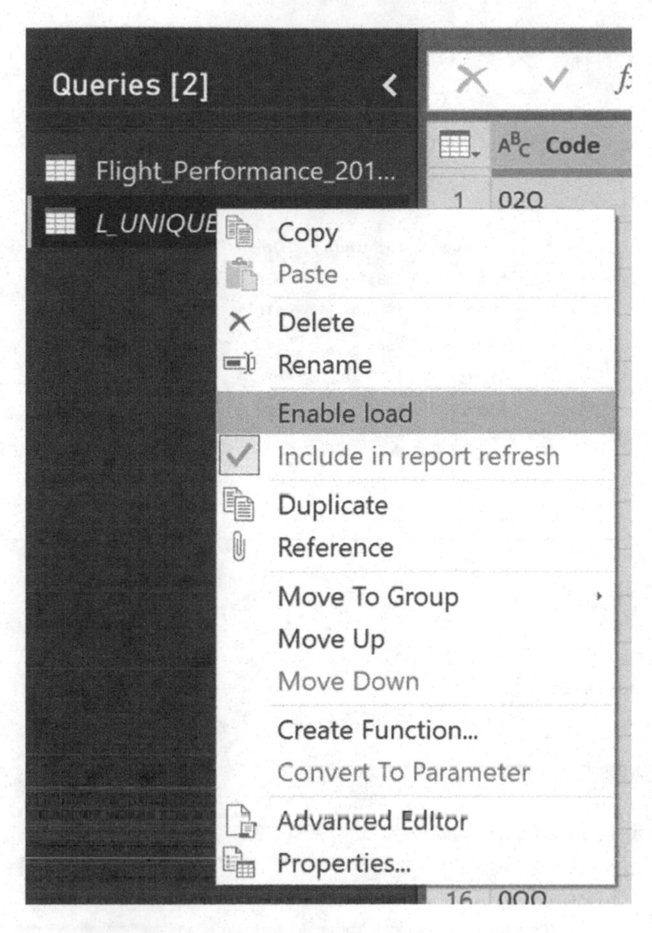

***Figure 3-12.** Disable loading into the Power BI model*

Next, select the main query and click the Merge Queries drop-down on the Home tab. This will launch the Merge window (see Figure 3-13), where you select the lookup query and the columns that link the data together.

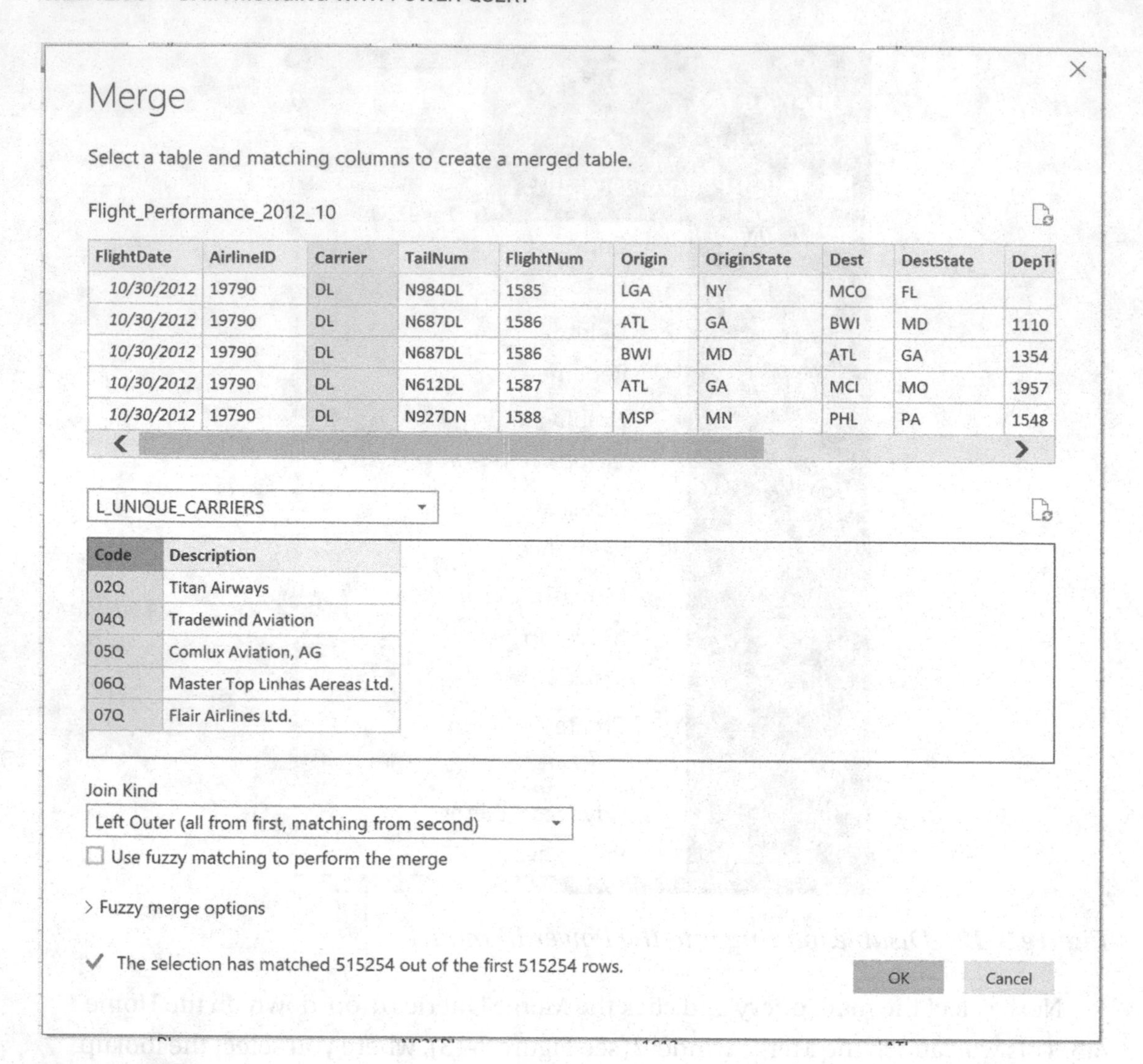

***Figure 3-13.** Merging data from two queries*

Once you merge the queries, you will see a new column containing a table type (see Figure 3-14).

ArrTime	ArrDelay	L_UNIQUE_CARRIERS
1911	9	Table
2330	-7	Table
2335	5	Table
1131	7	Table
1635	109	Table
2203	80	Table
859	-13	Table
1339	14	Table
821	-9	Table
930	-10	Table
721	-3	Table
2026	13	Table
649	21	Table
1740	-10	Table

***Figure 3-14.** Expanding the merged table*

By expanding this column, you can choose which columns to keep (see Figure 3-15).

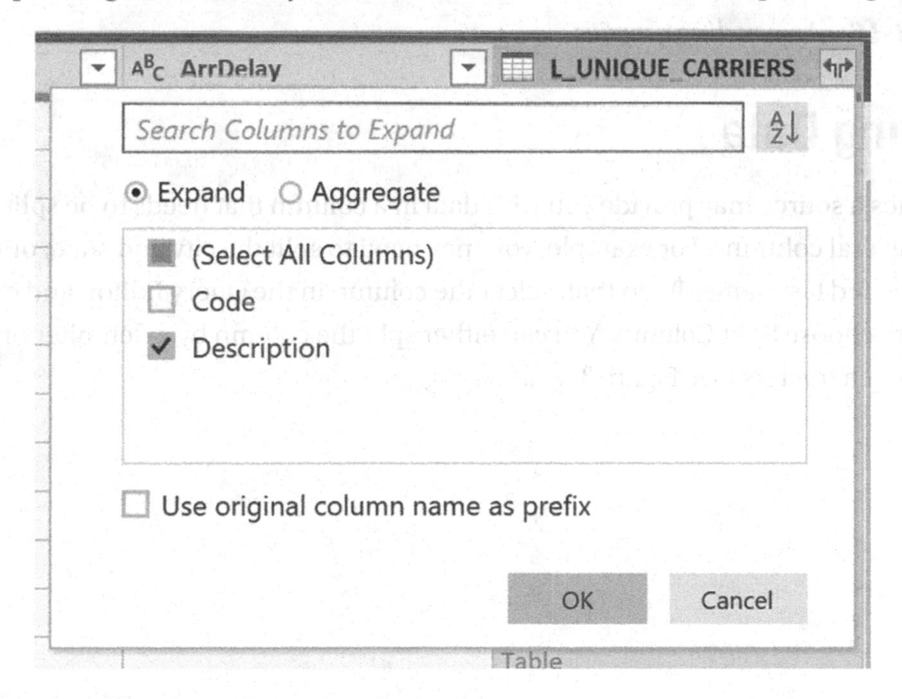

***Figure 3-15.** Choosing columns to keep*

Appending Data

Along with merging data from lookup tables, you may also need to append data from two different sources. For example, say you have flight data for each year separated into different source files or tables and want to combine multiple years into the same table. In this case, you would create two similar queries, each using a different source. First, open one of the queries in the Query Editor and select the Append Queries button on the Home tab. You can then select the other query as the table to append (see Figure 3-16).

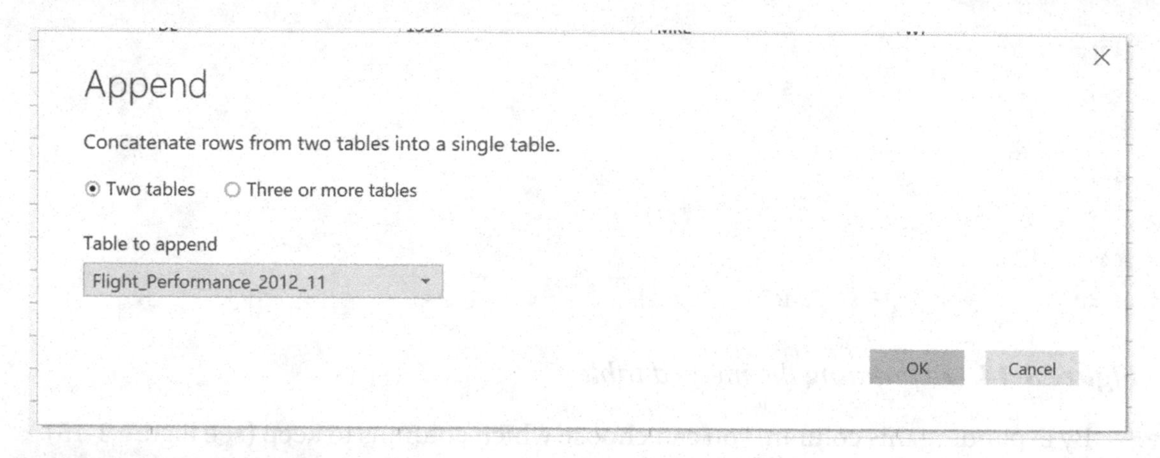

Figure 3-16. *Appending queries*

Splitting Data

Sometimes a source may provide you with data in a column that needs to be split up among several columns. For example, you may need to split the city and state, or the first name and last name. To do that, select the column in the Query Editor, and on the Home tab, choose Split Column. You can either split the column by a delimiter or by the number of characters (see Figure 3-17).

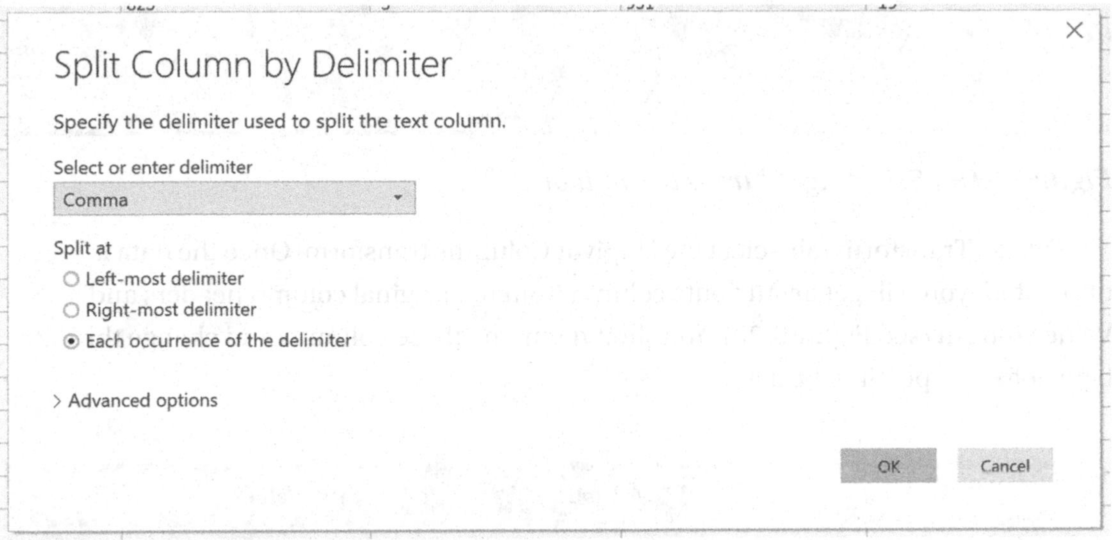

Split Column by Delimiter

Specify the delimiter used to split the text column.

Select or enter delimiter

Comma

Split at

- Left-most delimiter
- Right-most delimiter
- Each occurrence of the delimiter

> Advanced options

OK Cancel

Figure 3-17. *Splitting a column using a delimiter*

Unpivoting Data

Another scenario you may run into is when the data source contains data that is not in tabular form, but rather in a matrix, as in Figure 3-18.

Carrier	10/1/2012	10/2/2012	10/3/2012	10/4/2012	10/5/2012
AA	27579	18675	19041	21242	21115
DL	18763	11226	15112	16914	12681
UA	13784	11043	20777	20327	16497
US	3242	4083	7362	7567	4443

Figure 3-18. *Using a matrix as a data source*

To import this data into the data model, you will need to unpivot the data to get it in a tabular form. In the Query Editor, select the columns that need to be unpivoted (see Figure 3-19).

	Carrier	10/1/2012	10/2/2012	10/3/2012	10/4/2012	10/5/2012
1	AA	27579	18675	19041	21242	21115
2	DL	18763	11226	15112	16914	12681
3	UA	13784	11043	20777	20327	16497
4	US	3242	4083	7362	7567	4443

Figure 3-19. *Selecting columns to unpivot*

On the Transform tab, select the Unpivot Columns transform. Once the data is unpivoted, you will get an Attribute column from the original column headers and a Value column (see Figure 3-20). You should rename these columns and change the data type before importing the data.

	Carrier	Attribute	Value
1	AA	10/1/2012	27579
2	AA	10/2/2012	18675
3	AA	10/3/2012	19041
4	AA	10/4/2012	21242
5	AA	10/5/2012	21115
6	DL	10/1/2012	18763
7	DL	10/2/2012	11226
8	DL	10/3/2012	15112
9	DL	10/4/2012	16914
10	DL	10/5/2012	12681
11	UA	10/1/2012	13784
12	UA	10/2/2012	11043
13	UA	10/3/2012	20777
14	UA	10/4/2012	20327
15	UA	10/5/2012	16497
16	US	10/1/2012	3242
17	US	10/2/2012	4083
18	US	10/3/2012	7362
19	US	10/4/2012	7567
20	US	10/5/2012	4443

Figure 3-20. *Resulting rows from the unpivot transformation*

As you bring data into the model, you often don't need the detail-level data; instead, you need an aggregate value at a higher level—for example, product level sales or monthly sales. Using Power Query, you can easily group and aggregate the data before importing it.

Grouping and Aggregating Data

The need to group and aggregate data is a common scenario you may run into when importing raw data. For example, you may need to roll the data up by month or sales territory, depending on the analysis you want. To aggregate and group the data in the Query Editor, select the column you want to group by and select the Group By transform in the Home tab. You are presented with a Group By window (see Figure 3-21).

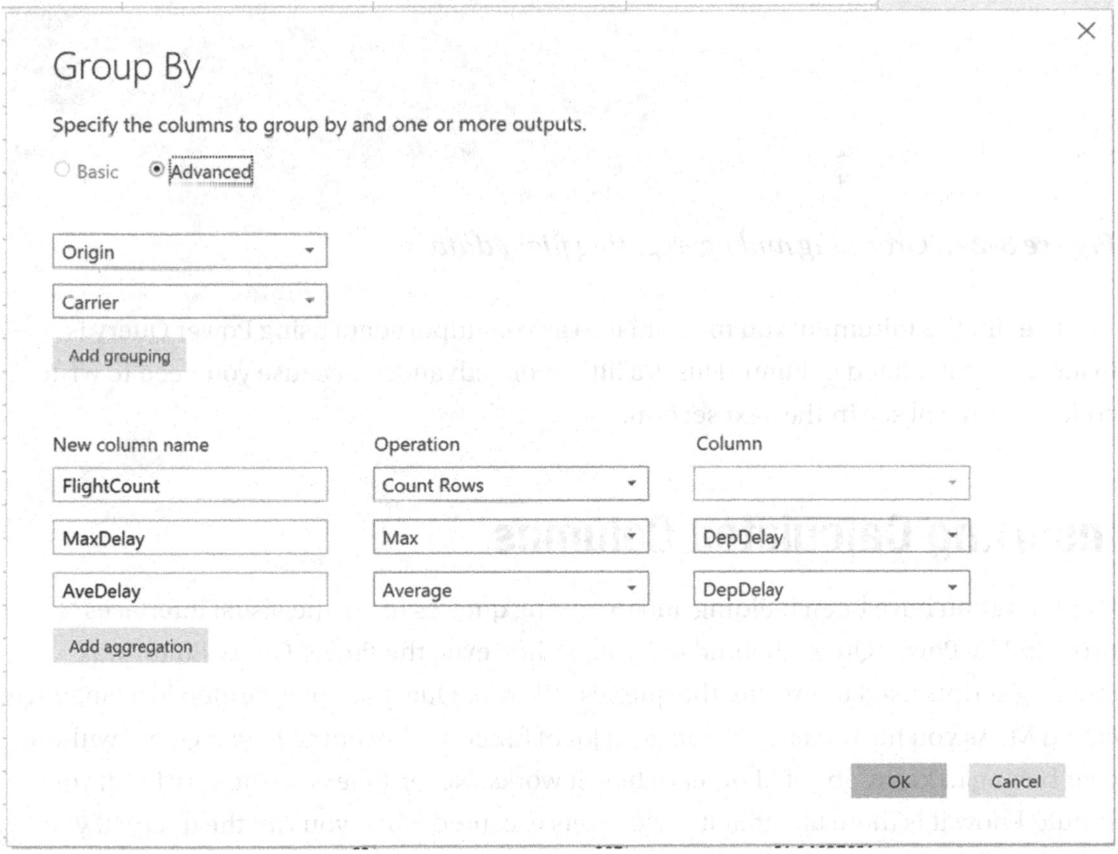

Figure 3-21. *Grouping data in Power Query*

You can group by multiple columns and aggregate multiple columns using the standard aggregate functions. Figure 3-22 shows some of the results from grouping by origin and carrier and aggregating the average and maximum departure delays.

	Origin	Carrier	FlightCount	MaxDelay	AveDelay
1	MSP	DL	1141	336	26.44785276
2	ATL	DL	6339	820	19.91717937
3	DEN	DL	233	488	34.63948498
4	JFK	DL	397	345	25.64735516
5	MOT	DL	11	40	12
6	DTW	DL	1090	433	25.95963303
7	RDU	DL	97	327	23.03092784
8	SFO	DL	226	307	31.02212389
9	LAS	DL	226	493	24.82300885
10	MIA	DL	170	333	26.50588235
11	PBI	DL	126	247	26.82539683
12	MKE	DL	92	162	20.06521739
13	PDX	DL	72	690	45.375
14	SLC	DL	644	479	21.32453416
15	SDF	DL	56	566	44.44642857
16	CVG	DL	183	327	20.96174863
17	SNA	DL	44	93	16.22727273
18	LAX	DL	586	447	20.34812287
19	ELP	DL	15	83	24.86666667
20	SAN	DL	86	479	28.40697674
21	JAX	DL	122	155	22.26229508
22	PWM	DL	39	67	10.97435897

Figure 3-22. *Grouping and aggregating flight data*

The final requirement you may run into as you import data using Power Query is inserting a calculated column. This is a little more advanced because you need to write code, as you will see in the next section.

Inserting Calculated Columns

Until now you have been building and executing queries using the visual interfaces provided by Power Query. Behind the scenes, however, the Power Query Editor was creating scripts used to execute the queries. A Power Query script is written in a language called M. As you have seen, you can get a lot of functionality out of Power Query without ever having to know about M or learn how it works. Nevertheless, at the very least, you should know it is there and that it is what gets executed when you run the query. If you navigate to the View tab in the Query Editor, you will see an option to open the Advanced Editor, which will display the M code used to build the query (see Figure 3-23).

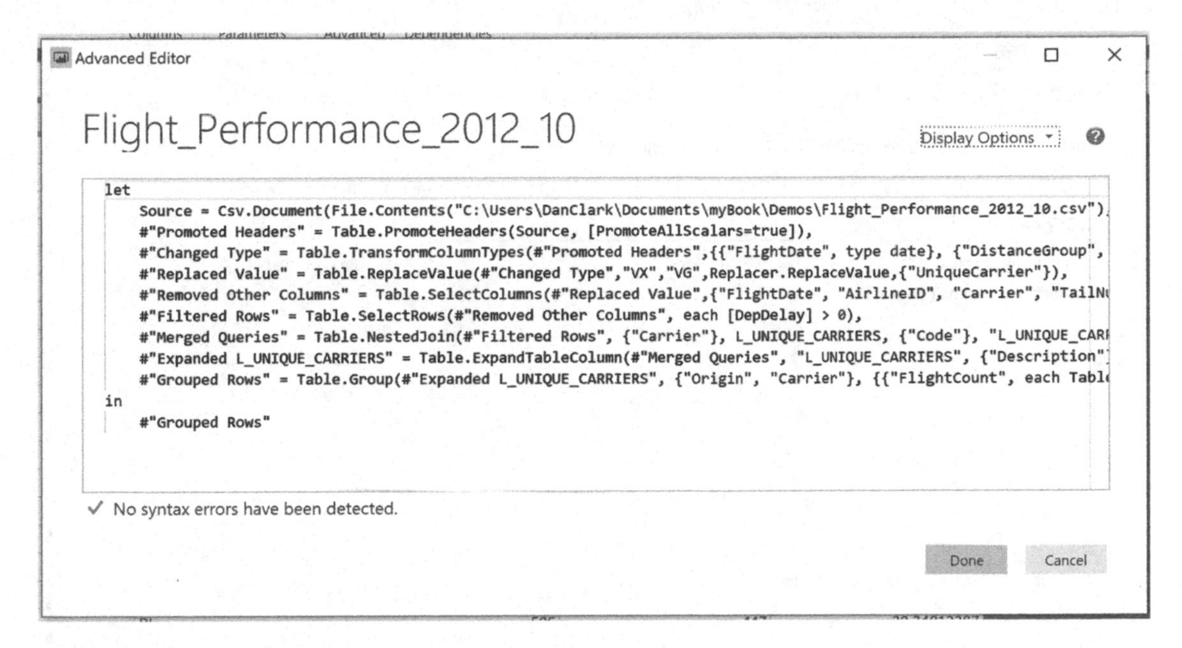
Advanced Editor
Flight_Performance_2012_10
Display Options
let
Source = Csv.Document(File.Contents("C:\Users\DanClark\Documents\myBook\Demos\Flight_Performance_2012_10.csv"),
#"Promoted Headers" = Table.PromoteHeaders(Source, [PromoteAllScalars=true]),
#"Changed Type" = Table.TransformColumnTypes(#"Promoted Headers",{{"FlightDate", type date}, {"DistanceGroup",
#"Replaced Value" = Table.ReplaceValue(#"Changed Type","VX","VG",Replacer.ReplaceValue,{"UniqueCarrier"}),
#"Removed Other Columns" = Table.SelectColumns(#"Replaced Value",{"FlightDate", "AirlineID", "Carrier", "TailN
#"Filtered Rows" = Table.SelectRows(#"Removed Other Columns", each [DepDelay] > 0),
#"Merged Queries" = Table.NestedJoin(#"Filtered Rows", {"Carrier"}, L_UNIQUE_CARRIERS, {"Code"}, "L_UNIQUE_CAR
#"Expanded L_UNIQUE_CARRIERS" = Table.ExpandTableColumn(#"Merged Queries", "L_UNIQUE_CARRIERS", {"Description"
#"Grouped Rows" = Table.Group(#"Expanded L_UNIQUE_CARRIERS", {"Origin", "Carrier"}, {{"FlightCount", each Tabl
in
#"Grouped Rows"
No syntax errors have been detected.
Done
Cancel

Figure 3-23. *Building a query with M*

You can use the Advanced Editor to write the query directly with M, thereby exposing some advanced data processing not available in the visual interface tools.

If you want to insert a calculated column into the query, you need to use the M functions. On the Add Column tab of the Query Editor, you can duplicate columns, insert an index column, merge columns, and insert a custom column. When you select the Add Custom Column option, you are presented with an Add Custom Column editor where you insert the M function used to create the column. For example, in Figure 3-24, we are checking to see if the Carrier column has a Q in the name.

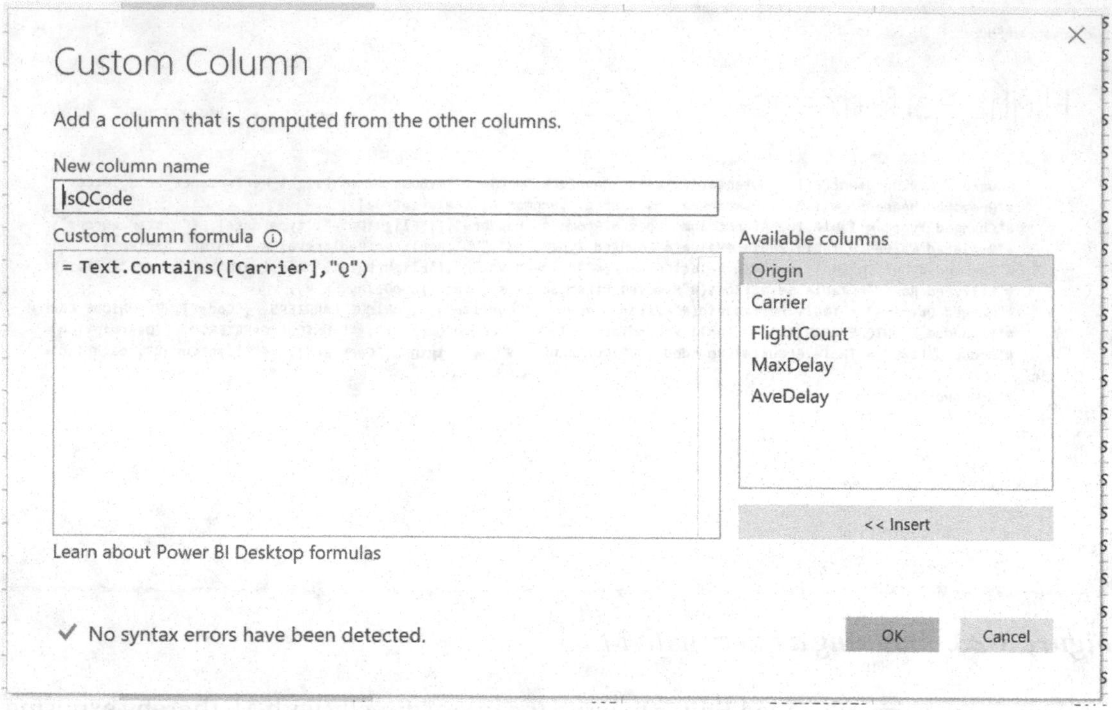

Custom Column

Add a column that is computed from the other columns.

New column name

IsQCode

Custom column formula

```
= Text.Contains([Carrier],"Q")
```

Available columns

Origin
Carrier
FlightCount
MaxDelay
AveDelay

<< Insert

Learn about Power BI Desktop formulas

✓ No syntax errors have been detected.

OK Cancel

Figure 3-24. *Creating columns using M formulas*

Figure 3-25 shows the results of the query with the custom column added.

	Origin	Carrier	IsQCode
1	ABE	F9	FALSE
2	ABE	YV	FALSE
3	ABE	EV	FALSE
4	ABI	MQ	TRUE
5	ABQ	EV	FALSE
6	ABQ	DL	FALSE
7	ABQ	AA	FALSE
8	ABQ	UA	FALSE
9	ABQ	WN	FALSE
10	ABQ	MQ	TRUE
11	ABQ	YV	FALSE
12	ABQ	OO	FALSE
13	ABQ	US	FALSE
14	ABR	OO	FALSE
15	ABY	EV	FALSE
16	ACK	B6	FALSE

***Figure 3-25.** Displaying the results of the query*

So, if you need to create columns using Power Query before inserting the data into the Power Pivot model, you use M code. If you create the columns after importing the data into the Power Pivot model, you use DAX code (covered in later chapters).

Now that you have seen how Power Query works, it is time to get some hands-on experience using it to import, cleanse, and shape data.

HANDS-ON LAB: IMPORTING AND SHAPING DATA WITH POWER QUERY

In the following lab, you will

- Create a query to import data
- Filter and transform data
- Append and shape data
- Group and aggregate data

1. In the LabStarterFiles\Chapter3Lab1 folder, create a Power BI Desktop file called Chapter3Lab1.pbix.

2. On the Home tab, select Edit Queries to launch the Power Query Editor.

3. On the Home tab, select the New Source drop-down and choose the Text/CSV option. Navigate to the FlightPerformance_2012_10.csv file in the LabStarterFiles\Chapter3Lab1 folder. Click OK to load the query into the editor. After loading the file, you should see the Query Editor window with airline delay data, as shown in Figure 3-26.

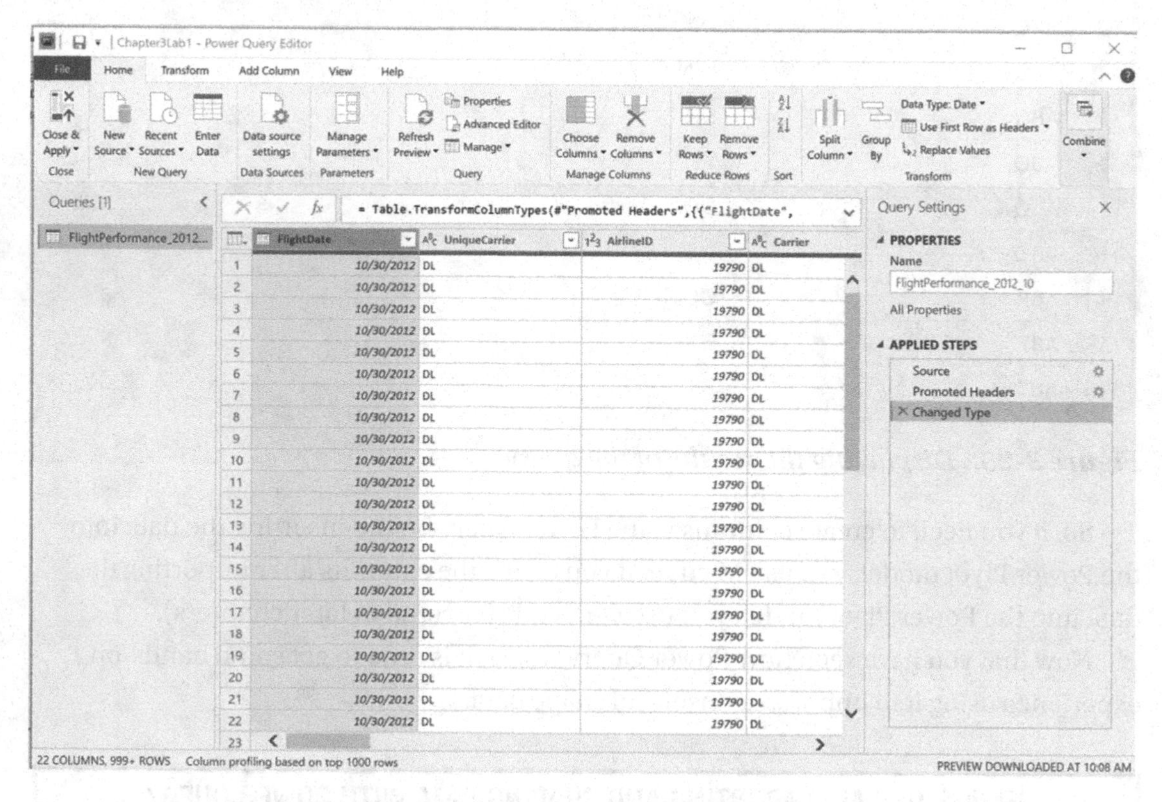

Figure 3-26. *The Query Editor window with delay data*

4. In the Query Settings pane, rename the query to `FlightDelays`.

5. In the Applied Steps list, if the Query Editor did not automatically add an item labeled "Promoted headers" that transformed the first row to headers, add it now.

6. Check the types of each column to see whether the Query Editor updated the FlightDate to a Date data type and the number type columns ("FlightNum" for example) to a Number data type. If it did not, change them now by right-clicking the column headers and choosing Change Type from the context menu.

7. On the Home tab, select Choose Columns. Clear all the selections and then select just the Carrier, Origin, OriginCityName, and DepDelay columns.

8. Use the DepDelay column and apply a filter so that the query only pulls rows that have a flight departure delay of greater than 15 minutes. Choose the drop-down arrow in the column heading and use the "Number filters" option.

9. To pull data from another month, complete steps 2–8 for the FlightPerformance_2012_11.csv file, except this time, name the query `FlightDelays2`.

10. In the Queries pane on the left side of the editor, right-click the query FlightDelays2 and uncheck Enable load (see Figure 3-27).

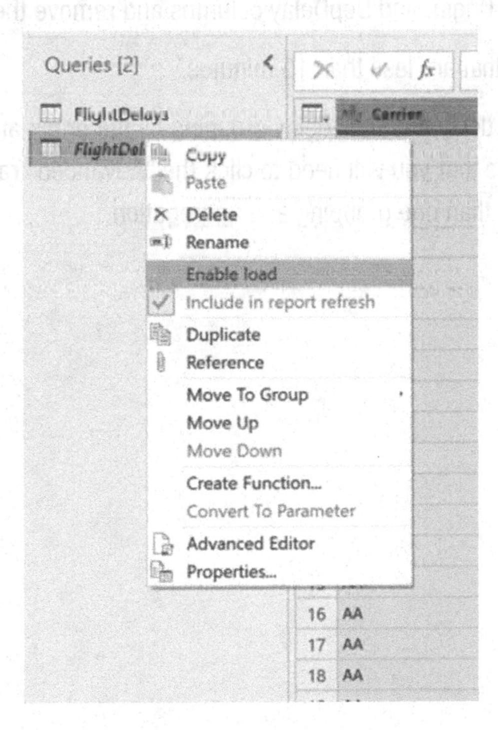

Figure 3-27. *Disabling loading the query to the model*

11. Select the `FlightDelays` query in the Queries list.
12. Select the Append Queries option in the Combine section of the Home tab. Append the `FlightDelays2` query to the `FlightDelays` query.
13. Note that the OriginCityName contains both a city and a state. To split that column into two, select the OriginCityName column, and on the Transform tab, select Split Column. Split it into two columns, using the comma as the delimiter.
14. Rename the two resulting columns OriginCity and OriginState.
15. By right-clicking the OriginCity column, replace West Palm Beach/Palm Beach in the OriginCity column with just Palm Beach by choosing Replace values.
16. On the Home tab, select the New Source drop-down and choose the Text/CSV option. Connect to the FlightPerformance_2012_12.csv in the LabStarterFiles\Chapter3Lab1 folder.
17. In the Query Settings pane, rename the query to `DelaySummary`.
18. Keep the Carrier, Origin, and DepDelay columns and remove the rest.
19. Filter out delays that are less than 15 minutes.
20. Find the average delay and max delay grouping by the origin and carrier (see Figure 3-28). Note that you will need to click the "Advanced" radio button to be able to add more than one grouping and aggregation.

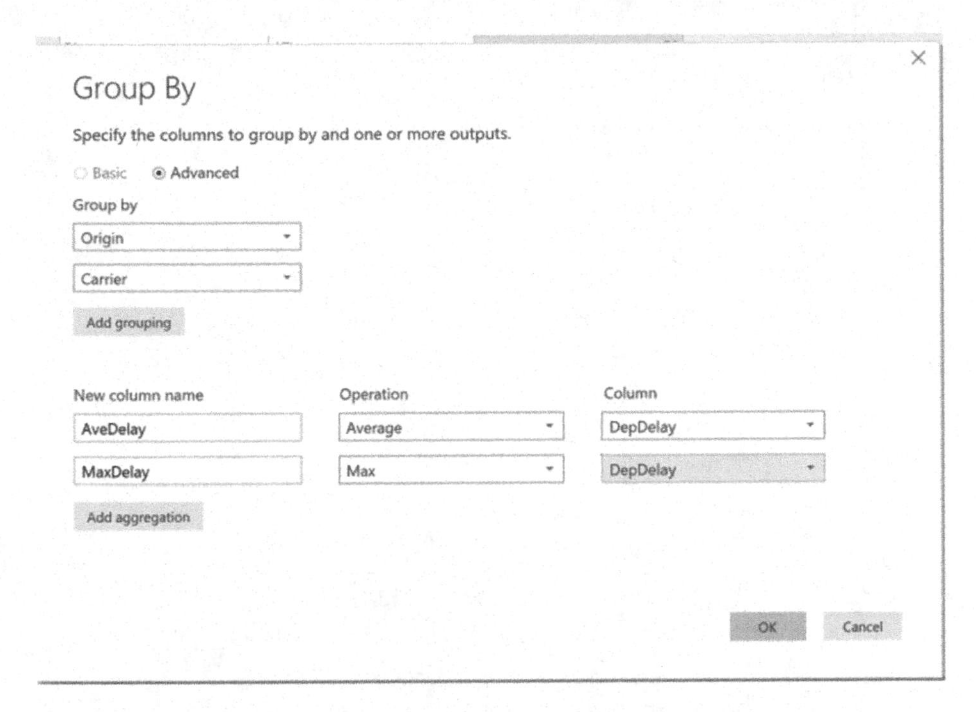

Figure 3-28. *Grouping by origin and carrier*

21. On the Transform tab, use rounding to round down the AveDelay to the nearest minute.

22. Add a query called `Carriers` that gets the code and description from the Carriers.csv file. Verify that the first row was promoted as a header for the data.

23. Select the `DelaySummary` query and select the Merge Queries option on the Home tab. Using the `Carriers` query, merge the Description column matching the Carrier column in the `DelaySummary` query with the Code column in the `Carriers` query (see Figure 3-29).

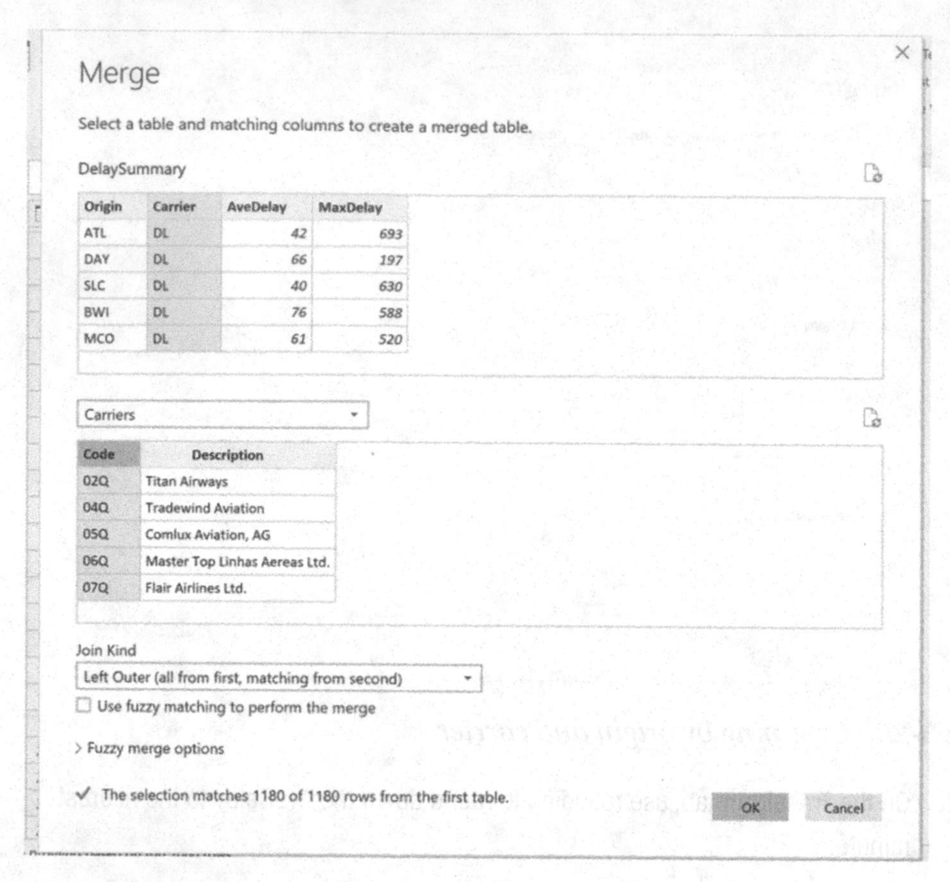

Figure 3-29. *Merging data from two queries*

24. Expand the Carriers column and select the Description field (see Figure 3-30).

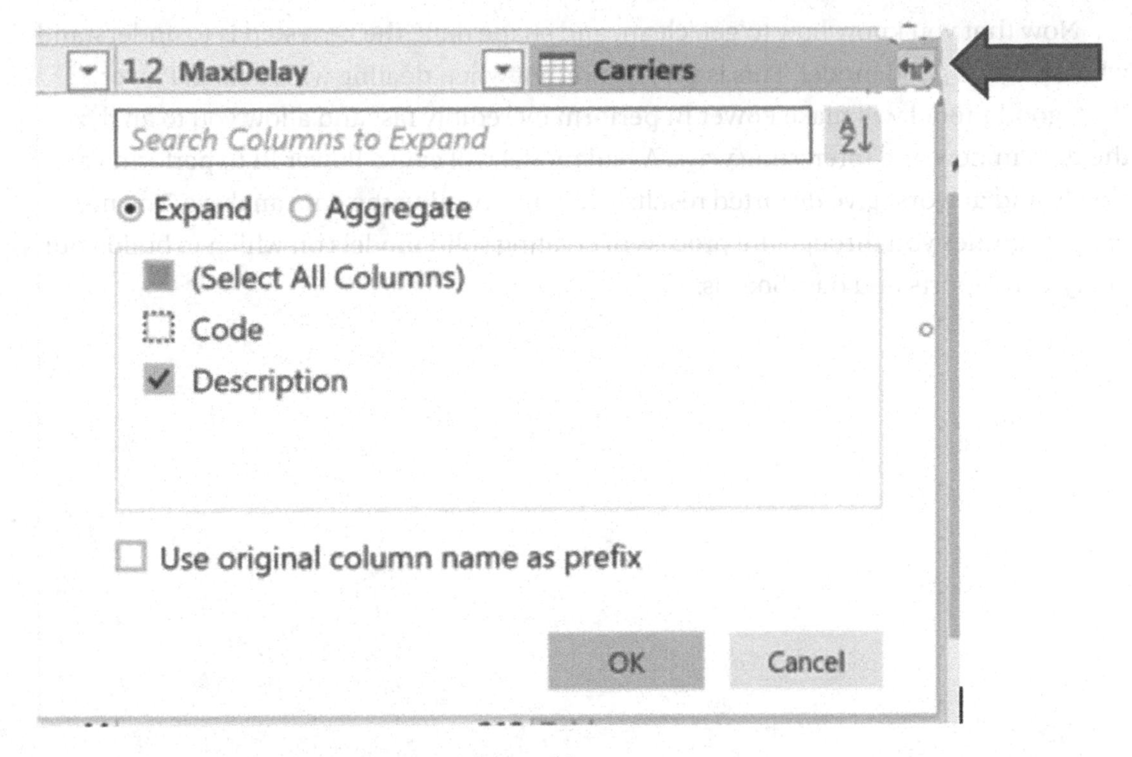

***Figure 3-30.** Expanding the Carriers table*

25. Rename the Carrier column to *CarrierCode* and the Description column to *Carrier*.
26. Select Close & Apply on the Home tab.
27. Verify that the Carriers, DelaySummary, and FlightDelays tables were added to the Power BI model.
28. Save the file and close Power BI Desktop.

Summary

Power Query is a very helpful tool you can use to get data from many different types of sources. In this chapter, you learned how to employ Power Query to do some data munging—shaping, cleansing, and transforming data, applying intuitive interfaces without having to use code. Power Query builds the code for you but doesn't hide it from you. If you need to alter or enhance the code, you can make use of the Advanced Editor view. Although this chapter only touched on the M query language, we will return to this topic in Chapter 13, where we will dive deeper into it.

Now that you know how to get, clean, and shape data, the next step is to understand what makes a good model. This is very important when dealing with data in Power BI. A good model will make Power BI perform incredibly fast and allow you to analyze the data in new and interesting ways. A bad model will cause Power BI to perform very slowly and at worst give distorted results when performing the data analysis. The next chapter guides you through the process of creating solid models on which to build your analytical reports and dashboards.

CHAPTER 4

Creating the Data Model

Now that you know how to get data into the Power BI model, the next step is to understand what makes a good model. This is very important when dealing with data in Power BI. A good model will make Power BI perform amazingly fast and allow you to analyze the data in new and interesting ways. A bad model will cause Power BI to perform very slowly and at worst give misleading results when performing the data analysis. This chapter guides you through the process of creating a solid model that will become the foundation for your data analysis. In addition, you will look at how to present a user-friendly model to ease the development of reports. This includes renaming tables and fields, presenting appropriate data types, and hiding extraneous fields.

After completing this chapter, you will be able to

- Explain what a data model is
- Create relationships between tables in the model
- Create and use a star schema
- Understand when and how to denormalize the data
- Create and use hierarchies
- Make a user-friendly model

What Is a Data Model?

Fundamentally a *data model* is made up of tables, columns, data types, and table relations. Typically, data tables are constructed to hold data for a business entity; for example, customer data is contained in a customer table and employee data is contained in an employee table. Tables consist of columns that define the attributes of the entity. For example, you may want to hold information about customers such as name, address,

D. Clark, *Beginning Microsoft Power BI*, https://doi.org/10.1007/978-1-4842-5620-6_4

birth date, household size, and so on. Each of these attributes has a data type that depends on what information the attribute holds—the name would be a string data type, the household size would be an integer, and the birth date would be a date. Each row in the table should be unique. Take a customer table, for example; if you had the same customer in multiple rows with different attributes, say birth date, you would not know which was correct.

In the previous example, you would know that one of the rows was incorrect because the same person could not have two different birthdays. There are many times, however, when you want to track changes in attribute values for an entity. For example, a product's list price will probably change over time. To track the change, you need to add a time stamp to make the row unique. Then each row can be identified by the product number and time stamp, as shown in Figure 4-1.

	ProductNumber	StandardCost	ListPrice	ProductLine	DealerPrice	ModelName	StartDate
1	FR-R38B-58	176.1997	297.6346	R	178.5808	LL Road Frame	2005-07-01 00:00:00.000
2	FR-R38B-58	170.1428	306.5636	R	183.9382	LL Road Frame	2006-07-01 00:00:00.000
3	FR-R38B-58	204.6251	337.22	R	202.332	LL Road Frame	2007-07-01 00:00:00.000

***Figure 4-1.** Using a time stamp to track changes*

Once you have the tables of the model identified, it is important that you recognize whether the tables are set up to perform efficiently. This process is called *normalizing* the model. Normalization is the process of organizing the data to make data querying easier and more efficient. For example, you should not mix attributes of unrelated entities together in the same table—you would not want product data and employee data in the same table. Another example of proper normalization is to not hold more than one attribute in a column. For example, instead of having one customer address column, you would break it up into street, city, state, and zip code. This would allow you to easily analyze the data by state or by city. The spreadsheet shown in Figure 4-2 shows a typical non-normalized table (this is one continuous table split to visualize it on the page). If you find that the data supplied to you is not sufficiently normalized, you can use Power Query to break the data up into multiple tables that you can then relate together in your model.

Customer Name	Address
Tom Smith	128 Elm St. Littleton, PA 12555
Jannet Jones	1399 Firestone Drive, San Francisco, CA 94109
Jon Yang	9539 Glenside Dr, Phoenix AZ 85004
Tom Smith	9707 Coldwater Drive, Orlando, FL 32804

Order Date	Item 1	Item 1 Description	Item 1 Price	Item 2	Item 2 Description	Item 2 Price
7/23/2013	FR-R38B-58	LL Road Frame	183	HB-R504	LL Road Handlebars	26.75
7/23/2013	FW-R623	Road Front Wheel	85.5			
8/1/2013	RB-9231	Rear Brakes	21.98			
8/1/2013	FW-R623	LL Road Front Wheel	85.5	RB-9231	Breaks	21.98

Figure 4-2. *A non-normalized table*

Once you are satisfied the tables in your model are adequately normalized, the next step is to determine how the tables are related. For example, you might need to relate the customer table, sales table, and product table to analyze how various products are selling by age group. The way you relate the different tables is by using *keys*. Each row in a table needs a column or a combination of columns that uniquely identifies the row. This is called the *primary key*. The key may be easily identified, such as a sales order number or a customer number that has been assigned by the business when the data was entered. Sometimes you will need to do some analysis of a table to find the primary key, especially if you get the data from an outside source. For example, you may get data that contains potential customers. The fields are name, city, state, zip, birth date, and so on. You cannot just use the name as the key because it is very likely that you have more than one customer with the same name. If you use the combination of name and city, you have less of a chance of having more than one customer identified by the same key. As you use more columns, such as zip and birth date, your odds get even better.

When you go to relate tables in the model, the primary key from one table becomes a *foreign* key in the related table. For example, to relate a customer to their sales, the customer key needs to be contained in the sales table where it is considered a foreign key. When extracting the data, the keys are used to get the related data. By far, the best type of key to use for performance reasons is a single column integer. For this reason, a lot of database tables are designed with a surrogate key. This key is an integer that gets assigned to the record when it is loaded. Instead of using the natural key, the surrogate is used to connect the tables. Figure 4-3 shows a typical database table containing both the surrogate key (CustomerKey) and the natural key (CustomerAlternateKey).

CustomerKey	GeographyKey	CustomerAlternateKey	Title	FirstName	MiddleName	LastName	NameStyle	BirthDate
11000	26	AW00011000	NULL	Jon	V	Yang	0	1966-04-08
11001	37	AW00011001	NULL	Eugene	L	Huang	0	1965-05-14
11002	31	AW00011002	NULL	Ruben	NULL	Torres	0	1965-08-12
11003	11	AW00011003	NULL	Christy	NULL	Zhu	0	1968-02-15
11004	19	AW00011004	NULL	Elizabeth	NULL	Johnson	0	1968-08-08
11005	22	AW00011005	NULL	Julio	NULL	Ruiz	0	1965-08-05
11006	8	AW00011006	NULL	Janet	G	Alvarez	0	1965-12-06
11007	40	AW00011007	NULL	Marco	NULL	Mehta	0	1964-05-09
11008	32	AW00011008	NULL	Rob	NULL	Verhoff	0	1964-07-07
11009	25	AW00011009	NULL	Shannon	C	Carlson	0	1964-04-01

Figure 4-3. *A table containing both a surrogate key and a natural key*

It is important that you are aware of the keys used in your sources of data. If you can retrieve the keys from the source, you are much better off. This is usually not a problem when you are retrieving data from a relational database, but if you are combining data from different sources, make sure you have the appropriate keys.

Once you have the keys between the tables identified, you are ready to create the relationships in the Power BI model.

Creating Table Relations

There are a few rules to remember when establishing table relationships in a Power BI model. First, you can't use composite keys in the model. If your table uses a composite key, you will need to create a new column by concatenating the composite columns together and using this column as the key. Second, you can only have one active relationship path between two tables, but you can have multiple inactive relationships. Third, relationships are usually one-to-many; in other words, creating a relationship between the customer table (one side) and the sales table (many sides) is common and works well. Sometimes however, you need to create a many-to-many relationship. For example, consider customers and products, the customer can buy many products, and the same product can be bought by many customers. In these cases, it is a best practice to create a junction table to connect the tables together.

To create a relationship between two tables in the Power BI model, you open Power BI Desktop and switch to the Model view. In the Home tab, select Manage relationships. This launches the Manage relationships window (see Figure 4-4).

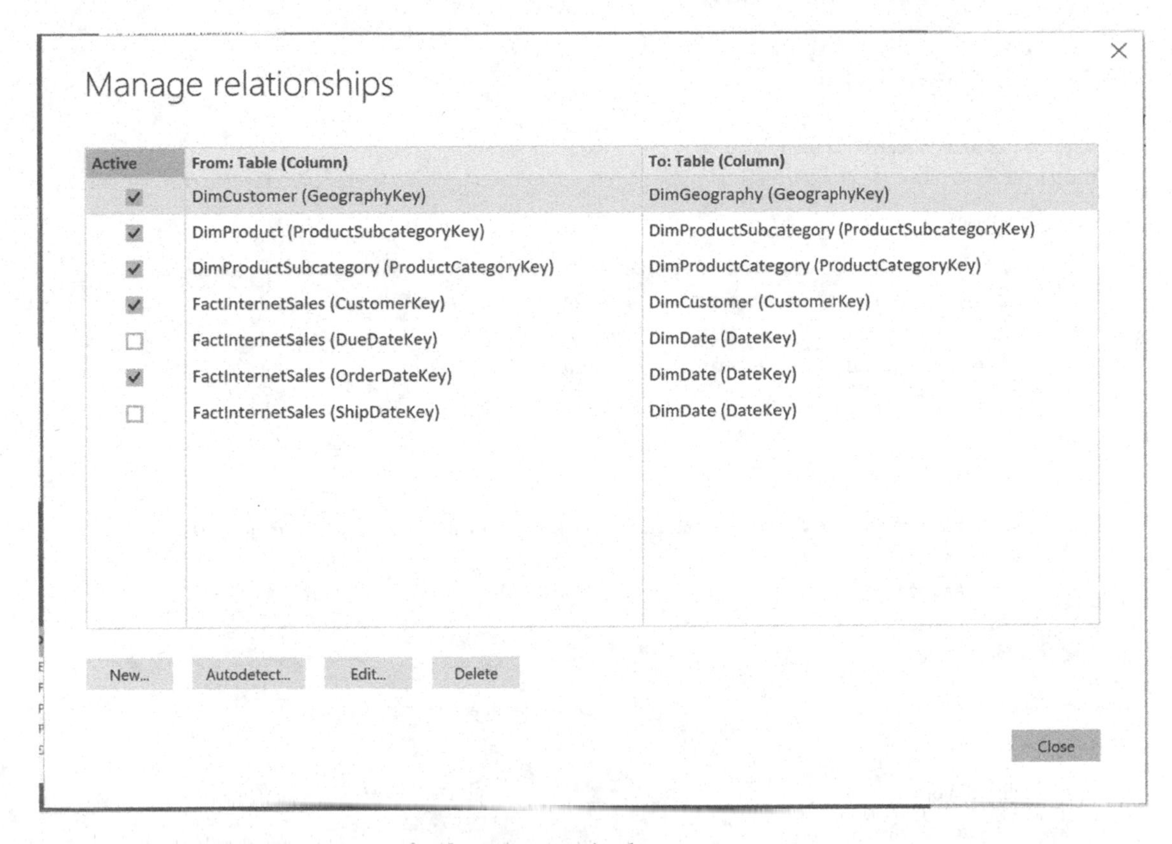

Figure 4-4. *The Manage relationships window*

This window shows the relationships currently defined in the model. You can create a new relationship, edit an existing relationship, and delete a relationship. Be careful selecting the Autodetect button; this may overwrite the existing relationships and could give erroneous results.

Selecting the New button in the Manage relationships window launches the Create relationship window (see Figure 4-5). Choose the related tables and the key column in each table. Figure 4-5 shows creating a relationship between the FactInternetSales table and the DimProduct table. Note that the "ProductKey" column in each table is highlighted to indicate the relationship. By default, the first relationship created between the tables is marked as the active relationship.

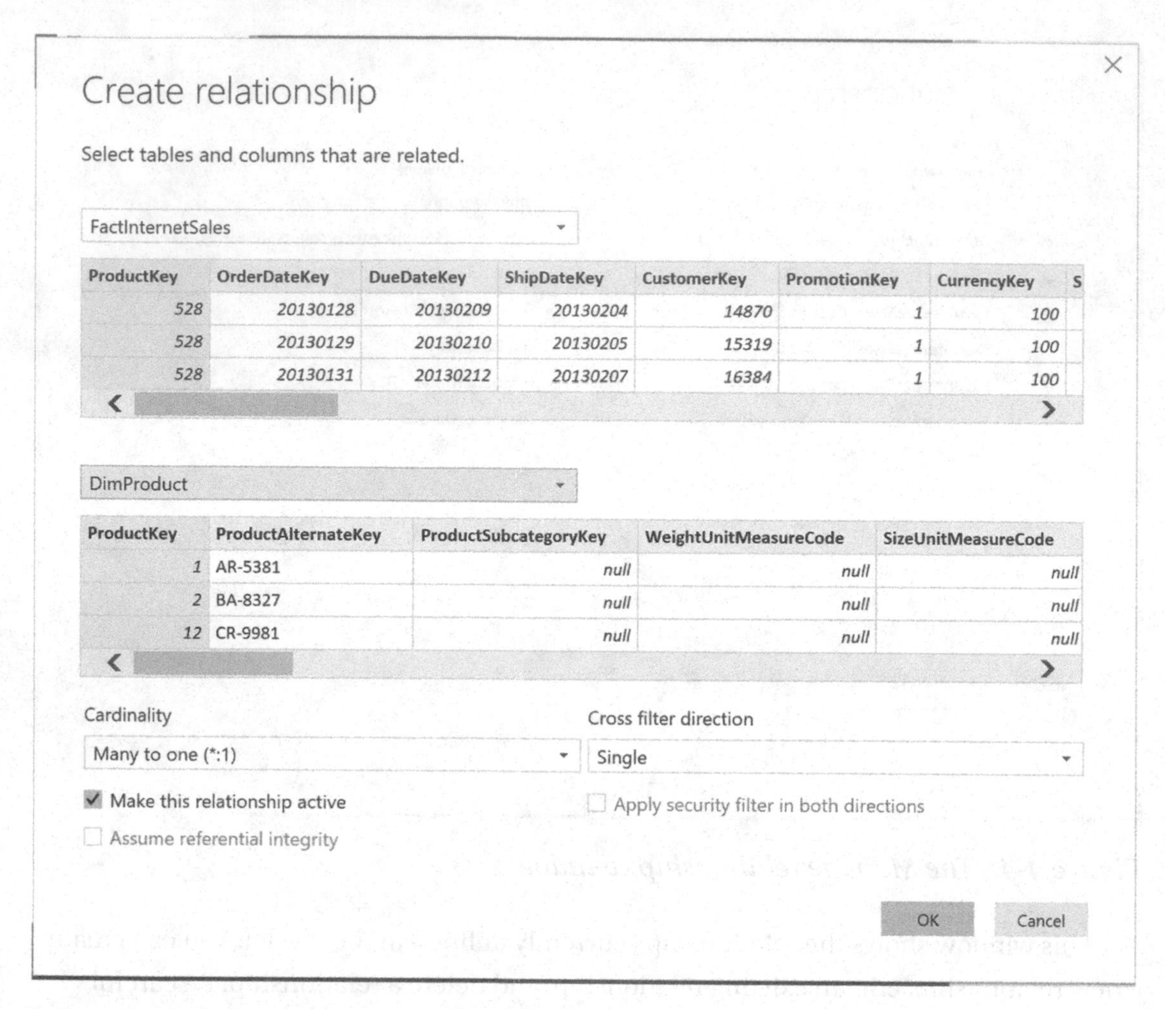

***Figure 4-5.** Creating a table relationship*

Figure 4-6 shows the resulting relationship in the Diagram View. If you hover the mouse over the relationship arrow, the two key columns of the relationship are highlighted. The * indicates the *many* side of the relationship. For each row in the DimProduct table, there can be many related rows in the FactInternetSales table.

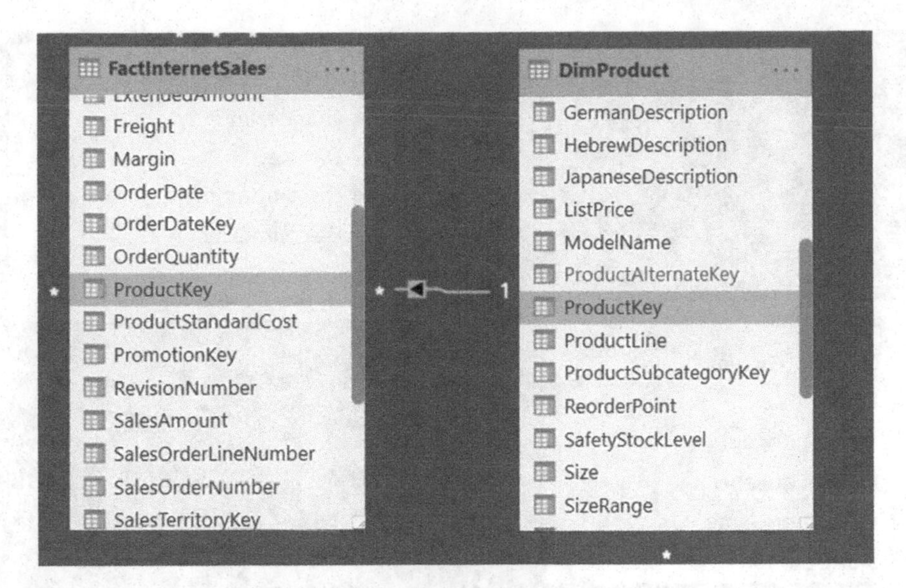

Figure 4-6. *Viewing a relationship in the Diagram View*

You can create more than one relationship between two tables, but remember, only one can be the active relationship. If you try to make two active relationships, you'll get an error like the one shown in Figure 4-7.

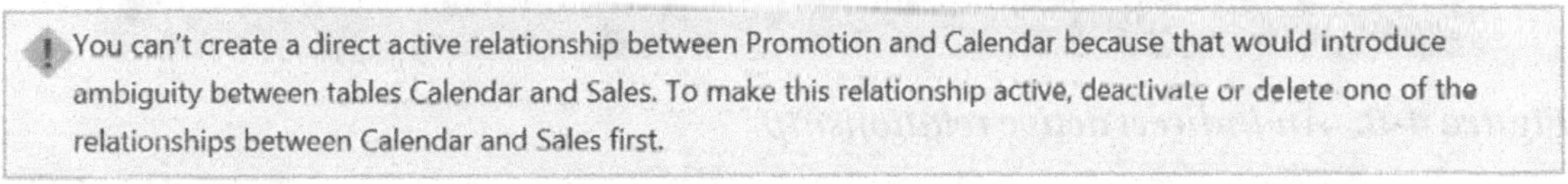

You can't create a direct active relationship between Promotion and Calendar because that would introduce ambiguity between tables Calendar and Sales. To make this relationship active, deactivate or delete one of the relationships between Calendar and Sales first.

Figure 4-7. *Trying to create a second active relationship between two tables*

Sometimes the active relationship is not so obvious. Figure 4-8 shows an active relationship between the Sales table and the Calendar table and another one between the Sales table and the Promotions table. There is also an inactive relationship between the Promotions table and the Calendar table. If you try to make this one active, you get the same error message shown in Figure 4-7. This is because you can trace an active path from the Calendar table to the Sales table to the Promotions table.

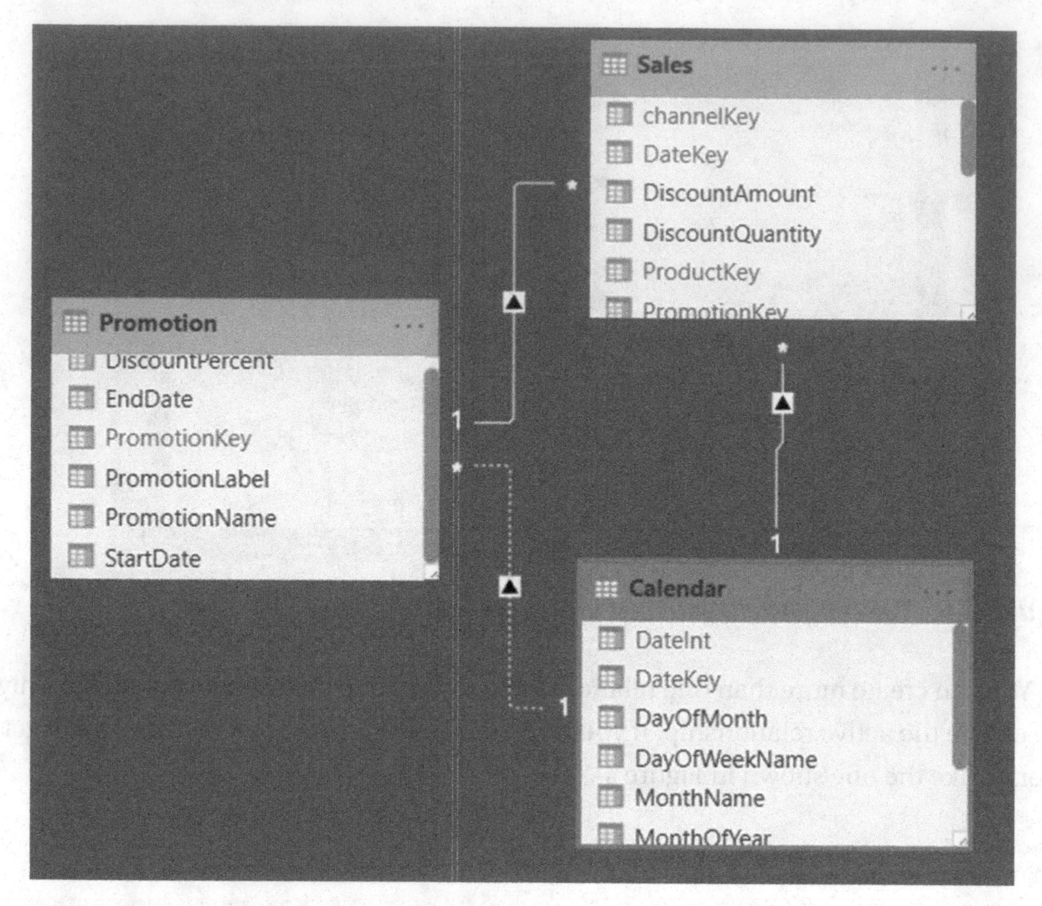

Figure 4-8. *An indirect active relationship*

Another common error you may run into is when you try to create a relationship between two tables and the key is not unique in at least one of the tables. Figure 4-9 shows a Flights table and a Carriers table that both contain duplicate Carrier Codes.

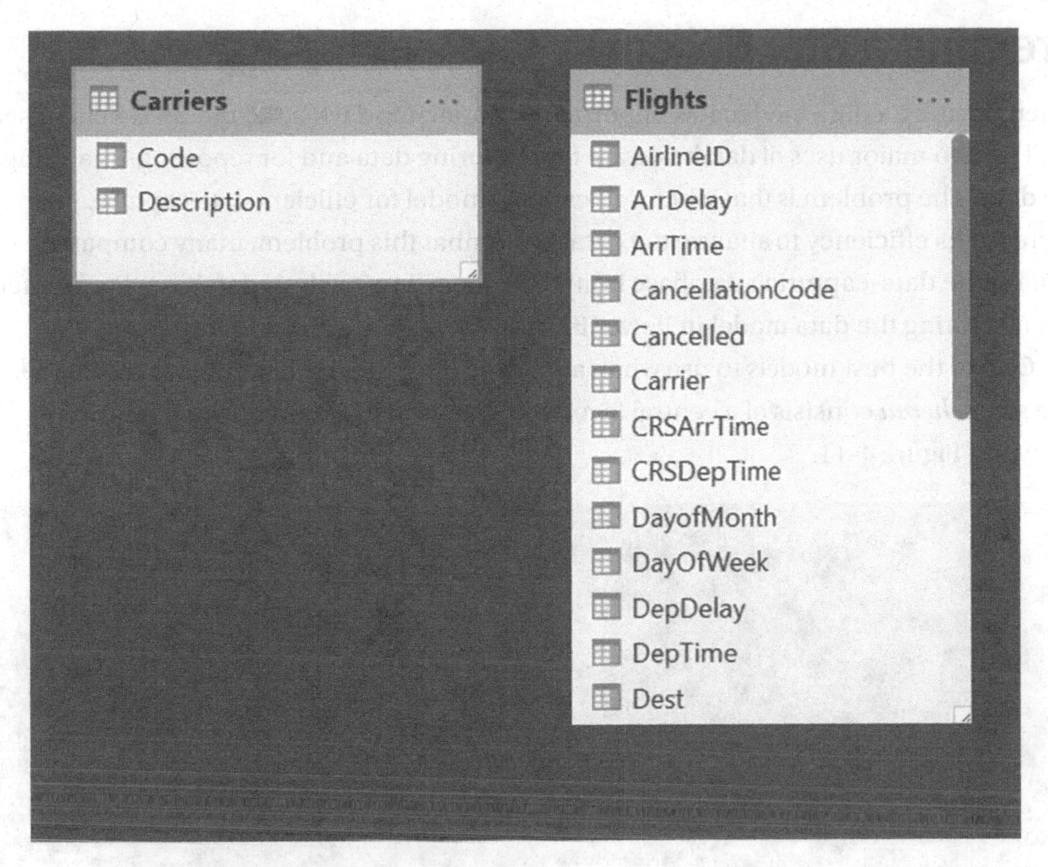

***Figure 4-9.** The Carriers and Flights tables*

When you try to create a relationship between the tables, you get the warning shown in Figure 4-10.

This relationship has cardinality Many-Many. This should only be used if it is expected that neither column (Code and Carrier) contains unique values, and that the significantly different behavior of Many-many relationships is understood. Learn more

***Figure 4-10.** Getting a duplicate value warning*

In this example, the Code column in the Carriers table is supposed to be unique, but it turns out there are duplicates. Even though Power BI can make this a many-to-many relationship, this is not what we want. To fix this, you would have to change the query for the Carriers table data to ensure that you are not getting duplicates.

Now that you know how to create table relationships in the model, you are ready to look at the benefits of using a star schema.

Creating a Star Schema

When creating a data model, it is important to understand what the model is being used for. The two major uses of databases are for capturing data and for reporting/analyzing the data. The problem is that when you create a model for efficient data capture, you decrease its efficiency to analyze the data. To combat this problem, many companies split off the data-capturing database from their reporting/analysis database. Fortunately, when creating the data model in Power BI, we only need to tune it for reporting.

One of the best models to use when analyzing large sets of data is the star schema. The *star schema* consists of a central fact table surrounded by dimension tables, as shown in Figure 4-11.

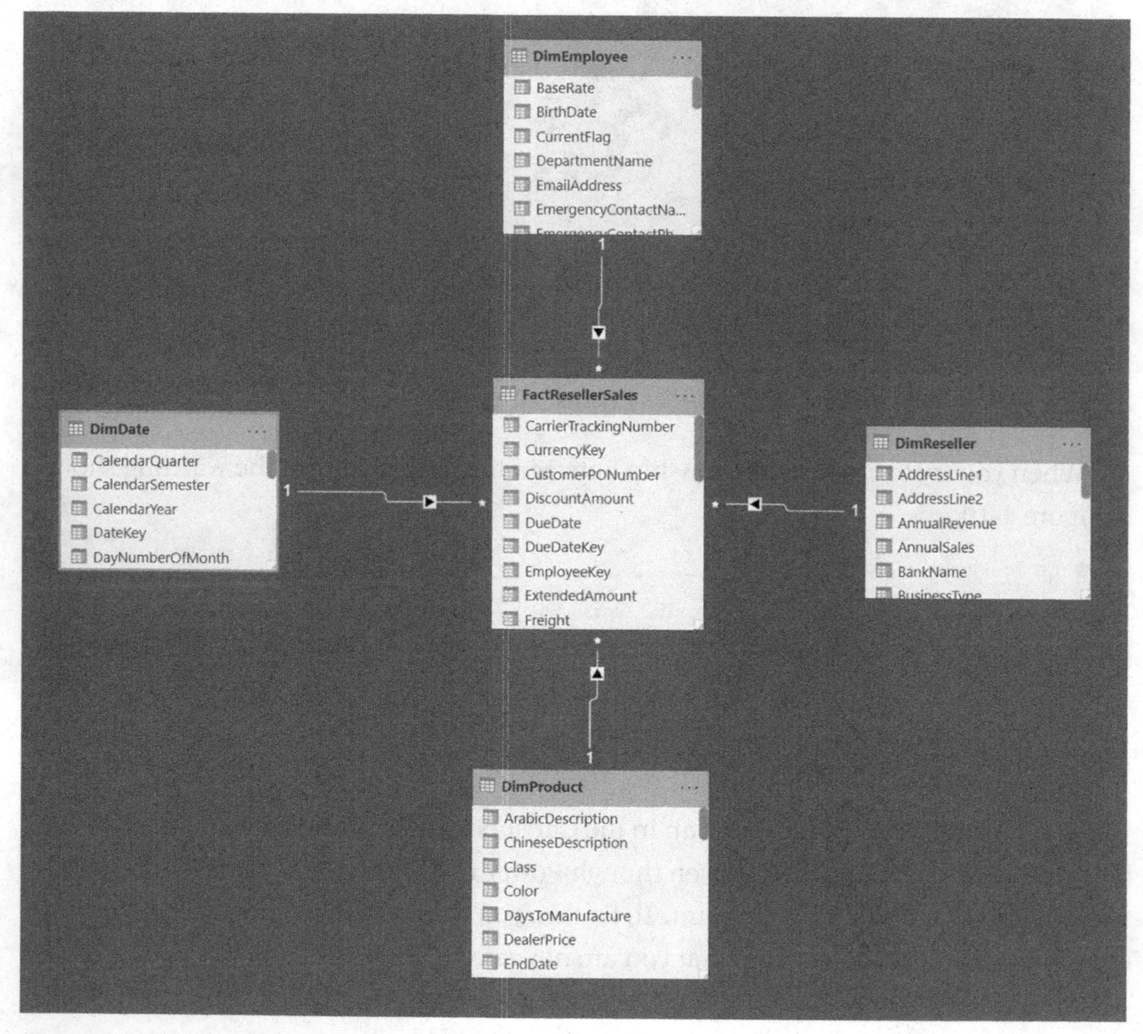

Figure 4-11. *A typical star schema*

The fact table contains quantitative data related to the business or process. For example, the Sales table in Figure 4-11 contains measurable aspects of a sale, such as total costs, sales amount, and quantity. Fact tables usually contain many rows and have a date or time component that records the time point at which the event occurred. The dimension tables contain attributes about the event. For example, the Date table can tell you when the sale occurred and allows you to roll up the data to the month, quarter, or year level. The Product table contains attributes about the product sold, and you can look at sales by product line, color, and brand. The Reseller table contains attributes about the store involved in the sale. Dimension tables usually don't contain as many rows as fact tables but can contain quite a few columns. When you ask a question like "Which bikes are selling the best in the various age groups?", the measures (sales dollar values) come from the fact table, whereas the categorizations (age and bike model) come from the dimension tables.

The main advantage of the star schema is that it provides fast query performance and aggregation processing. The disadvantage is that it usually requires a lot of preprocessing to move the data from a highly normalized transactional system to a more denormalized reporting system. The good news is that your business may have a reporting system feeding a traditional online analytical processing (OLAP) database such as Microsoft's Analysis Server or IBM's Cognos. If you can gain access to these systems, they are probably the best source for your core business data.

To create a star schema from your source data systems, you may have to perform some data denormalization, which is covered in the following section.

Understanding When to Denormalize the Data

Although transactional database systems tend to be highly normalized, reporting systems are denormalized into the star schema. If you don't have access to a reporting system where the denormalization is done for you, you will have to denormalize the data as you load it into the Power BI model. As an example, Figure 4-12 shows the tables that contain customer data in the Adventureworks transactional sales database.

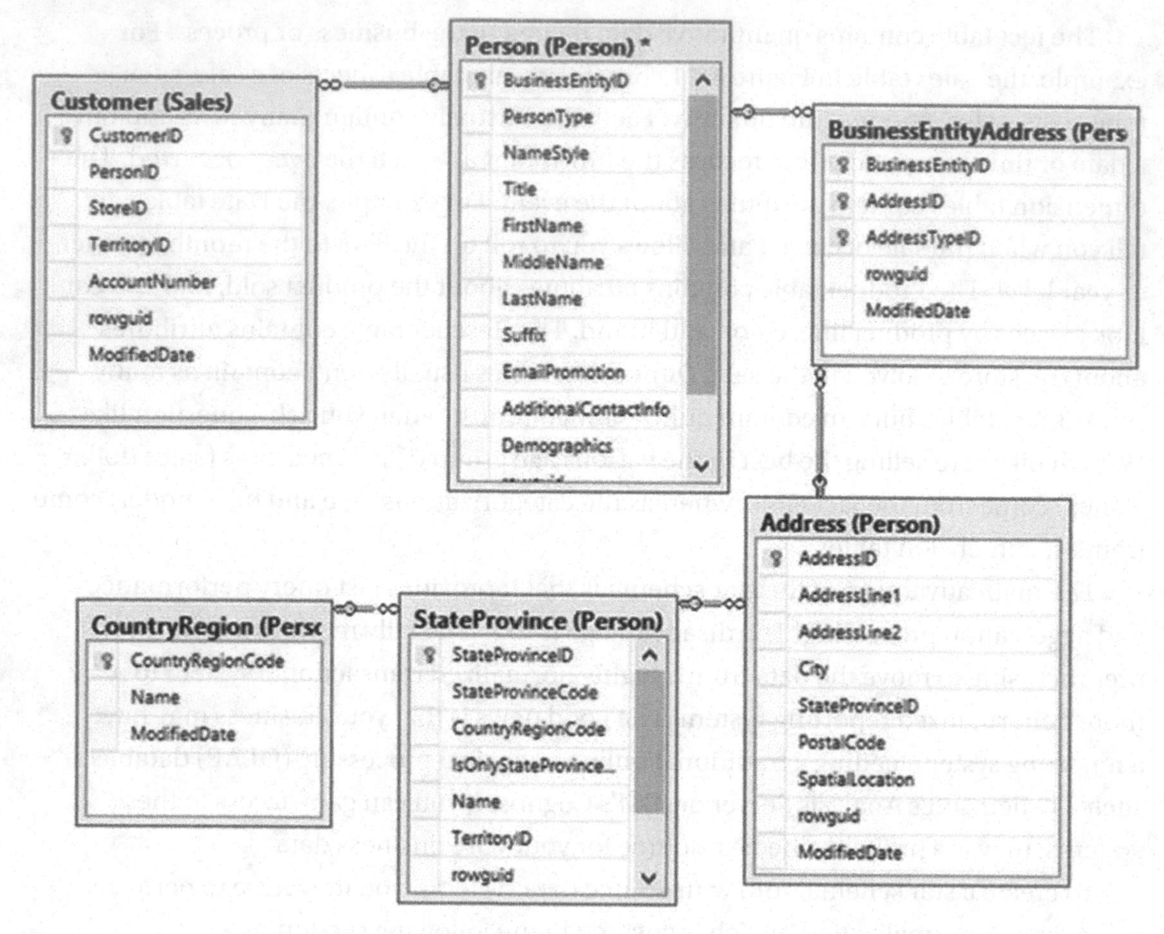

***Figure 4-12.** A highly normalized schema*

To denormalize the customer data into your model, you need to create a query that combines the data into a single customer dimension table. If you are not familiar with creating complex queries, the easiest way to do this is to have the database developers create a view you can pull from that combines the tables for you. If the query isn't too complex, you can create it yourself using Power Query. For example, Figure 4-13 shows a Customer table and a Geography table.

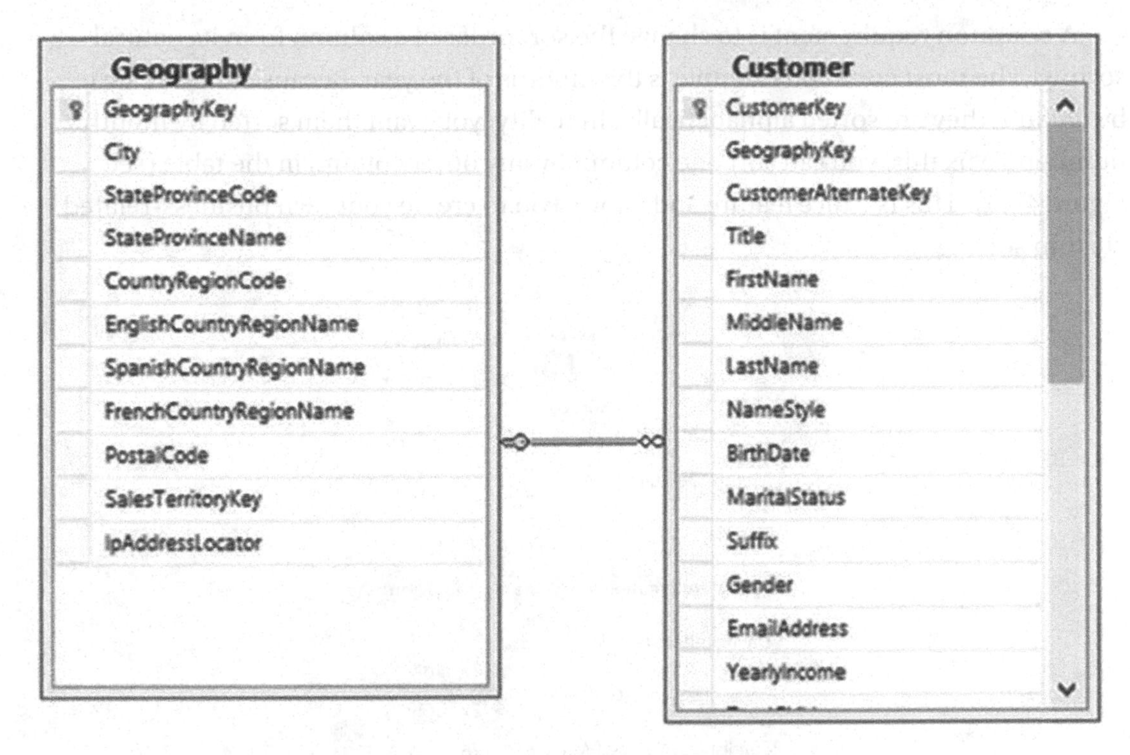

Figure 4-13. *Combining customer and customer location data*

You can combine these into one customer dimension table using the merge function in Power Query. Although you do not have to be a query expert to get data into your Power BI model, it is very beneficial to know the basics of querying the data sources, even if it just helps to ask the right questions when you talk to the database developers.

Making a User-Friendly Model

As you are creating your model, one thing to keep in mind is making the model easy and intuitive to use. Chances are that the model may get used by others for analysis. There are properties and settings you can use that will make your models more user-friendly.

One of the most effective adjustments you can make is to rename the tables and columns. Use names that make sense for business users instead of a cryptic naming convention that only makes sense to the database developers. Another good practice is to make sure the data types and formats of the columns are set correctly. A field from a text file may come in typed as a string when in reality it is numeric data. In addition, you can hide fields that are of no use to the user, such as the surrogate keys used for linking the tables.

A common requirement is to change the sort order of a column from its natural sorting. The most common example is the months of the year. Because they are text by default, they are sorted alphabetically. In reality, you want them sorted by month number. To fix this, you can sort one column by any other column in the table (see Figure 4-14). This is a nice feature and allows you to create your own business-related custom sorting.

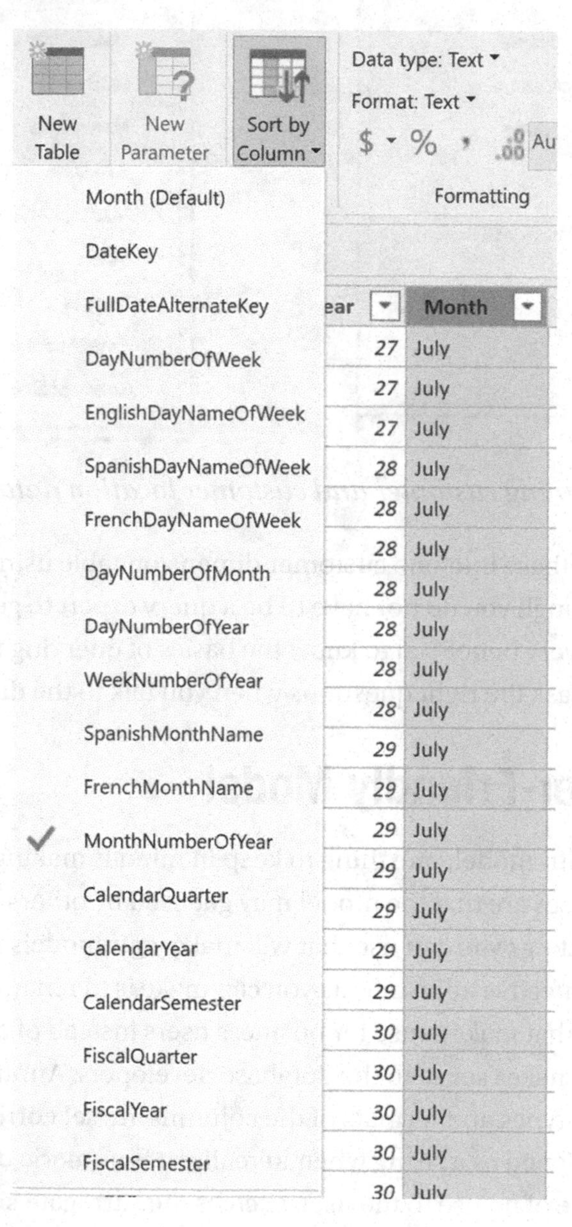

***Figure 4-14.** Sorting one column by another column*

Another useful feature when analyzing data is using hierarchies. Hierarchies define various levels of aggregations. For example, it is common to have a calendar-based hierarchy based on year, quarter, and month levels. An aggregate like sales amount is then rolled up from month to quarter to year. Another common hierarchy might be from department to building to region. You could then roll cost up through the different levels. Creating hierarchies in a Power Pivot model is very easy. In the Field list window, select the column that you want as the top level of the hierarchy. Right-click the column and select New hierarchy (see Figure 4-15).

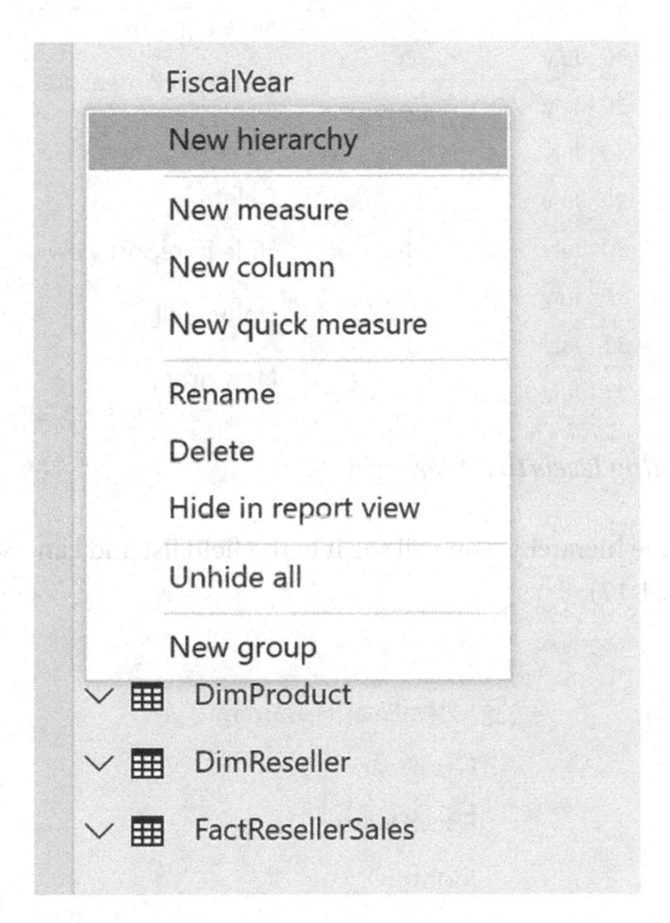

***Figure 4-15.** Creating a hierarchy*

Next in the Field list, right-click the column that will be the next level in the hierarchy. Select Add to hierarchy and select the hierarchy in the list (see Figure 4-16).

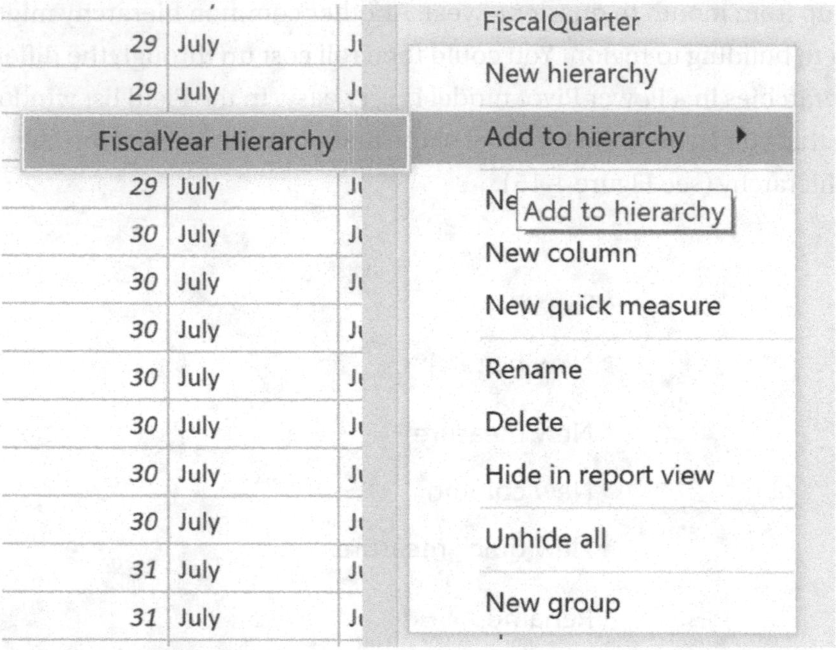

***Figure 4-16.** Adding levels to a hierarchy*

After creating the hierarchy, you will see it in the field list and can use it on your reports (see Figure 4-17).

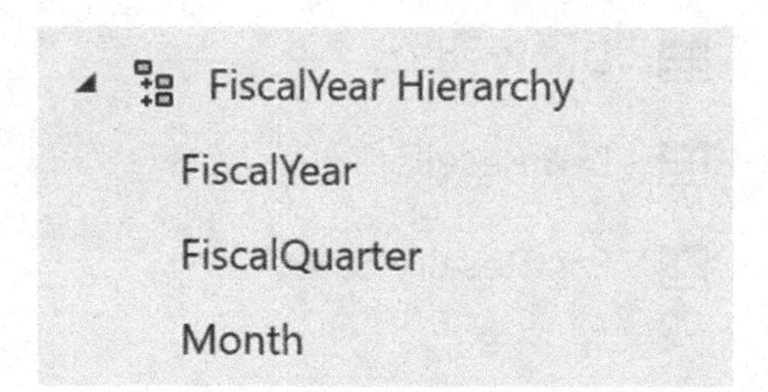

***Figure 4-17.** Selecting a hierarchy in the field list*

Creating hierarchies is one way to increase the usability of your model and help users instinctively gain more value in their data analysis. There are many other techniques you can use to create good models for the various client tools. We will revisit this topic in more detail in Chapter 9.

The following hands-on lab will help you solidify the topics covered in this chapter.

HANDS-ON LAB: CREATING A DATA MODEL IN POWER PIVOT

In the following lab, you will

- Create table relations
- Denormalize data
- Set up a hierarchy
- Make a user-friendly model

1. Open Power BI Desktop and create a new file called Chapter4Lab1.pbix.
2. Connect to the Adventureworks.accdb file in the LabStarterFiles\Chapter4Lab1 folder. (If you get an error, refer to this page to troubleshoot.)
3. Once connected, use the "Edit" button to launch Power Query.
4. In the Power Query window, select the tables and fields listed and import the data. Tip: You can use the "Recent Sources" button to pull the second and third tables from the same source file.

Source Table Name	Friendly Name	Fields (Columns)
DimDate	Date	DateKey FullDateAlternateKey EnglishMonthName MonthNumberOfYear CalendarQuarter CalendarYear
DimCustomer	Customer	CustomerKey BirthDate MaritalStatus Gender YearlyIncome TotalChildren HouseOwnerFlag NumberOfCarsOwned
FactInternetSales	Internet Sales	ProductKey OrderDateKey ShipDateKey CustomerKey SalesTerritoryKey SalesOrderNumber SalesOrderLineNumber OrderQuantity UnitPrice TotalProductCost SalesAmount

5. Select Close & Apply to load the data into the model.
6. Switch to the Data view in Power BI Desktop; rename and/or hide the columns indicated in the table. To rename or hide a column, right-click it and choose Rename or Hide From Report view in the context menu. (Note: You could change the column names in Power Query instead.)

Table	Column	Friendly Name	Hide
Date	DateKey		X
	FullDateAlternateKey	Date	
	EnglishMonthName	Month	
	MonthNumberOfYear	Month No	X
	CalendarQuarter	Quarter	
	CalendarYear	Year	
Customer	CustomerKey		X
	BirthDate	Birth Date	
	MaritalStatus	Marital Status	
	Gender	Gender	
	YearlyIncome	Income	
	TotalChildren	Children	
	HouseOwnerFlag	Home Owner	
	NumberOfCars	Cars	
Internet Sales	ProductKey		X
	OrderDateKey		X
	ShipDateKey		X
	CustomerKey		X
	SalesTerritoryKey		X
	SalesOrderNumber	Order Number	
	SalesOrderLineNumber	Order Line Number	
	OrderQuantity	Quantity	
	UnitPrice	Unit Price	
	TotalProductCost	Product Cost	
	SalesAmount	Sales Amount	

7. Switch to the Model view. You should see the Date, Customer, and Internet Sales tables. The Customer table and the Internet Sales table have a relationship defined between them. This was inferred by Power BI based on the column names and data type.

8. Drag the DateKey from the Date table and drop it on the OrderDateKey in the Internet Sales table. Similarly, create a relationship between the DateKey and the ShipDateKey. Double-click this second relationship to launch the Edit relationship window (see Figure 4-18). Try to make it an active relationship. You should get an error because you can only have one active relationship between two tables in the model. After reviewing the error, click "Cancel."

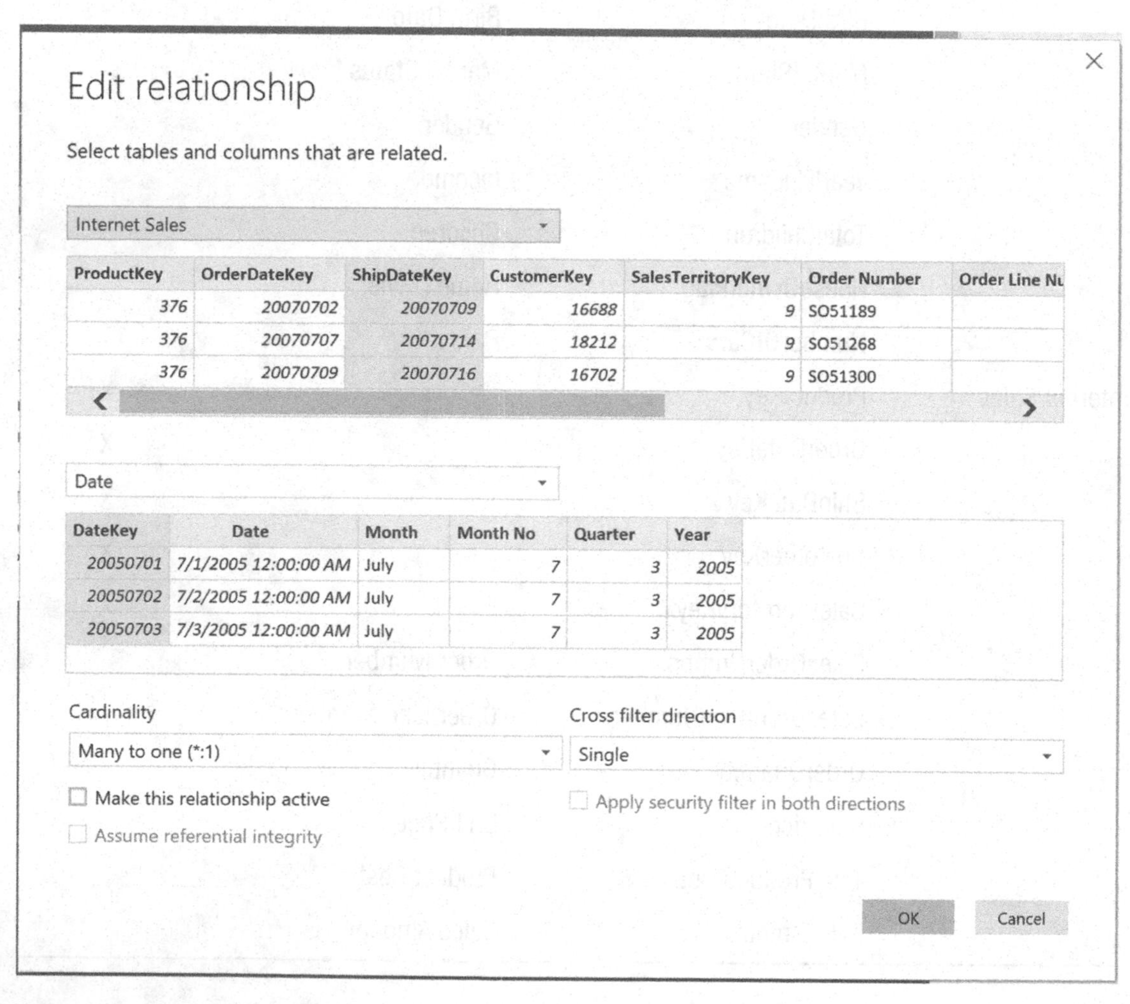

Figure 4-18. *Creating the table relationship*

9. Now it's time to add some product information to our data set. On the Home tab, select Recent Sources and choose the Adventureworks database (see Figure 4-19).

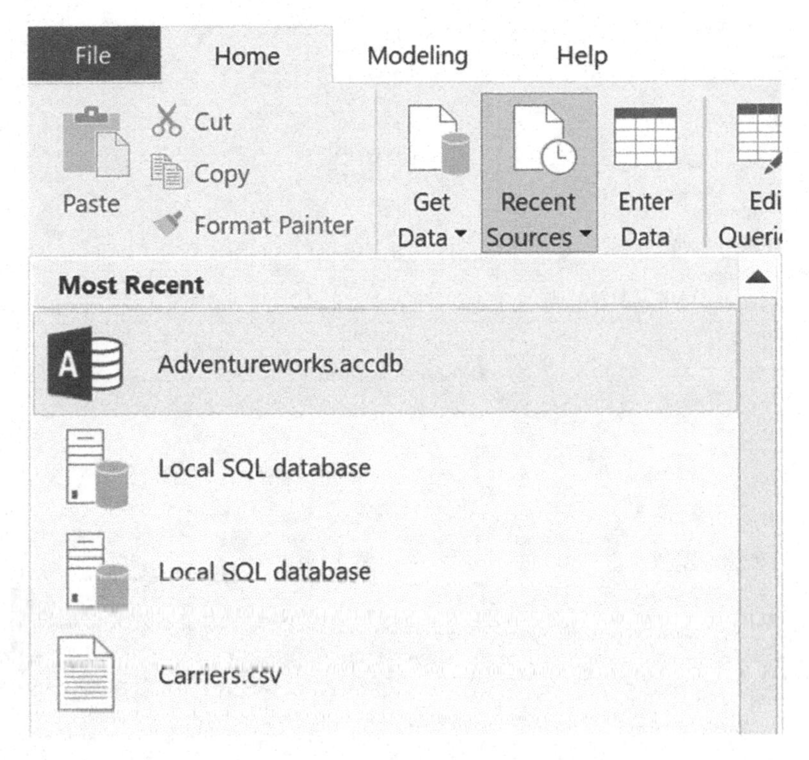

***Figure 4-19.** Connecting to the Adventureworks database*

10. In the Navigator window, select the DimProduct, DimProductCategory, and the DimProductSubcategory tables and click the Edit button (see Figure 4-20).

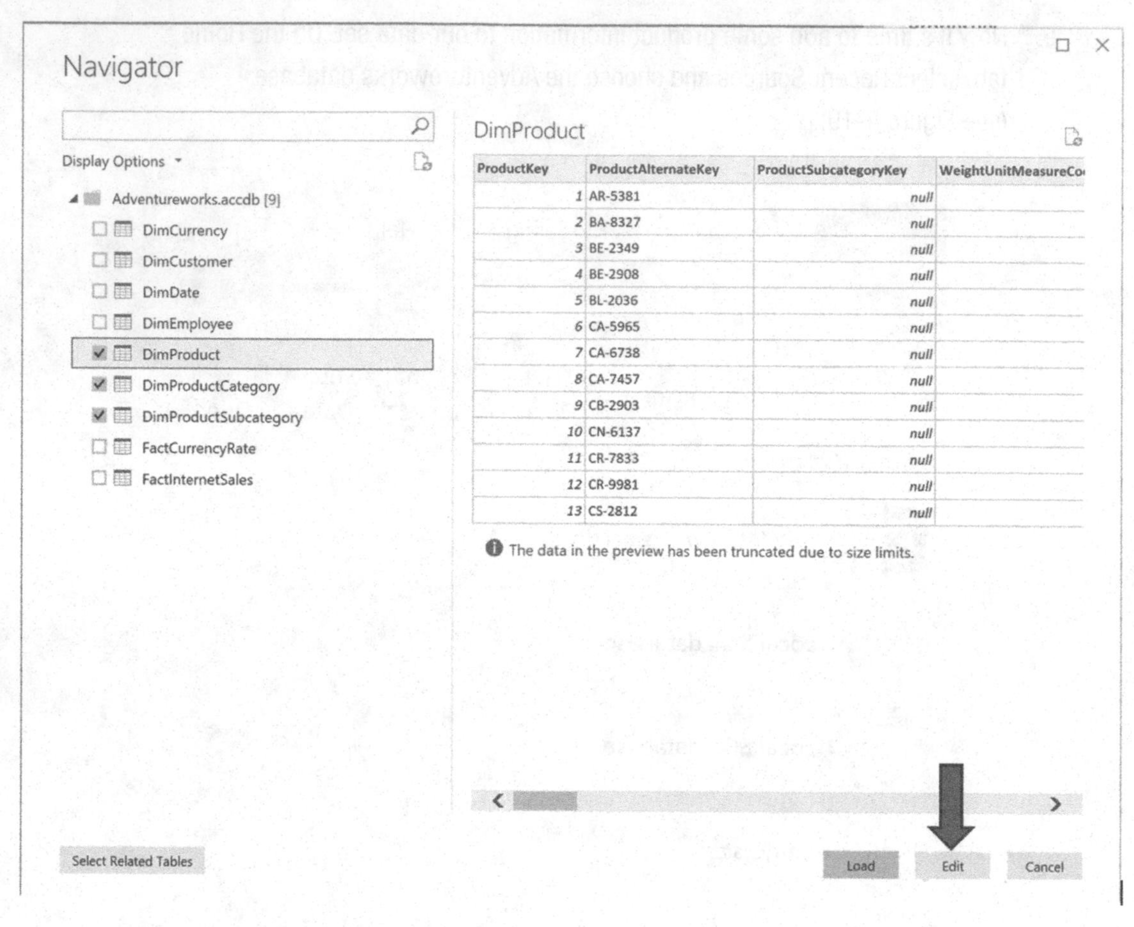

Figure 4-20. *Selecting additional tables*

11. Using the Power Query designer, select the ProductKey, ProductAlternateKey, ProductSubCategoryKey, WeightUnitMeasureCode, SizeUnitMeasureCode, EnglishProductName, ListPrice, Size, SizeRange, Weight, and Color columns from the DimProduct query. Change the Name of the query to `Product`.

12. Merge the DimProductSubcategory to the `Product` query using a Left Outer join (see Figure 4-21). Start by selecting the Product query and then clicking "Merge queries" on the toolbar.

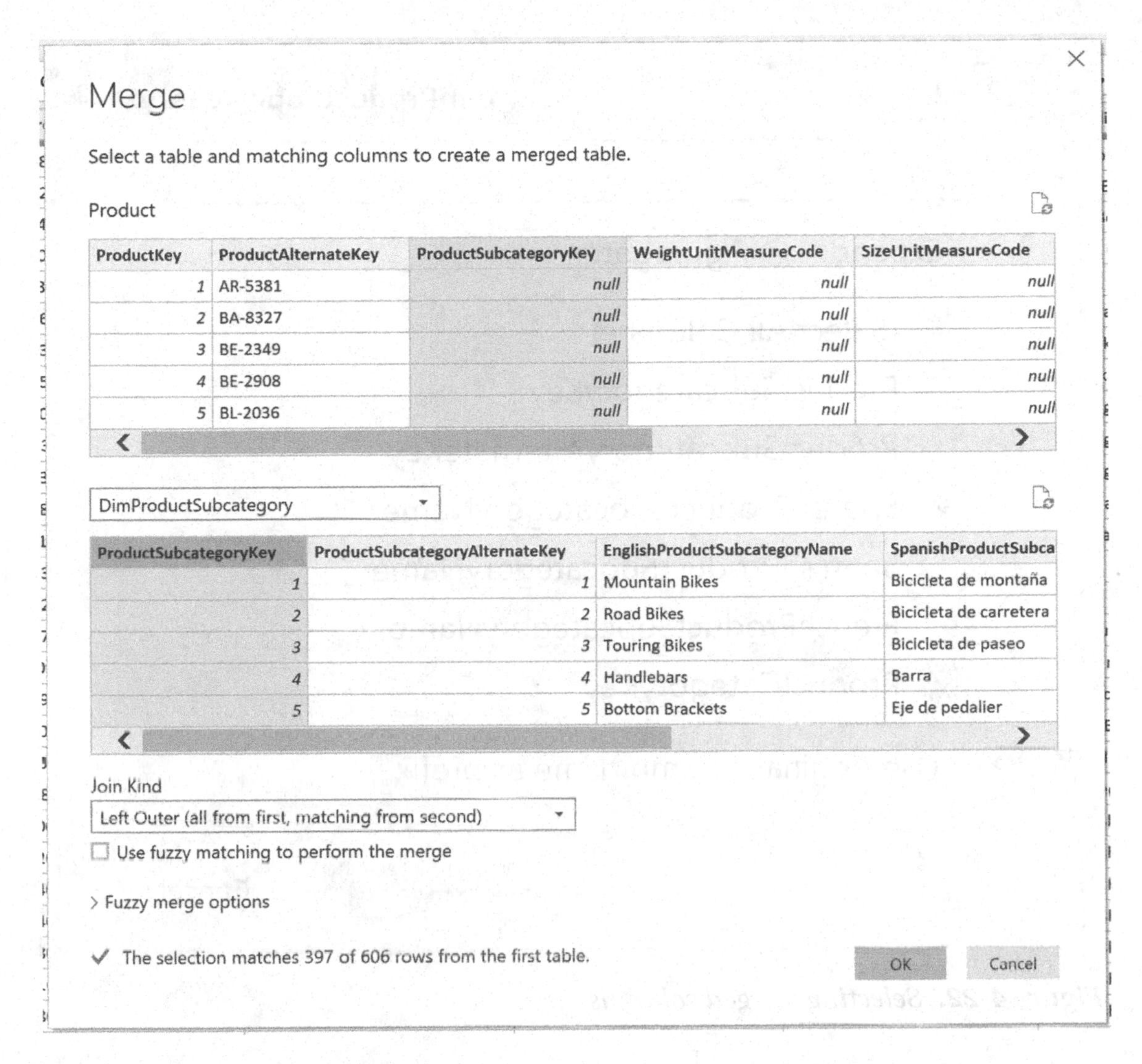

Figure 4-21. *Merging queries*

13. Expand the resulting column and select the EnglishProductSubcategoryName and the ProductCategoryKey columns. Uncheck the Use original column name as prefix check box (see Figure 4-22).

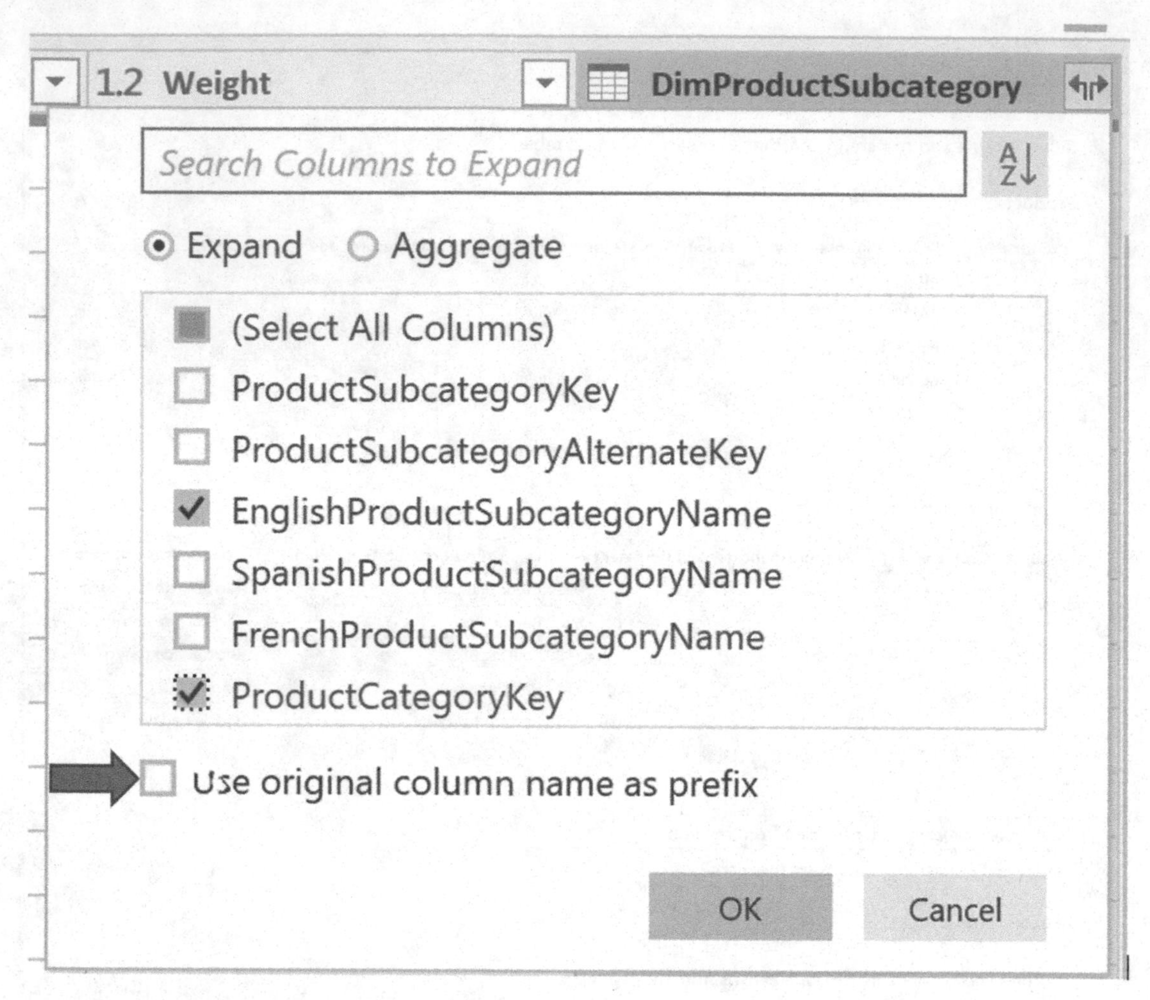

Figure 4-22. *Selecting merged columns*

14. Repeat the previous steps to merge the EnglishProductCategoryName from the DimProductCategory query into the `Product` query.

15. Now that the queries have been merged, you can remove the ProductSubcategoryKey and the ProductCategoryKey columns from the `Product` Query.

16. Rename the columns as follows:

Table	Column	Friendly Name
Product	ProductKey	
	ProductAlternateKey	Product Code
	WeightUnitMeasureCode	Weight UofM Code
	SizeUnitMeasureCode	Size UofM Code
	EnglishProductName	Product Name
	ListPrice	List Price
	Size	
	SizeRange	Size Range
	Weight	
	Color	
	EnglishProductCategoryName	Category
	EnglishProductSubcategoryName	Subcategory

17. In the Queries list, right-click the DimProductCategory query and uncheck the Enable load (see Figure 4-23).

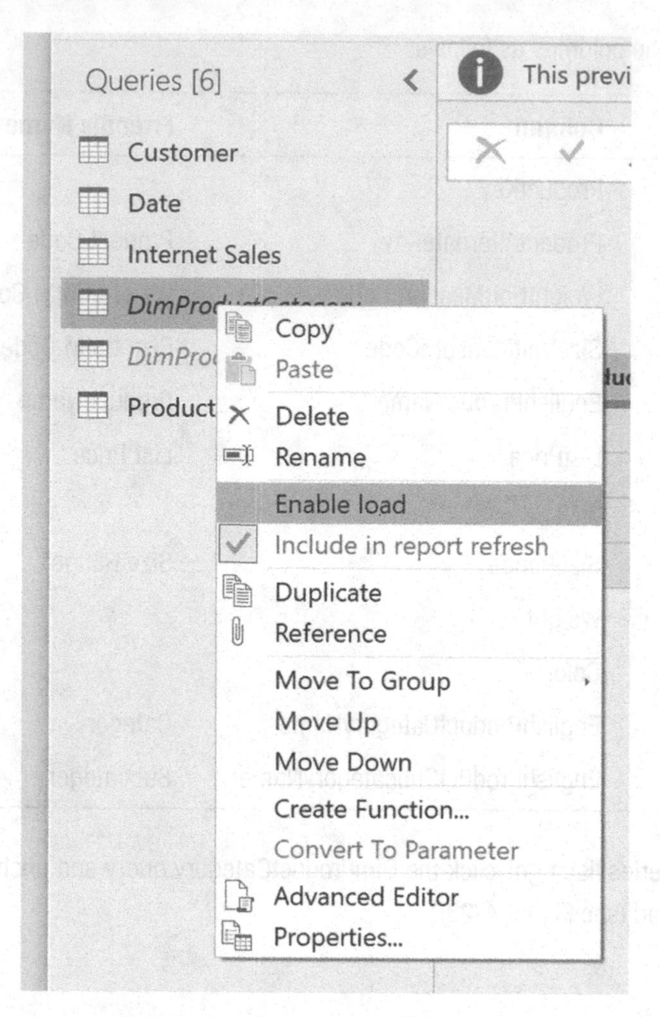

***Figure 4-23.** Disable query loading*

18. Repeat step 16 for the DimProductSubcategory query.
19. Choose "Close & Apply."
20. After loading the product data into the Power BI model, hide the ProductKey column in the Report view.
21. Create a relationship between the Internet Sales and the Product tables using the ProductKey (Power BI may have already added this automatically). Your final diagram should look like Figure 4-24.

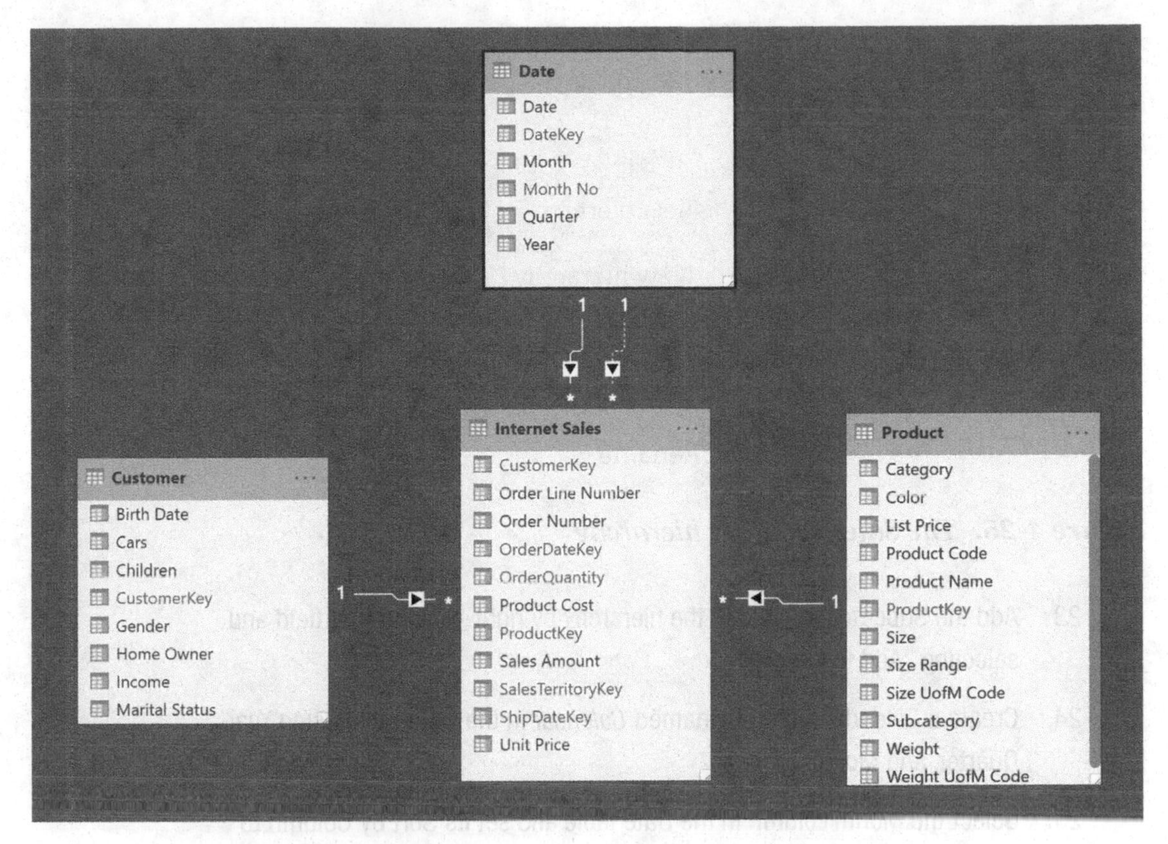

***Figure 4-24.** Viewing the data model relationships*

22. To create a hierarchy, switch back to the Data view and select the Category column in the Product table. Right-click the column and select New hierarchy (see Figure 4-25).

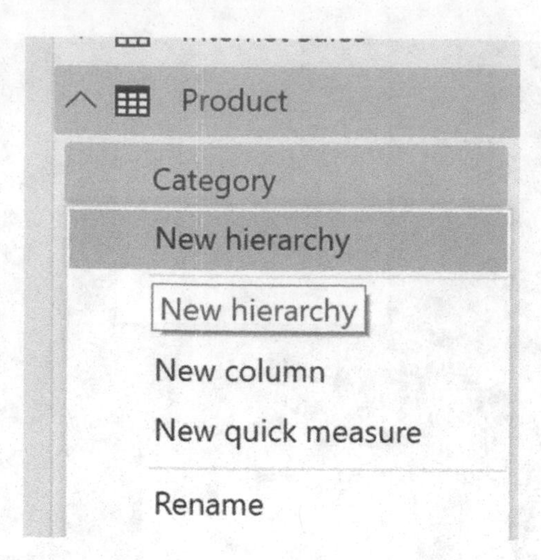

Figure 4-25. *The Sales Territory hierarchy*

23. Add the Subcategory field to the hierarchy by right-clicking that field and selecting "Add to hierarchy."
24. Create a Calendar hierarchy named *Calendar* in the Date table using Year, Quarter, and Month.
25. Select the Month column in the Date table and set its Sort by Column to the Month No column. Remember, the "Sort by Column" button is on the "Modeling" tab.
26. When done, save and close Power BI Desktop.

Summary

When working in Power BI, it is very important to understand what makes a good model. A good model will make Power BI perform incredibly fast and allow you to easily analyze large amounts of data. This chapter guided you through the process of creating a solid model that will become the foundation for your data analysis. In addition, you saw how to present a user-friendly model to client tools.

Now that you have a solid foundation for your model, you are ready to extend the model with custom calculations. The next chapter introduces the Data Analysis Expressions (DAX) language and explains how to create calculated columns in the data model. It includes plenty of examples to help you create common calculations in the model.

CHAPTER 5

Creating Calculations with DAX

Now that you know how to create a robust data model to base your analysis on, the next step is to add to the model any calculations required to aid your exploration of the data. For example, you may have to translate code values into meaningful descriptions or parse out a string to obtain key information. This is where Data Analysis Expressions (DAX) comes into play. This chapter introduces you to DAX and shows you how to use DAX to create calculated columns to add to the functionality of your model.

After completing this chapter, you will be able to

- Use DAX to add calculated columns to a table
- Implement DAX operators
- Work with text functions in DAX
- Use DAX date and time functions
- Use conditional and logical functions
- Get data from a related table
- Use math, trig, and statistical functions

What Is DAX?

DAX is a formula language used to create calculated columns and measures in the Power BI model. It is a language developed specifically for the tabular data model Power BI is based on. If you are familiar with Excel's formula syntax, you will find that the DAX syntax is very familiar. In fact, some of the DAX formulas have the same syntax and functionality as their Excel counterparts. The major difference—and one that you need

D. Clark, *Beginning Microsoft Power BI*, https://doi.org/10.1007/978-1-4842-5620-6_5

to wrap your head around—is that Excel formulas are cell based, whereas DAX is column based. For example, if you want to concatenate two values in Excel, you would use a formula like the following:

```
=A1 & " " & B1
```

where `A1` is the cell in the first row and first column and `B1` is the cell in the second column of the first row (see Figure 5-1).

C1 =A1 & " " & B1

	A	B	C	D	E	F
1	Dan	Clark	Dan Clark			
2	Al	Smith				
3	Jane	Doe				

***Figure 5-1.** Entering a formula in Excel*

This is very similar to the DAX formula:

```
=[First Name] & " " & [Last Name]
```

where `First Name` and `Last Name` are columns in a table in the model (see Figure 5-2).

```
1 FullName = [FirstName] & " " & [LastName]
```

EmployeeKey	FirstName	LastName	FullName
1	Guy	Gilbert	Guy Gilbert
2	Kevin	Brown	Kevin Brown
3	Roberto	Tamburello	Roberto Tamburello
4	Rob	Walters	Rob Walters
5	Rob	Walters	Rob Walters
6	Thierry	D'Hers	Thierry D'Hers
7	David	Bradley	David Bradley
8	David	Bradley	David Bradley
9	JoLynn	Dobney	JoLynn Dobney
10	Ruth	Ellerbrock	Ruth Ellerbrock
11	Gail	Erickson	Gail Erickson
12	Barry	Johnson	Barry Johnson
13	Jossef	Goldberg	Jossef Goldberg
14	Terri	Duffy	Terri Duffy

***Figure 5-2.** Entering a DAX formula in Power BI*

The difference is that the DAX formula is applied to all rows in the table, whereas the Excel formula only works on the specific cells. In Excel you need to re-create the formula in each row.

What this means is that although you can do something like this in Excel

```
=A1 & " " & B2
```

where you are taking a cell from the first row and concatenating a cell from the second row (see Figure 5-3), you can't do that in DAX.

C2 =A1 & " " & B2

	A	B	C	D	E
1	Dan	Clark	Dan Clark		
2	Al	Smith	Dan Smith		
3	Jane	Doe			

Figure 5-3. *Using cells in different rows*

When creating DAX formulas, it is important to consider the data types and any conversions that may take place during the calculations. If you don't take these into account, you may experience errors in the formula or unexpected results. The supported data types in the model are whole number, decimal number, currency, Boolean, text, and date. DAX also has a table data type that is used in many functions that take a table as an input value and return a table.

When you try to add a numeric data type with a text data type, you get an implicit conversion. If DAX can convert the text to a numeric value, it will add them as numbers; if it can't, you will get an error. On the other hand, if you try to concatenate a numeric data type with a text data type, DAX will implicitly convert the numeric data type to text. Although most of the time implicit conversions give you the results you are looking for, they come at a performance cost and should be avoided if possible. For example, if you import data from a text file and the column is set to a text data type, but you know it is in fact numeric, you should change the data type in the model.

When creating calculations in DAX, you need to reference tables and columns. If the table name doesn't contain spaces, you can just refer to it by name. If the table name contains spaces, you need to enclose it in single quotes. Columns and measures are enclosed in brackets. If you just list the column name in the formula, it is assumed that the column exists in the same table. If you are referring to a column in another table, you need to use the fully qualified name, which is the table name followed by the column name. The following code demonstrates the syntax:

```
=[SalesAmount] - [TotalCost]
=Sales[SalesAmount] - Sales[TotalCost]
='Internet Sales'[SalesAmount] - 'Internet Sales'[TotalCost]
```

Here are some other points to keep in mind when working with DAX:

- DAX formulas and expressions can't modify or insert individual values in tables.
- You can't create calculated rows using DAX. You can create only calculated columns and measures.
- When defining calculated columns, you can nest functions to any level.

The first thing to understand when creating a calculation is which operators are supported and what the syntax to use them is. In the next section, you will investigate the various DAX operators.

Implementing DAX Operators

DAX contains a robust set of operators, including arithmetic, comparison, logic, and text concatenation. Most of these should be familiar to you and are listed in Table 5-1.

Table 5-1. DAX Operators

Category	Symbol	Use
Arithmetic operators		
	+	Addition
	-	Subtraction
	*	Multiplication
	/	Division
	^	Exponentiation
Comparison operators		
	=	Equal to
	>	Greater than
	<	Less than
	>=	Greater than or equal to
	<=	Less than or equal to
		Not equal to
Text concatenation operator		
	&	Concatenation
Logic operators		
	&&	And
	||	Or

As an example of the arithmetic operator, the following code is used to divide the Margin column by the Total Cost column to create a new column, the Margin Percentage.

```
=[Margin]/[TotalCost]
```

It is very common to have several arithmetic operations in the same calculation. In this case, you must be aware of the order of operations. Exponents are evaluated first, followed by multiplication/division, and then addition/subtraction. You can control order of operations by using parentheses to group calculations; for example, the following formula will perform the subtraction before the division:

```
=([Sales Amount]-[Total Cost])/[Total Cost]
```

The comparison operators are primarily for `IF` statements. For example, the following calculation checks to see whether a store's selling area size is greater than 1000. If it is, it is classified as a large store; if not, it is classified as small:

```
=IF([Selling Area Size]>1000,"Large","Small")
```

The logical operators are used to create multiple comparison logic. The following code checks to see whether the store size area is greater than 1000 or if it has more than 35 employees to classify it as large:

```
=IF([Selling Area Size]> 1000 || [Employee Count] > 35,"Large","Small")
```

When you start stringing together a series of logical conditions, it is a good idea to use parentheses to control the order of operations. The following code checks to see whether the store size area is greater than 1000 and if it has more than 35 employees to classify it as large. It will also classify it as large if it has annual sales of more than $1,000,000 regardless of its size area or number of employees:

```
=IF(([Selling Area Size]> 1000 && [Employee Count] > 35) || [Annual Sales]
> 1000000,"Large","Small")
```

When working with DAX calculations, you may need to nest one formula inside another. For example, the following code nests an `IF` statement inside the false part of another `IF` statement. If the employee count is not greater than 35, it jumps to the next `IF` statement to check whether it is greater than 20:

```
=IF(Store[EmployeeCount]>35,"Large",IF(Store[EmployeeCount]>20,"Medium",
"Small"))
```

DAX contains many useful functions for creating calculations and measures. These functions include text functions, date and time functions, statistical functions, math functions, and informational functions. The next few sections look at using the various function types in your calculations.

Working with Text Functions

A lot of calculations involve text manipulation. You may need to truncate, parse, search, or format the text values that you load from the source systems. DAX contains many useful functions for working with text. The functions are listed in Table 5-2 along with a description of what they are used for.

Table 5-2. *DAX Text Functions*

Function	Description
BLANK	Returns a blank
CONCATENATE	Joins two text strings into one text string
EXACT	Compares two text strings and returns TRUE if they are exactly the same, and FALSE otherwise
FIND	Returns the starting position of one text string within another text string
FIXED	Rounds a number to the specified number of decimals and returns the result as text
FORMAT	Converts a value to text according to the specified format
LEFT	Returns the specified number of characters from the start of a text string
LEN	Returns the number of characters in a text string
LOWER	Converts all letters in a text string to lowercase
MID	Returns a string of characters from a text string, given a starting position and length
REPLACE	Replaces part of a text string with a different text string
REPT	Repeats text a given number of times. Use REPT to fill a cell with a number of instances of a text string
RIGHT	Returns the last character or characters in a text string, based on the number of characters you specify
SEARCH	Returns the number of the character at which a specific character or text string is first found, reading left to right
SUBSTITUTE	Replaces existing text with new text in a text string
TRIM	Removes all spaces from text except for single spaces between words
UPPER	Converts a text string to all uppercase letters
VALUE	Converts a text string that represents a number to a number

As an example of using a text function in a calculation, let's say you have a product code column in a products table where the first two characters represent the product family. To create the product family column, you would use the Left function as follows:

```
=Left([Product Code],2)
```

You can use the FIND function to search a text for a subtext. You can use a (?) to match any single character and a (*) to match any sequence of characters. You have the option of indicating the starting position for the search. The FIND function returns the starting position of the substring found. If it doesn't find the substring, it can return a 0, -1, or a blank value. The following code searches the product description column for the word *mountain*:

```
=FIND("mountain",[Description],1,-1)
```

The FORMAT function converts a value to text based on the format provided. For example, you may need to convert a date to a specific format. The following code converts a date data type to a string with a format like "Mon - Dec 02, 2019":

```
=FORMAT([StartDate],"ddd - MMM dd, yyyy")
```

Along with the ability to create your own format, there are also predefined formats you can use. The following code demonstrates using the Long Date format "Monday, December 2, 2019":

```
=FORMAT([StartDate],"Long Date")
```

Now that you have seen how to use some of the text functions, the next types of functions to look at are the built-in date and time functions.

Using DAX Date and Time Functions

Most likely you will find that your data analysis has a date component associated with it. You may need to look at sales or energy consumption and need to know the day of the week the event occurred. You may have to calculate age or maturity dates. DAX has quite a few date and time functions to help create these types of calculations. Table 5-3 summarizes the various date and time functions available.

Table 5-3. *DAX Date and Time Functions*

Function	Description
DATE	Returns the specified date in datetime format
DATEVALUE	Converts a date in the form of text to a date in datetime format
DAY	Returns the day of the month
EDATE	Returns the date that is the indicated number of months before or after the start date
EOMONTH	Returns the date in datetime format of the last day of the month, before or after a specified number of months
HOUR	Returns the hour as a number from 0 (12:00 a.m.) to 23 (11:00 p.m.)
MINUTE	Returns the minute as a number from 0 to 59
MONTH	Returns the month as a number from 1 (January) to 12 (December)
NOW	Returns the current date and time in datetime format
SECOND	Returns the seconds of a time value, as a number from 0 to 59
TIME	Converts hours, minutes, and seconds given as numbers to a time in datetime format
TIMEVALUE	Converts a time in text format to a time in datetime format
TODAY	Returns the current date
WEEKDAY	Returns a number from 1 to 7 identifying the day of the week of a date
WEEKNUM	Returns the week number for the given date
YEAR	Returns the year of a date as a four-digit integer
YEARFRAC	Calculates the fraction of the year represented by the number of whole days between two dates

As an example of using the date functions, say you need to calculate years of service for employees. The first thing to do is find the difference between the current year and the year they were hired. The following code gets the year from today's date:

```
=YEAR(Today())
```

Now you can subtract the year of their hire date. Notice we are nesting one function inside another. Nesting functions is a common requirement for many calculations:

```
=YEAR(TODAY()) - YEAR([HireDate])
```

Astute readers will realize that the result of this calculation is only correct if the current month is greater than or equal to the month they were hired. You can adjust for this using a conditional If statement as follows:

```
= If (MONTH(TODAY())>=MONTH([HireDate]) ,YEAR(TODAY()) -
YEAR([HireDate]),YEAR(TODAY()) - YEAR([HireDate])-1)
```

As you can see, calculations can get quite complicated. The challenge is making sure that the open and closing parentheses of each function line up correctly. One way to organize the code is to use multiple lines and indenting. To get a new line in the formula editor bar, you need to hold down the Shift key while you press Enter. I find the following easier to understand:

```
= If (MONTH(TODAY())>=MONTH([HireDate]),
       YEAR(TODAY()) - YEAR([HireDate]),
       YEAR(TODAY()) - YEAR([HireDate])-1
   )
```

There is often more than one way to create a calculation. You may find an easier way to make the calculation or one that performs better. The following calculates the years of service using the YEARFRAC function and the TRUNC function (one of the math functions) to drop the decimal part of the number:

```
=TRUNC(YEARFRAC([HireDate],TODAY()))
```

The next group of functions you are going to investigate are the informational and logical functions. These functions are important when you want to determine whether a condition exists such as a blank value, or whether an error is occurring due to a calculation. These functions allow you to trap for conditions and respond to them in an appropriate way.

Using Informational and Logical Functions

As you start building more complex calculations, you often need to use informational and logical functions to check for conditions and respond to various conditions. One common example is the need to check for blank values. The ISBLANK function returns TRUE if the value is blank and FALSE if it is not. The following code uses a different calculation depending on whether the middle name is blank:

```
=IF(ISBLANK([MiddleName]),
[FirstName] & " " & [LastName],
[FirstName] & " " & [MiddleName] & " " & [LastName]
)
```

The ISERROR function is used to check whether a calculation or function returns an error. The following calculation checks to see if a divide by zero error occurs during a division:

```
=IF(ISERROR([TotalProductCost]/[SalesAmount]),
   BLANK(),
     [TotalProductCost]/[SalesAmount]
   )
```

Another way to create this calculation is to use the IFERROR function, which returns the value if no error occurs and an alternate value if an error occurs:

```
=IFERROR([TotalProductCost]/[SalesAmount],BLANK())
```

Tables 5-4 and 5-5 list the logical and informational functions available in DAX.

Table 5-4. *The DAX Logical Functions*

Function	Description
AND	Checks whether both arguments are TRUE
FALSE	Returns the logical value FALSE
IF	Checks whether a condition provided as the first argument is met. Returns one value if the condition is TRUE and another value if the condition is FALSE
IFERROR	Evaluates an expression and returns a specified value if the expression returns an error; otherwise, returns the value of the expression itself
NOT	Changes FALSE to TRUE, or TRUE to FALSE
OR	Checks whether one of the arguments is TRUE to return TRUE
SWITCH	Evaluates an expression against a list of values and returns one of multiple possible result expressions
TRUE	Returns the logical value TRUE

Table 5-5. *The DAX Informational Functions*

Function	Description
CONTAINS	Returns TRUE if values for all referred columns exist, or are contained, in those columns
ISBLANK	Checks whether a value is blank
ISERROR	Checks whether a value is an error
ISLOGICAL	Checks whether a value is a Boolean value
ISNONTEXT	Checks whether a value is not text (blank cells are not text)
ISNUMBER	Checks whether a value is a number
ISTEXT	Checks whether a value is text
LOOKUPVALUE	Returns the value in the column for the row that meets all criteria specified by a search

When you are analyzing data, you often need to look up corresponding data from a related table. You may need to obtain descriptions from a related code or summarize data and import it into a table, such as lifetime sales. The following section looks at how you go about looking up related data using DAX.

Getting Data from Related Tables

There are times when you need to look up values in other tables to complete a calculation. If a relationship is established between the tables, you can use the RELATED function. This allows you to denormalize the tables and make it easier for users to navigate. For example, you may have a Customer table related to a Geography table (see Figure 5-4).

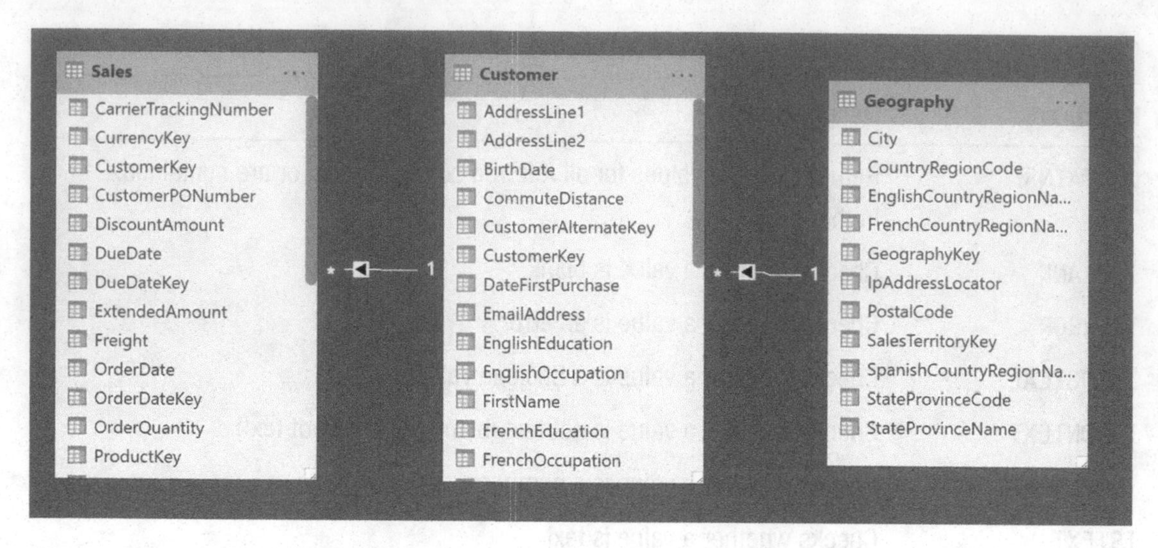

***Figure 5-4.** Related tables*

If you need to look at sales by customer's country, you can use the RELATED function to create a Country column in the Customers table:

```
=RELATED(Geography[CountryRegionName])
```

You can then hide the Geography table from client tools to keep the model cleaner and less confusing to users.

Although the related table returns a single value, there are times when you want to look at a set of related data and aggregate it before displaying the value in the column. For example, you may want to add a column to the Customers table that lists their lifetime sales amount. In this case, you would use the RELATEDTABLE function to get the related sales and then sum them up for each customer:

```
=SUMX(RELATEDTABLE(Sales),[SalesAmount])
```

Note The previous code uses the SUMX function, which is used instead of the SUM function because you are applying a filter. Chapter 6 discusses this in more detail.

The final set of functions to look at are the math, trig, and statistical functions. These functions allow you to perform common analysis such as logs, standard deviation, rounding, and truncation.

Using Math, Trig, and Statistical Functions

Along with the functions discussed thus far, DAX also includes quite a few math, trig, and statistical functions. The math functions (see Table 5-6) are used for rounding, truncating, and summing up the data. They also contain functions you may use in scientific, engineering, and financial calculations; for example, you may need to calculate the volume of a sphere given the radius. This is calculated in DAX as follows:

```
=4*PI()*POWER([Radius],3)/3
```

As another example, say you want to calculate compounding interest on an investment. The following DAX calculation determines the compounding rate of return for an investment:

```
=[Principal]*POWER(1+([IntRate]/[CompoundRate]),[CompoundRate]*[Years])
```

Table 5-6. *Some of the Math and Trig Functions Available in DAX*

Function	Description
ABS	Returns the absolute value of a number
CEILING	Rounds a number up to the nearest integer or to the nearest multiple of significance
EXP	Returns *e* raised to the power of a given number
FACT	Returns the factorial of a number
FLOOR	Rounds a number down, toward zero, to the nearest multiple of significance
LOG	Returns the logarithm of a number to the base you specify
PI	Returns the value of pi, 3.14159265358979, accurate to 15 digits
POWER	Returns the result of a number raised to a power
ROUND	Rounds a number to the specified number of digits
SQRT	Returns the square root of a number
SUM	Adds all the numbers in a column
TRUNC	Truncates a number to an integer by removing the decimal, or fractional, part of the number

When you are analyzing data, you often want to look at not only the relationship between the data but also the quality of the data and how well you can trust your predictions. This is where the statistical analysis of the data comes into play. With statistics, you can do things like determine and account for outliers in the data, examine the volatility of the data, and detect fraud. As an example, you can use DAX to determine and filter out the outliers in your data using the standard deviation. The following DAX function calculates the standard deviation of the sales amount:

```
=STDEVX.P(RELATEDTABLE(Sales),Sales[SalesAmount])
```

Table 5-7 lists some of the statistical functions available in DAX.

Table 5-7. *Some Statistical Functions Available in DAX*

Function	Description
AVERAGE	Returns the average of all the numbers in a column
COUNT	Counts the number of cells in a column that contain numbers
COUNTA	Counts the number of cells in a column that are not empty
COUNTBLANK	Counts the number of blank cells in a column
COUNTROWS	Counts the number of rows in the specified table
DISTINCTCOUNT	Counts the number of different cells in a column of numbers
MAX	Returns the largest numeric value in a column
MIN	Returns the smallest numeric value in a column
RANK.EQ	Returns the ranking of a number in a list of numbers
RANKX	Returns the ranking of a number in a list of numbers for each row in the table argument
STDEV.S	Returns the standard deviation of a sample population
TOPN	Returns the top *N* rows of the specified table
VAR.S	Returns the variance of a sample population

Now that you have seen what functions you have available in Power BI through DAX, I want to review some tips on creating functions in general.

Tips for Creating Calculations in Power BI

Before turning you loose on a hands-on lab, I want to give you a few pointers on creating calculations in Power BI. To create a calculated column, right-click the table you want to add the column to in the Field list window. In the context menu, select New Column (see Figure 5-5).

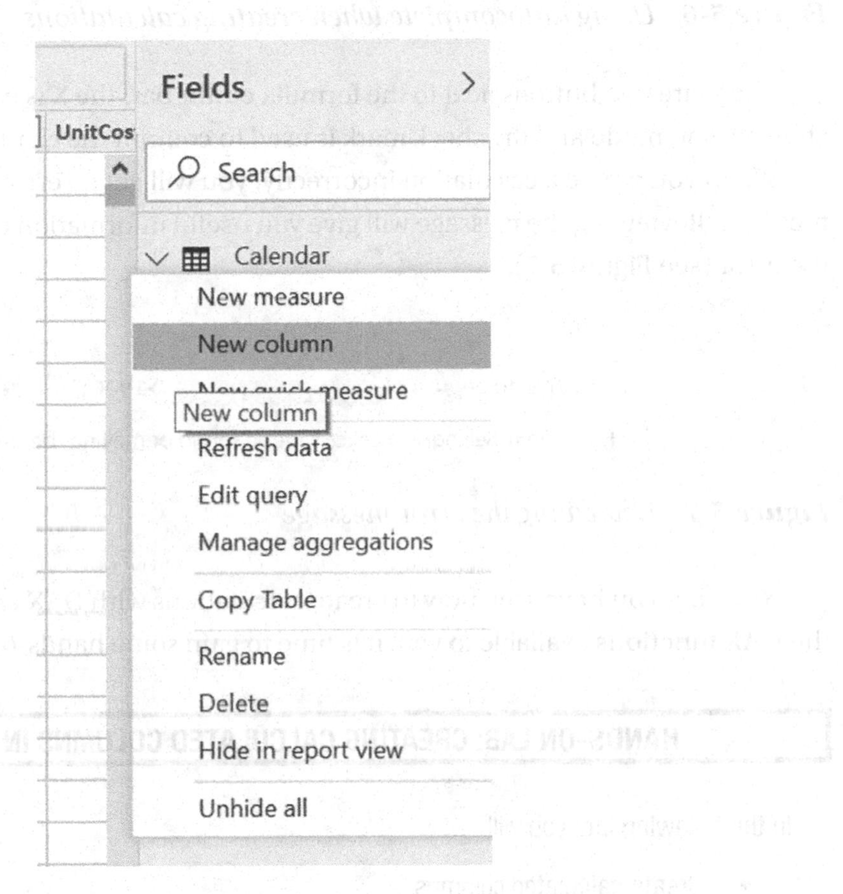

Figure 5-5. *Adding a calculated column*

Enter the formula in the formula editor bar. Calculated columns start with the name of the column followed by an equal sign (=) and then the formula. The formula editor bar supplies an autocomplete feature that you should take advantage of (see Figure 5-6). Select the function, table, or column from the drop-down list and press the Tab key to insert it into the formula. If you don't see the autocomplete drop-down, chances are there is an error in your formula.

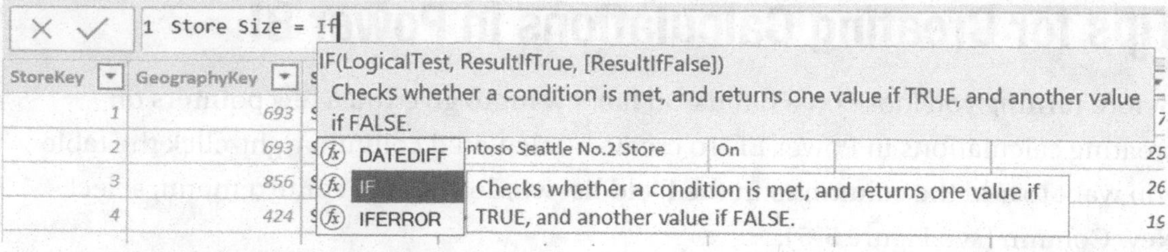

Figure 5-6. Using autocomplete when creating calculations

There are two buttons next to the formula editor bar: the X is used to cancel the changes you made and the check mark is used to commit the changes.

When you create a calculation incorrectly, you will get an error indicator and a message. Reviewing the message will give you useful information that can help you fix the error (see Figure 5-7).

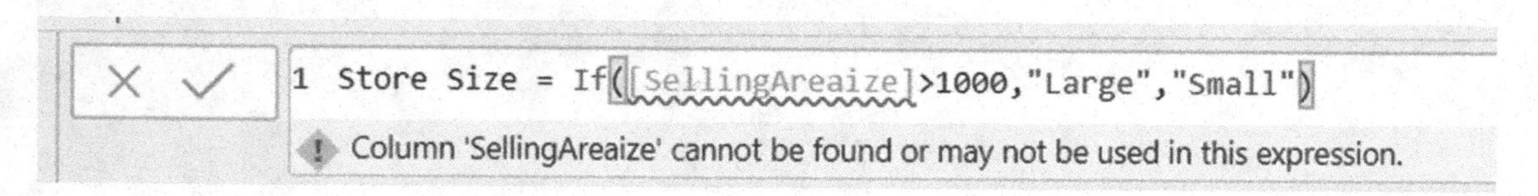
1 Store Size = If([SellingAreaize]>1000,"Large","Small")
Column 'SellingAreaize' cannot be found or may not be used in this expression.

Figure 5-7. Reviewing the error message

Now that you have seen how to create calculations with DAX and are familiar with the DAX functions available to you, it is time to gain some hands-on experience.

HANDS-ON LAB: CREATING CALCULATED COLUMNS IN POWER BI

In the following lab, you will

- Create calculated columns
- Use DAX text functions
- Use date functions in a DAX expression
- Use data from a related table in an expression
- Implement conditional logic in an expression

1. In the LabStarterFiles\Chapter5Lab1 folder, open the Chapter5Lab1.pbix file. This file contains a data model consisting of sales data, product data, and store data.
2. View the model diagram (see Figure 5-8).

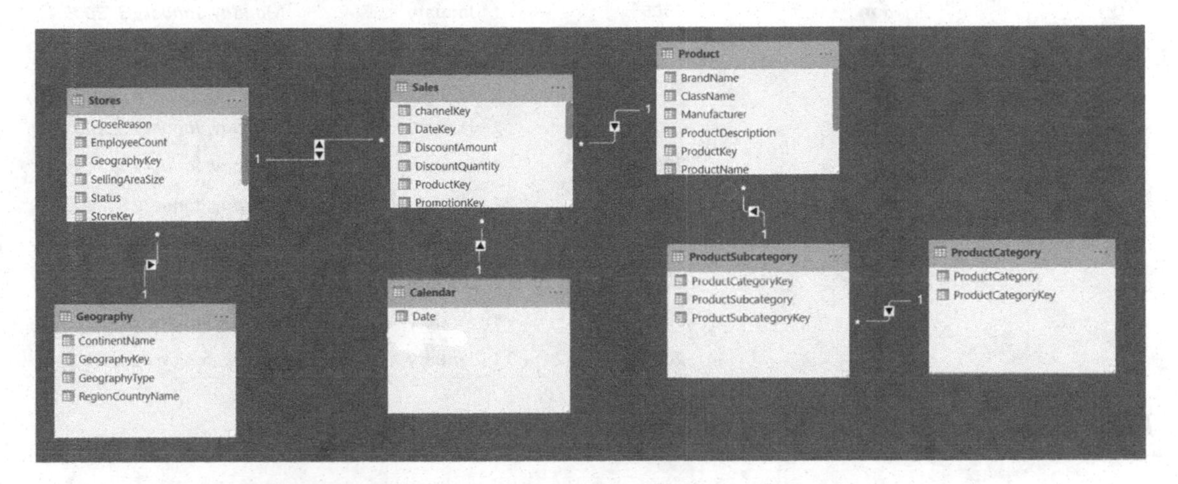

Figure 5-8. *The data model for store sales*

3. Switch to the Data view and select the Sales table.
4. Right-click the Sales table in the Fields list and select New Column. In the formula bar above the table, enter the following to calculate margin:

```
Margin = [SalesAmount] - [TotalCost]
```

5. Repeat this procedure to create a Margin Percent column with the following formula:

```
Margin Percent = DIVIDE([Margin],[SalesAmount],BLANK())
```

6. In the Field list window, select the Calendar table.
7. Use the WeekDay function to create a DayOfWeek column.
8. Use the Format function to get the Weekday (name) column.
9. Create a Year, Month No, and Month column. Your Calendar table should look like Figure 5-9. (Hint: Use the Format function to get the Month column.)

```
1 Weekday = FORMAT([Date],"ddd")
```

DayOfWeek	Weekday	Year	Month No	Month	Date
7	Sat	2005	1	January	Saturday, January 1, 2005
1	Sun	2005	1	January	Sunday, January 2, 2005
2	Mon	2005	1	January	Monday, January 3, 2005
3	Tue	2005	1	January	Tuesday, January 4, 2005
4	Wed	2005	1	January	Wednesday, January 5, 2005
5	Thu	2005	1	January	Thursday, January 6, 2005
6	Fri	2005	1	January	Friday, January 7, 2005
7	Sat	2005	1	January	Saturday, January 8, 2005
1	Sun	2005	1	January	Sunday, January 9, 2005
2	Mon	2005	1	January	Monday, January 10, 2005
3	Tue	2005	1	January	Tuesday, January 11, 2005
4	Wed	2005	1	January	Wednesday, January 12, 2005

***Figure 5-9.** Adding calculated columns to the Calendar table*

10. Using the Year and Month columns, create a Calendar hierarchy in the Date table.

11. In the Products table, insert a Weight Label column with the following formula:

    ```
    =if(ISBLANK([Weight]),BLANK(), [Weight] & " " &
    [WeightUnitMeasureID])
    ```

12. Create a Product Category column using the `RELATED` function:

    ```
    = RELATED('ProductCategory'[ProductCategory])
    ```

13. Using the `RELATED` function, create a ProductSubcategory column.

14. Hide the ProductCategory and ProductSubcategory tables from the client tools.

15. Switch to the Store table and create a Years Open column with the following formula:

    ```
    =TRUNC(YEARFRAC([OpenDate],
            If(ISBLANK([CloseDate]),TODAY(),[CloseDate])),0)
    ```

16. Create a Lifetime Sales column using the following formula:

    ```
    =SUMX(RELATEDTABLE(Sales),Sales[SalesAmount])
    ```

17. Save and close the file.

Summary

This chapter introduced you to the DAX language and the built-in functions that you can use to create calculations. At this point, you should be comfortable with creating calculated columns and using the DAX functions. I strongly recommend that you become familiar with the various functions available and how to use them in your analysis.

In the next chapter, you will continue working with DAX to create measures. *Measures* are one of the most important parts of building your model in Power BI; the measures are the reason you are looking at your data. You want to answer questions such as how sales are doing or what influences energy consumption. Along with creating measures, you will also see how filter context affects measures. Filter context is one of the most important concepts you need to master to create powerful Power BI models and reports.

CHAPTER 6

Creating Measures with DAX

Creating measures in DAX is the most important skill necessary to create solid data models. This chapter covers the common functions used to create measures in the data model. It also covers the important topic of data context and how to alter or override the context when creating measures.

After completing this chapter, you will be able to

- Understand the difference between measures and attributes
- Understand how context affects measurements
- Create common aggregates
- Know how and when to alter the filter context

Measures vs. Attributes

If you look at a typical star model for a data warehouse, you have a fact table surrounded by dimension tables. For example, Figure 6-1 shows a financial fact table surrounded by several dimension tables.

D. Clark, *Beginning Microsoft Power BI*, https://doi.org/10.1007/978-1-4842-5620-6_6

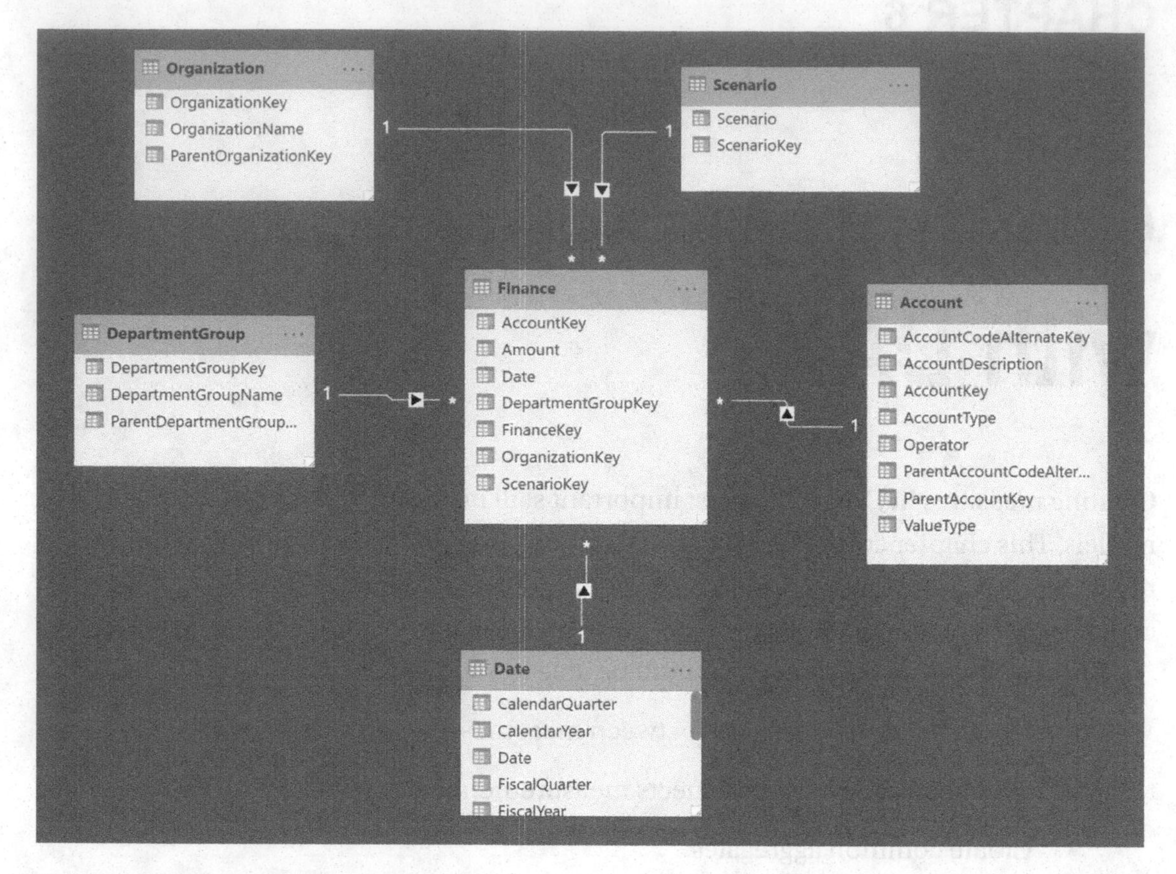

Figure 6-1. *Typical star schema*

Remember, the fact table contains numbers that you need to aggregate; for example, in the finance table, you have the amount, which is a monetary value that needs to be aggregated. In a sales fact table, you may have a sales amount and item counts. In a human resources system, you might have hours worked. The dimension tables contain the attributes that you are using to categorize and roll up the measures. For example, the financial measures are classified as profit, loss, and forecasted. You want to roll the values up to the department and organization level and you want to compare values between months and years.

When you start slicing and dicing the data in a matrix, the attributes become the row and column headers, whereas the measures are the values in the cells. Attributes are also commonly used as filters either in the filter pane or in a slicer. Figure 6-2 shows a matrix containing research and development spending, actual and budgeted, for the months in the fiscal year 2006.

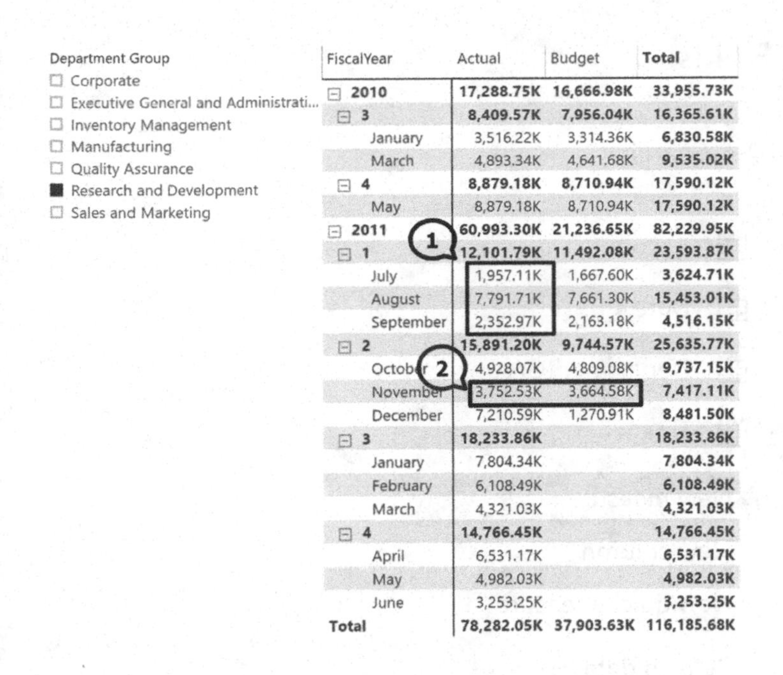

FiscalYear	Actual	Budget	Total
2010	17,288.75K	16,666.98K	33,955.73K
3	8,409.57K	7,956.04K	16,365.61K
January	3,516.22K	3,314.36K	6,830.58K
March	4,893.34K	4,641.68K	9,535.02K
4	8,879.18K	8,710.94K	17,590.12K
May	8,879.18K	8,710.94K	17,590.12K
2011	60,993.30K	21,236.65K	82,229.95K
1	12,101.79K	11,492.08K	23,593.87K
July	1,957.11K	1,667.60K	3,624.71K
August	7,791.71K	7,661.30K	15,453.01K
September	2,352.97K	2,163.18K	4,516.15K
2	15,891.20K	9,744.57K	25,635.77K
October	4,928.07K	4,809.08K	9,737.15K
November	3,752.53K	3,664.58K	7,417.11K
December	7,210.59K	1,270.91K	8,481.50K
3	18,233.86K		18,233.86K
January	7,804.34K		7,804.34K
February	6,108.49K		6,108.49K
March	4,321.03K		4,321.03K
4	14,766.45K		14,766.45K
April	6,531.17K		6,531.17K
May	4,982.03K		4,982.03K
June	3,253.25K		3,253.25K
Total	78,282.05K	37,903.63K	116,185.68K

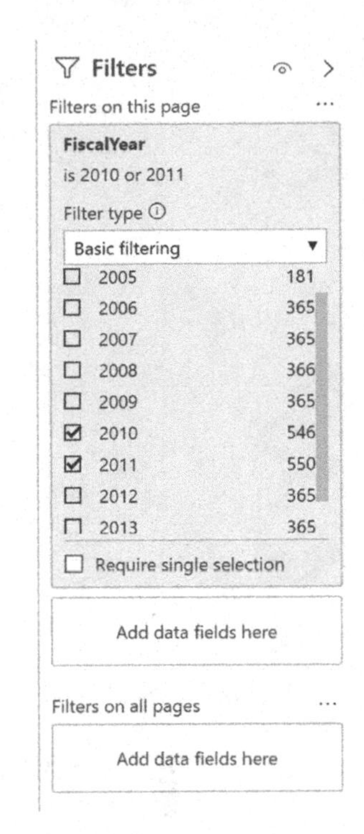

Figure 6-2. *Analyzing data in a matrix*

If you look at the filtering for each cell, you should realize they are all filtered a little differently. The three measures indicated by the first box differ by month, whereas the measures indicated by the second box differ by actual vs. budgeted amount. As you change the fiscal year, department, or organization, the values for the measures must be recalculated because the query context has changed.

In the following section, you will see how you can create some common aggregation measures in your Power BI model.

Creating Common Aggregates

It is very easy to create common aggregates such as sum, count, or average in Power BI. First, you need to determine which table you want to associate the measure with. If you follow the star schema model, this will most likely be the fact table, but it doesn't have to be.

To add the measure, in the Data view tab, right-click the table and select New measure (see Figure 6-3).

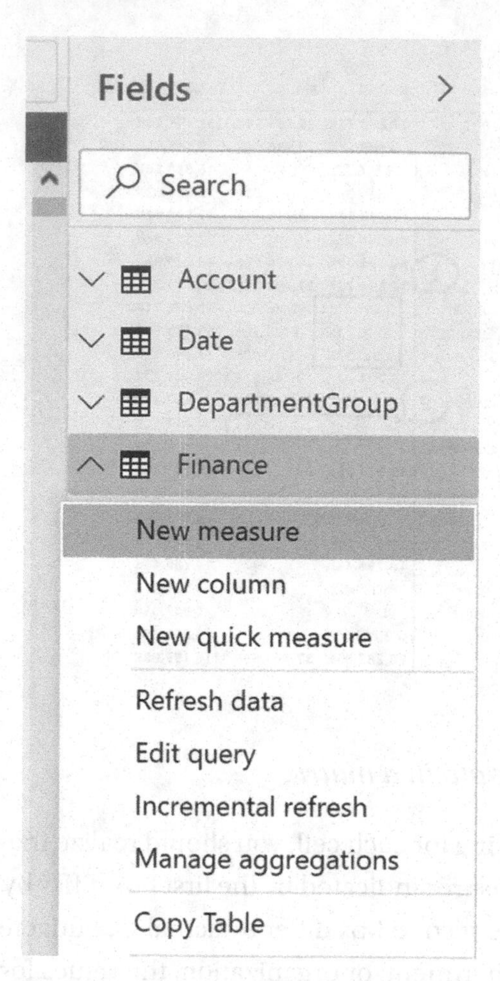

***Figure 6-3.** Adding a measure to the model*

Once you select New measure, you enter the measure into the formula bar (see Figure 6-4).

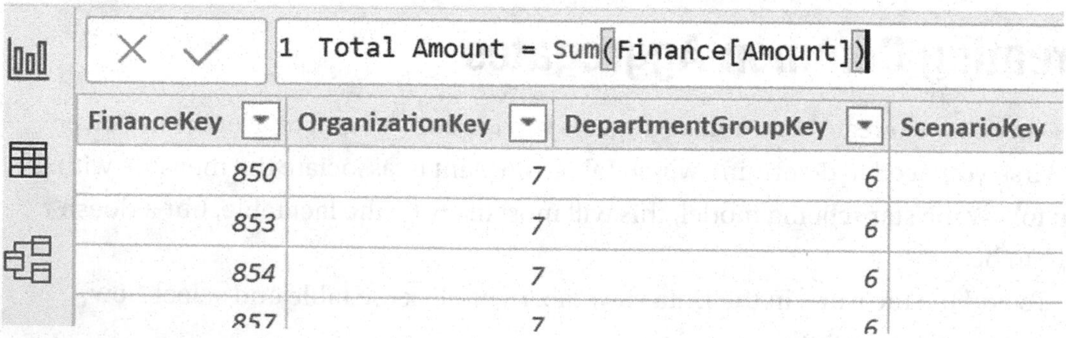

```
1 Total Amount = Sum(Finance[Amount])
```

FinanceKey	OrganizationKey	DepartmentGroupKey	ScenarioKey
850	7	6	
853	7	6	
854	7	6	
857	7	6	

***Figure 6-4.** Entering the DAX formula*

While creating a calculated column and a measure seem similar, they are very different. A calculated column is calculated when the model is loaded and is contained in a table. A measure is calculated on the fly as you change the filter context in the report or visual. The measure is not contained in a table but is merely associated with it. This determines where it shows up in the Field list. You can change the table association on the model (see Figure 6-5). This is also why you need to reference the table and the column when using a column in a measure formula.

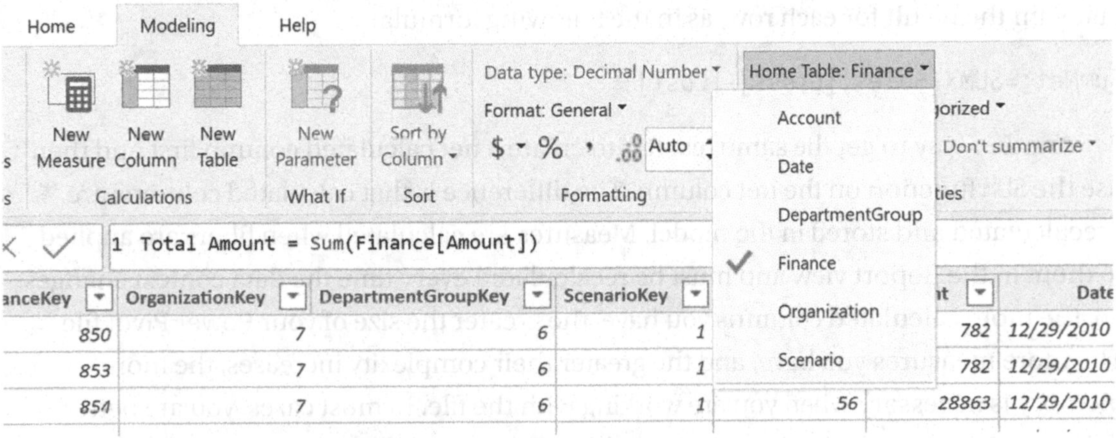

Figure 6-5. Entering the DAX formula

You may have noticed that the aggregate functions such as SUM, AVE, MIN, and MAX have corresponding SUMX, AVEX, MINX, and MAXX functions (see Figure 6-6).

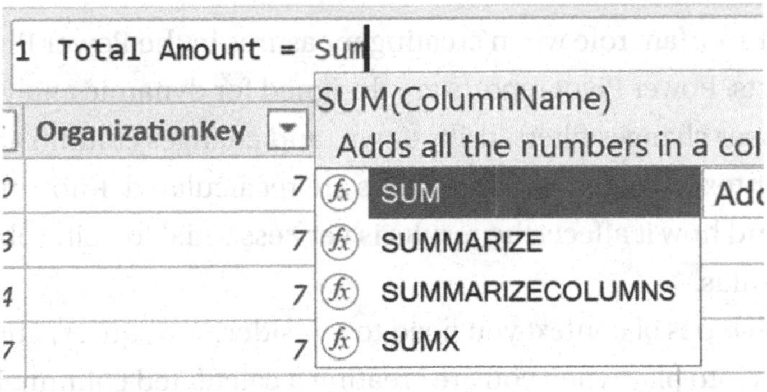

Figure 6-6. SUM and SUMX functions

The X functions are used when you are evaluating an expression for each row in the table and not just a single column. As an example, the SUMX function is defined as follows:

```
SUMX(<table>, <expression>)
```

where the table is the table containing the rows to be evaluated and the expression is what will be evaluated for each row.

As an example, say you have a sales table that contains a Cost and a Gross column. To figure out the total net sales amount, you can take the gross amount minus the cost and sum the result for each row, as in the following formula:

```
SumNet:=SUMX(Sales,[Gross]-[Cost])
```

Another way to get the same result is to create a net calculated column first and then use the SUM function on the net column. The difference is that calculated columns are precalculated and stored in the model. Measures are calculated when filters are applied to them in the Report view and must be recalculated every time the data context changes. So, the more calculated columns you have, the greater the size of your Power Pivot file. The more measures you have, and the greater their complexity increases, the more memory is necessary when you are working with the file. In most cases, you are better off doing as much calculation in memory rather than creating lots of calculated columns.

Understanding how data context changes the measurement value is very important when creating measures and is explored in the next section.

Mastering Data Context

Context plays an important role when creating measures in the Power Pivot model. Unlike static reports, Power Pivot reports are designed for dynamic analysis by the client. When the user changes filters, drills down, and changes column and row headers in a matrix, the context changes, and the values are recalculated. Knowing how the context changes and how it affects the results is very essential to being able to build and troubleshoot formulas.

There are three types of context you need to consider: row, query, and filter. The row context comes into play when you are creating a calculated column. It includes the values from all the other columns of the current row as well as the values of any table related to the row. If you create a calculated column, say margin

```
=[Gross] - [Cost]
```

DAX uses the row context to look up the values from the same row to complete the calculation. If you create a calculated column, such as lifetime sales

```
=SUMX(RELATEDTABLE(Sales),[SalesAmount])
```

DAX automatically looks up the related values using the row context of the current row. The row context is set once the model is loaded and will not change until new data is loaded. This is why calculated columns are precalculated and only need to be recalculated when data is refreshed.

Query context is the filtering applied to a cell in the matrix. When you drop a measure into a matrix, the DAX query engine examines the row and column headers and any filters applied. Each cell has a different query context applied to it (see Figure 6-7) and returns the value associated with the context. Because you can change the query context on the fly by changing row or column headers and filter values, the cell values are calculated dynamically, and the values are not held in the Power BI model.

FiscalYear: 2010

OrganizationName: Southeast Division

DepartmentGroup	Actual	Budget	Total
Corporate	137.83K	128.94K	**266.77K**
Executive General and Administration	583.30K	56.83K	**640.13K**
Inventory Management	7,512.78K	155.00K	**7,667.78K**
Manufacturing	115.09K	110.72K	**225.81K**
Quality Assurance	66.89K	65.18K	**132.07K**
Research and Development	2,425.49K	2,488.93K	**4,914.42K**
Sales and Marketing	20,261.75K		**20,261.75K**
Total	**31,103.12K**	**3,005.60K**	**34,108.72K**

***Figure 6-7.** The query context of a measure*

Filter context is added to the measure using filter constraints as part of the formula. The filter context is applied in addition to the row and query contexts. You can alter the context by adding to it, replacing it, or selectively changing it using filter expressions. For example, if you used the following formula to calculate sales,

```
AllStoreSales=CALCULATE(SUM(Sales[SalesAmount]),ALL(Store[StoreType]))
```

the filter context would clear any `StoreType` filter implemented by the query context.

In the next section, you will see why knowing when and how to alter the query context is an important aspect of creating measures.

Altering the Query Context

When creating calculations, you often need to alter the filter context being applied to the measure. For example, say you want to calculate the sales of a product category compared to the sales of all products (see Figure 6-8).

Category	ProductSales	ProductSalesRatio
Accessories	$571,297.9278	0.71%
Bikes	$66,302,381.557	82.41%
Clothing	$1,777,840.8391	2.21%
Components	$11,799,076.6584	14.67%
Total	**$80,450,596.9823**	**100.00%**

Figure 6-8. *Viewing the product sales ratio*

To calculate the sales ratio, you need to take the sales filtered by the query context (in this case, categories) and divide it by the sales of all products regardless of the product query context. To calculate sales, you just use the `SUM` function. To calculate the sum of all product sales, you need to override any product filtering applied to the cell. To do that, you use the `CALCULATE` function, which evaluates an expression in a context that is modified by the specified filters and has the following syntax:

```
CALCULATE(<expression>,<filter1>,<filter2>...)
```

where `expression` is essentially a measure to be evaluated and the `filters` are Boolean expressions or a table expression that defines the filters.

So, to override any product filters, you use the following code:

```
TotalProductSales=CALCULATE(SUM([SalesAmount]), ALL(Product))
```

This uses the `ALL` function, which returns all the rows in a table or all the values in a column, ignoring any filters that might have been applied. In this case, it clears all filters placed on the Product table. Figure 6-9 shows the measures in a matrix.

Category	ProductSales	TotalProductSales	ProductSalesRatio
Accessories	$571,297.9278	$80,450,596.9823	0.71%
Bikes	$66,302,381.557	$80,450,596.9823	82.41%
Clothing	$1,777,840.8391	$80,450,596.9823	2.21%
Components	$11,799,076.6584	$80,450,596.9823	14.67%
Total	**$80,450,596.9823**	**$80,450,596.9823**	**100.00%**

Figure 6-9. *Verifying the AllProductSales measure*

Notice the ProductSales measure is affected by the product filter (category), whereas the AllProductSales is not. The final step to calculate the ProductSalesRatio measure is to divide the ProductSales by the AllProductSales. You can use a measure inside another measure if you don't have a circular reference. So, the ProductSalesRatio is calculated as follows:

```
ProductSalesRatio = Divide([ProductSales],[TotalProductSales],0)
```

You can hide the AllProductSales measure from the client tools because, in this case, it is used as an intermediate measure and is not useful on its own.

In this section, you saw how to use the `CALCULATE` function and a filter function to alter the filters applied to a measure. There are many filter functions available in DAX, and it is important that you understand when to use them. The next section looks at several more important filter functions you can use.

Using Filter Functions

The filter functions in DAX allow you to create complex calculations that require you to interrogate and manipulate the data context of a row or cell in a matrix. Table 6-1 lists and describes some of the filter functions available in DAX.

Table 6-1. *Some DAX Filter Functions*

Function	Description
ALL	Returns all the rows in a table, or all the values in a column, ignoring any filters that might have been applied
ALLEXCEPT	Removes all context filters in the table except filters that have been applied to the specified columns
ALLNONBLANKROW	Returns all rows but the blank row and disregards any context filters that might exist
ALLSELECTED	Removes context filters from columns and rows, while retaining all other context filters or explicit filters
CALCULATE	Evaluates an expression in a context that is modified by the specified filters
CALCULATETABLE	Evaluates a table expression in a context modified by the given filters
DISTINCT	Returns a one-column table that contains the distinct values from the specified column
FILTER	Returns a table that represents a subset of another table or expression
FILTERS	Returns the values that are directly applied as filters
HASONEVALUE	Returns TRUE when the context has been filtered down to one distinct value
ISFILTERED	Returns TRUE when a direct filter is being applied
ISCROSSFILTERED	Returns TRUE when the column or another column in the same or related table is being filtered
KEEPFILTERS	Modifies how filters are applied while evaluating a CALCULATE or CALCULATETABLE function. Keeps applied filters and adds additional filters
RELATED	Returns a related value from another table
USERELATIONSHIP	Specifies the relationship to be used in a specific calculation
VALUES	Returns a one-column table that contains the distinct values from the specified column

You have already seen how you can use the CALCULATE function in combination with the ALL function to calculate the total product sales, ignoring any product filtering applied. Let's look at a few more examples.

Figure 6-10 shows a Power Pivot model for reseller sales. In the model, there is an inactive relationship between the Employee and the SalesTerritory tables. You can use this relationship to calculate the number of salespeople in each country.

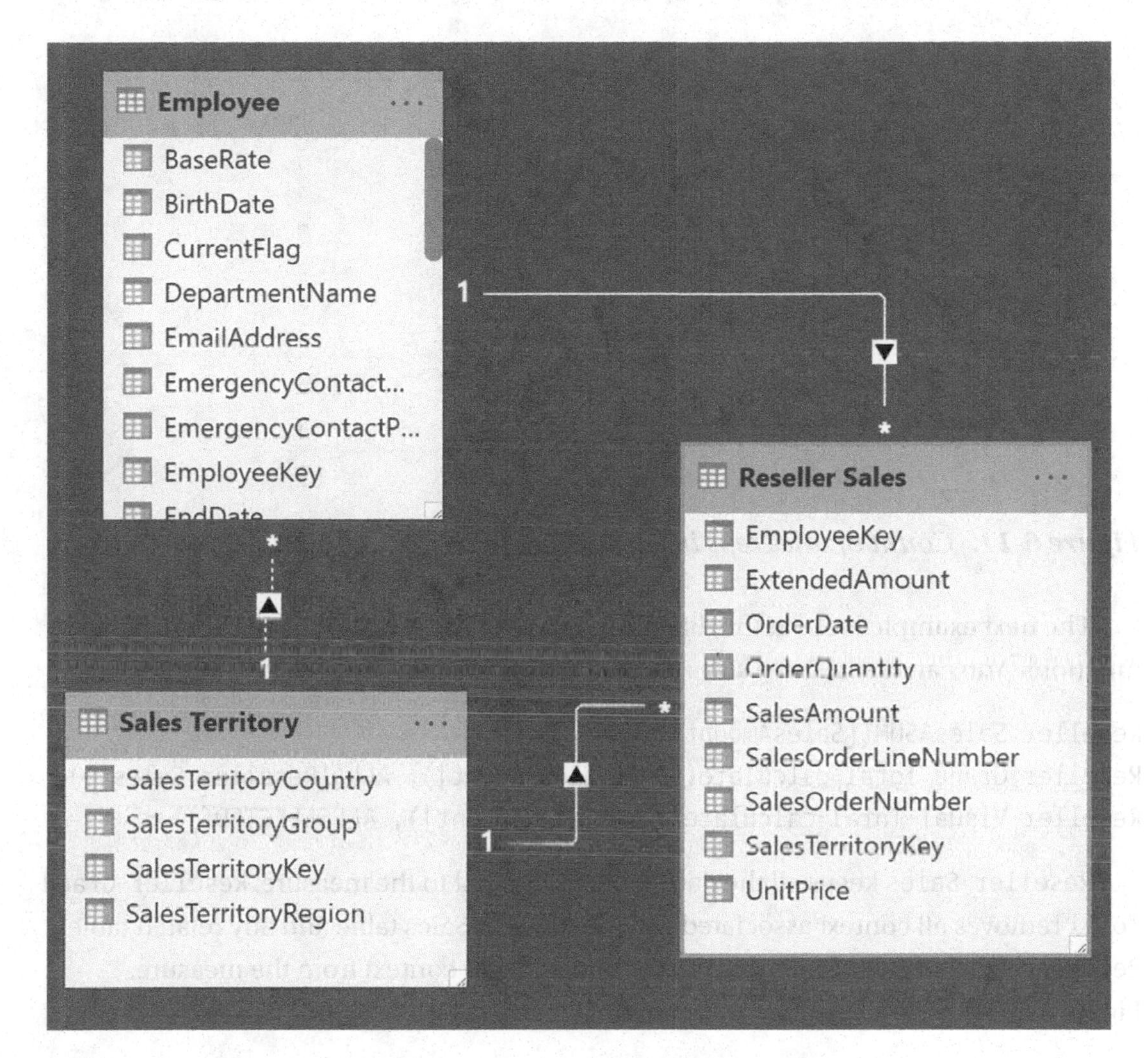

***Figure 6-10.** An inactive relationship in the Power BI model*

You calculate the number of sales reps in each country using the following code:

```
Sales Rep Cnt=CALCULATE(
DISTINCTCOUNT(Employee[EmployeeNationalIDAlternateKey]),
USERELATIONSHIP(Employee[SalesTerritoryKey],
'Sales Territory'[SalesTerritoryKey]))
```

In this case, you need to use the CALCULATE function so that you can apply the filter function USERELATIONSHIP to tell the DAX query engine which relationship to use. Figure 6-11 shows the resulting matrix.

SalesTerritoryCountry	Sales Rep Cnt
United States	8
NA	3
Canada	2
Australia	1
France	1
Germany	1
United Kingdom	1
Total	**17**

Figure 6-11. *Count of sales reps in each country*

The next example looks at the difference between the ALL and the ALLSELECTED filter functions. You can create three sales amount measures as follows:

```
Reseller Sales=SUM([SalesAmount])
Reseller Grand Total=calculate(sum([SalesAmount]), ALL('Reseller Sales'))
Reseller Visual Total=calculate(sum([SalesAmount]), ALLSELECTED())
```

`Reseller Sales` keeps all the data contexts applied to the measure. `Reseller Grand Total` removes all context associated with the ResellerSales table and any related table. `Reseller Visual Total` removes the column and row context from the measure. Figure 6-12 shows the resulting measures in a matrix.

StoreType
- ☐ (Blank)
- ■ Catalog
- ■ Online
- ■ Reseller
- ☐ Store

StoreType	Reseller Sales	Reseller Grand Total	Reseller Visual Total
Catalog	$13,137.3691	$80,450,596.9823	$1,153,619.7582
Online	$303,047.0449	$80,450,596.9823	$1,153,619.7582
Reseller	$837,435.3442	$80,450,596.9823	$1,153,619.7582
Total	**$1,153,619.7582**	**$80,450,596.9823**	**$1,153,619.7582**

Figure 6-12. *Results of using different filters*

Now let's look at a more complex example. In this example, you want to determine the best single-order customers in a particular time period. The final matrix is shown in Figure 6-13.

LastName	Large Sales	Top Sale	Date of Top Sale
Adams	$44,492.75	$2,456.41	12/20/2013
Alexander	$47,167.02	$2,414.99	3/2/2013
Allen	$34,450.42	$2,543.04	11/6/2013
Anderson	$31,996.05	$4,791.33	8/5/2013
Bailey	$41,863.83	$2,453.04	6/6/2013
Baker	$39,086.32	$4,085.95	9/9/2013
Barnes	$39,436.02	$3,548.54	7/29/2013
Brooks	$57,883.48	$3,028.02	5/30/2013
Brown	$31,362.58	$2,420.34	11/9/2013
Bryant	$64,567.71	$3,036.95	5/9/2013
Butler	$71,548.83	$3,706.36	11/2/2013
Campbell	$36,600	$3,535.95	9/10/2013
Carter	$31,622.15	$3,018.53	11/20/2013
Clark	$43,175.52	$4,892.1	12/14/2013
Coleman	$63,955.4	$2,945.55	7/24/2013
Collins	$41,487.32	$2,478.24	8/12/2013
Cook	$47,298.69	$4,170.04	12/4/2013
Cooper	$37,243.8	$4,142.94	9/9/2013
Cox	$51,341.14	$3,517.24	7/31/2013

CalendarYear
- ☐ 2005
- ☐ 2006
- ☐ 2007
- ☐ 2008
- ☐ 2009
- ☐ 2010
- ☐ 2011
- ☐ 2012
- ■ 2013
- ☐ 2014

CountryRegionName
- ☐ Australia
- ☐ Canada
- ☐ France
- ☐ Germany
- ☐ United Kingdom
- ■ United States

Figure 6-13. *Finding best single-order customers*

The first step is to find the customers who spent a lot of money during the time period. To calculate customer sales, you use the following measure:

```
Sum Sales=SUM([SalesAmount])
```

Next, you want to only look at large spenders (over $30,000 spent during the period) so you can filter out smaller values:

```
Large Sales=IF([Sum Sales]>=30000,[Sum Sales],Blank())
```

The next step is to find the order amounts for the customer and take the maximum value:

```
Top Sale=MAXX(VALUES(Date[Date]),[Sum Sales])
```

Because you only want to list the top sales for top customers, you can add an IF statement to make sure the customer has large sales:

```
Top Sale=
IF(ISBLANK([Large Sales]),Blank(),MAXX(VALUES(Date[Date]),[Sum Sales]))
```

As a final example, say you are working with the HR department and you want to create a matrix that will allow them to list employee counts for the departments at a particular date. There is an EmployeeDepartmentHistory table that lists employee, department, start date, and end date. There is also a Dates table that has a row for every date spanning the department histories. Figure 6-14 shows the matrix containing employee counts for each department.

As of Date

1/1/2009

Department	Emp Cnt
Document Control	1
Engineering	5
Facilities and Maintenance	1
Finance	2
Human Resources	4
Information Services	3
Marketing	2
Production	55
Production Control	2
Purchasing	1
Quality Assurance	2
Research and Development	1
Shipping and Receiving	3
Tool Design	1
Total	**83**

Figure 6-14. *Employee counts as of the selected date*

As of Date is used as a filter, and Emp Cnt is the measure. When the As of Date is changed, the Emp Cnt is recalculated to show the employee counts on that date. Figure 6-15 shows new counts after the date is changed.

As of Date

1/1/2010

Department	Emp Cnt
Document Control	5
Engineering	5
Executive	1
Facilities and Maintenance	3
Finance	11
Human Resources	6
Information Services	10
Marketing	6
Production	159
Production Control	5
Purchasing	3
Quality Assurance	5
Research and Development	4
Shipping and Receiving	6
Tool Design	1
Total	**230**

Figure 6-15. *Changing the As of Date*

The first step to creating the Emp Cnt is to use the COUNT function because you want to count the EmployeeID in the table:

```
Emp Cnt=COUNT(EmpDepHist[BusinessEntityID])
```

Because you need to filter the table to only active employees at the date chosen, you need to change this to the COUNTX function:

```
Emp Cnt=COUNTX(EmpDepHist, EmpDepHist[BusinessEntityID])
```

To filter the EmpDepHist table, you use the FILTER function:

```
FILTER(<table>,<filter>)
```

The FILTER function is a Boolean expression that evaluates to TRUE. In this case, you need to have the date that the employee started in the department less than or equal to the As of Date:

```
EmpDepHist[StartDate]<=Dates[As of Date]
```

Now, because the matrix user can select more than one date, and you want to make sure you only compare it to a single date, you can use the MAX function:

```
EmpDepHist[StartDate]<=MAX(Dates[As of Date])
```

You also want to make sure the date the employee left the department is greater than the As of Date:

```
EmpDepHist[EndDate] > Max(Dates[As of Date]
```

If the employee is currently in the department, the EndDate will be blank:

```
ISBLANK(EmpDepHist[EndDate])
```

When you combine these filter conditions, you get the following filter condition:

```
EmpDepHist[StartDate]<=MAX(Dates[As of Date])
&& (ISBLANK(EmpDepHist[EndDate]) || EmpDepHist[EndDate] > Max(Dates[As of Date]))
```

The final FILTER function then becomes the following:

```
FILTER(EmpDepHist,
EmpDepHist[StartDate]<=MAX(Dates[As of Date])
&& (ISBLANK(EmpDepHist[EndDate]) || EmpDepHist[EndDate] >
Max(Dates[As of Date])))
```

And the final employee count measure becomes the following:

```
Emp Cnt=COUNTX(FILTER(EmpDepHist,
EmpDepHist[StartDate]<=MAX(Dates[As of Date])
&& (ISBLANK(EmpDepHist[EndDate]) || EmpDepHist[EndDate] >
Max(Dates[As of Date]))),
EmpDepHist[BusinessEntityID])
```

As you can see, creating a measure can be quite complex, but if you break it up into steps, it becomes very manageable. A good practice to make the measures more manageable is to use variables. In the next section, you will learn to use variables in your measures.

Using Variables in DAX

As your measures become increasingly complex, it is a good idea to start using variables in your measure definitions. Using variables increases both the performance and readability of your code.

Variables can be both scalar values and tables. You define variables using the VAR keyword and use them in the return clause. For example, the following measure calculates the sales growth over the previous year:

```
 Sales Growth =
Var
    CurrentSales = Sum('Reseller Sales'[SalesAmount])
Var
    PrevYearSales = Calculate(
        Sum('Reseller Sales'[SalesAmount]),
        SAMEPERIODLASTYEAR('DimDate'[Datekey])
    )
Return
    Divide(
        CurrentSales - PrevYearSales,
        PrevYearSales,
        BLANK()
    )
```

As you can see, using variables and indenting can greatly increase the readability of your code. Although writing complex DAX measures can be daunting at first, rest assured the more you work with DAX and creating measures, the more intuitive and easier it becomes.

Now that you have seen how to create measures and alter the data context using DAX, it is time to get your hands dirty and create some measures in the following lab.

HANDS-ON LAB: CREATING MEASURES IN POWER PIVOT

In the following lab, you will

- Create aggregate measures
- Alter the data context in a measure
- Use a nonactive relationship in a measure
- Create a complex measure
- Create a KPI

1. In the LabStarterFiles\Chapter6Lab1 folder, open the Chapter6Lab1.pbix file. This file contains a data model consisting of sales data, product data, and store data.
2. View the model in the Power BI model using the Model view (Figure 6-16).

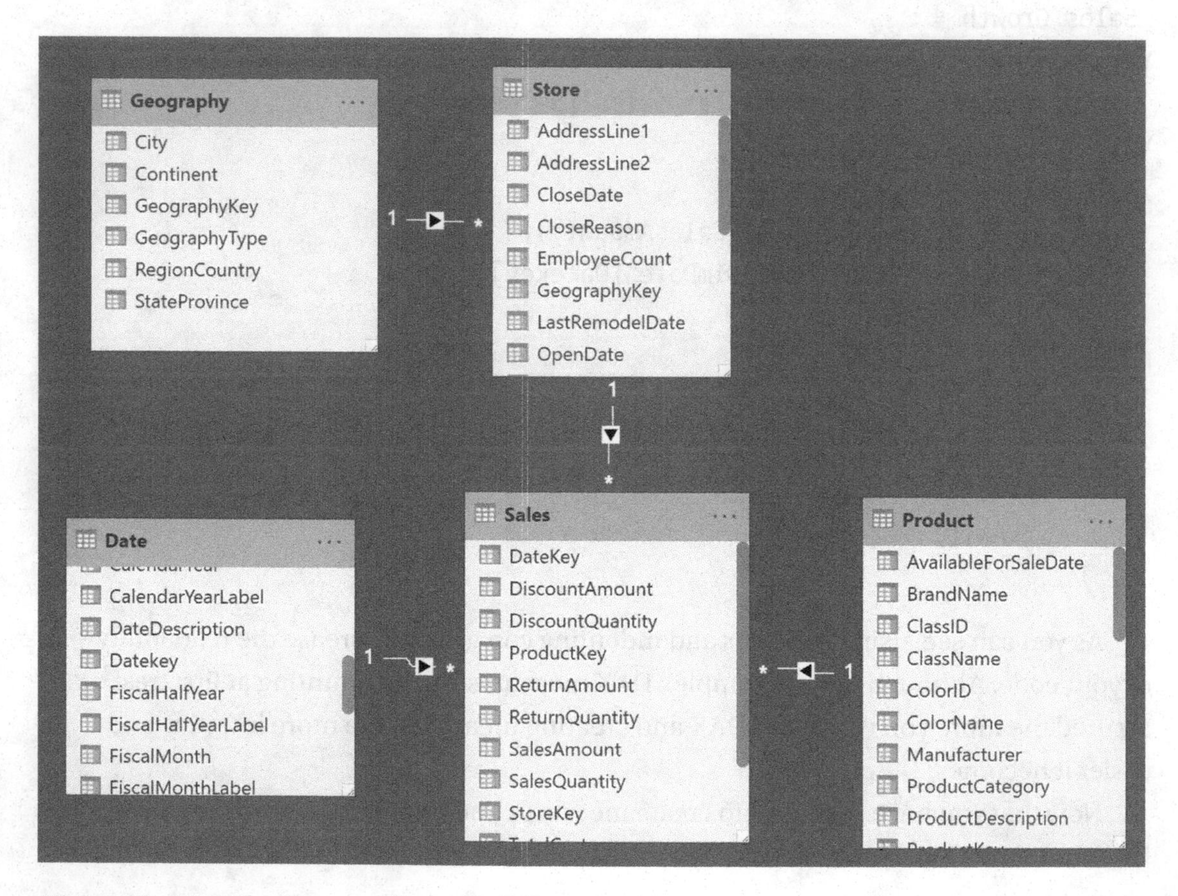

***Figure 6-16.** The Power Pivot model*

3. Switch to Data view in the Power BI window and select the Sales table.

4. Under the Sales table, create a Sum Sales measure by right-clicking the table name and choosing "New measure":

```
Sum Sales = Sum(Sales[SalesAmount])
```

5. Also create a Max Sales Quantity, a Min Sales Quantity, and an Ave Sales Quantity measure under the Sales table.

6. To test how the measures are recalculated as the filter context changes, switch to the Report tab and add a Matrix visual to the report (see Figure 6-17).

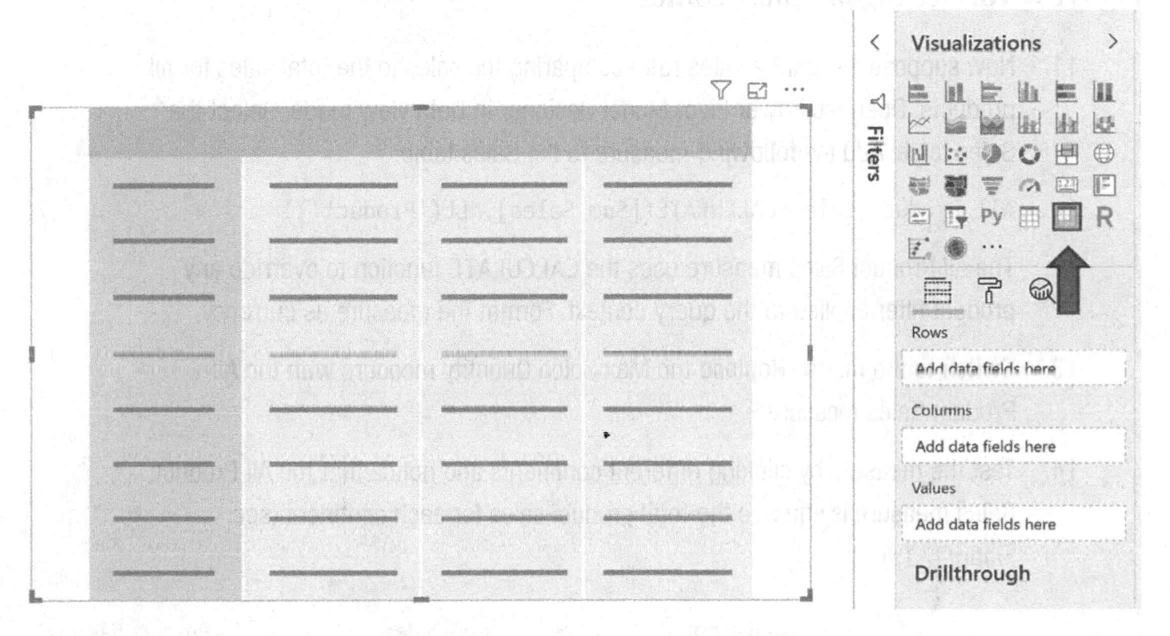

***Figure 6-17.** Creating a matrix*

7. If you don't see the field wells, click the matrix to show them.

8. Add the Sum Sales and Max Sales Quantity measures to the Values drop area. Under the Product table, add the ProductCategory to the Rows drop area.

9. Add a Slicer visual to the report that allows you to slice by the Continent field in the Geography table.

10. The report should look like the one shown in Figure 6-18. Test the measures by clicking the different continents. This changes the query context. Notice how the measure values are recalculated as the query context changes.

Continent

☐ Asia
■ Europe
☐ North America

ProductCategory	Sum Sales	Max Sales Quantity
Audio	$28,327,054.1609	100
Cameras and camcorders	$497,254,487.8651	100
Cell phones	$171,965,337.0499	2880
Computers	$635,370,758.1587	180
Games and Toys	$31,323,167.2138	400
Home Appliances	$771,257,471.64	96
Music, Movies and Audio Books	$32,047,454.4966	72
TV and Video	$258,602,375.7187	100
Total	**$2,426,148,106.3037**	**2880**

***Figure 6-18.** Testing the query context*

11. Now suppose we want a sales ratio comparing the sales to the total sales for all products. Open the Power Pivot Model Designer in Data view mode. Select the Sales table. Add the following measure to the Sales table:

```
All Product Sales=CALCULATE([Sum Sales],ALL('Product'))
```

12. The All Product Sales measure uses the `CALCULATE` function to override any product filter applied to the query context. Format the measure as currency.

13. Switch to the matrix. Replace the Max Sales Quantity measure with the All Product Sales measure.

14. Test the measure by clicking different continents and notice that the All Product Sales measure is equal to the total product sales for each continent (see Figure 6-19).

Continent

☐ Asia
■ Europe
☐ North America

ProductCategory	Sum Sales	All Product Sales
Audio	$28,327,054.1609	$2,426,148,106.30
Cameras and camcorders	$497,254,487.8651	$2,426,148,106.30
Cell phones	$171,965,337.0499	$2,426,148,106.30
Computers	$635,370,758.1587	$2,426,148,106.30
Games and Toys	$31,323,167.2138	$2,426,148,106.30
Home Appliances	$771,257,471.64	$2,426,148,106.30
Music, Movies and Audio Books	$32,047,454.4966	$2,426,148,106.30
TV and Video	$258,602,375.7187	$2,426,148,106.30
Total	**$2,426,148,106.3037**	**$2,426,148,106.30**

***Figure 6-19.** Testing the measure*

15. Switch back to the Data view tab and add the following measure to the Sales table (format the measure as percentage):

```
Product Sales Ratio=Divide([Sum Sales],[All Product Sales],Blank())
```

16. Return to the Report view tab and add the Product Sales Ratio as a value in the matrix you just created.

17. Open the Model view tab. To create a relationship between the Date table and the Store table, drag the OpenDate field from the Store table and drop it on top of the DateKey in the Date table (see Figure 6-20). Notice that this is not the active relationship between the Store and the Date tables as indicated by the dashed line. This is because the active relationship goes from the Store table, through the Sales table, and then to the Date table.

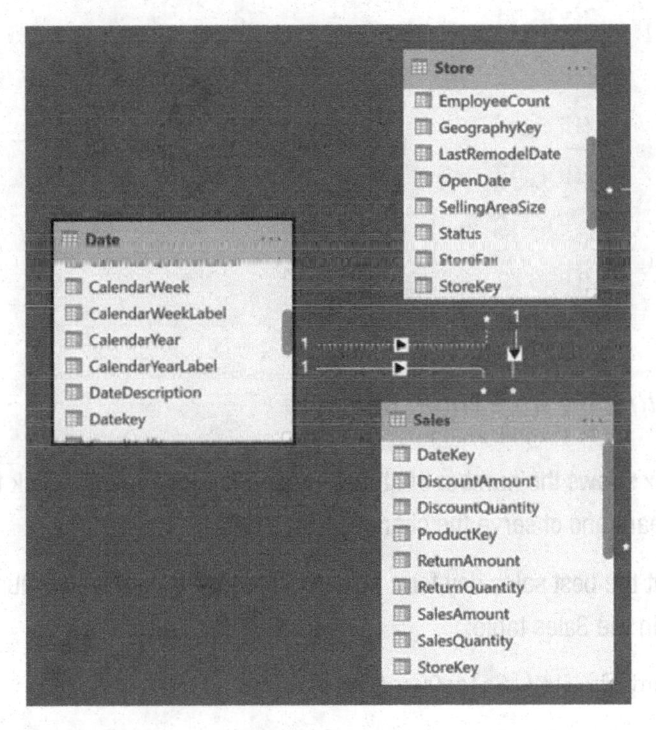

Figure 6-20. *Creating an inactive relationship*

18. Switch to the Data view tab and add the following measure to the Store table. Because you are using a nonactive relationship, you need to use the USERELATIONSHIP function:

```
Store Count=CALCULATE(DISTINCTCOUNT([StoreKey]),
USERELATIONSHIP(Store[OpenDate],'Date'[Datekey]))
```

19. To test the store count measure, create a matrix on a new report page. Use the Continent as the column labels and the CalendarMonth as the row labels. Use the Store Count measure as the value. Insert a slicer using CalendarYear. Change it from a slider to a list. Your report should look like Figure 6-21.

CalendarYear
- ☐ (Blank)
- ☐ 2005
- ■ 2006
- ☐ 2007
- ☐ 2008
- ☐ 2009
- ☐ 2010

CalendarMonth	Asia	Europe	North America	Total
200601			2	2
200602	1	2		3
200604	1			1
200607	1			1
200608	1			1
200611	1			1
Total	**5**	**2**	**2**	**9**

Figure 6-21. *Testing the store count measure*

20. The matrix shows the number of stores opened during a month. Click the various years and observe the changes in the data.

21. To find out the best sales day for a product category, create a Sale Quantity measure in the Sales table:

```
Sale Quantity=SUM([SalesQuantity])
```

22. Use the Sale Quantity measure to create a Top Sale Day Quantity measure. The MAXX function is used to break any ties and returns the most recent DateKey:

```
Top Sale Day Quantity=MAXX(values('Date'[Datekey]),[Sale Quantity])
```

23. To figure out the date of the top sales day, you first create a filter function that returns the dates when the Sale Quantity equals the Top Sale Day Quantity for the period:

```
Filter(VALUES('Date'[Datekey]),
[Sale Quantity]=CALCULATE([Top Sale Day Quantity],
VALUES('Date'[Datekey])))
```

24. This filter is then inserted into a CALCULATE function that returns the most recent date. Format the measure as a short date (select More Formats in the Format drop-down):

```
Top Sale Day=CALCULATE(MAX('Date'[Datekey]),
Filter(VALUES('Date'[Datekey]),[Sale Quantity]=
CALCULATE([Top Sale Day Quantity],VALUES('Date'[Datekey]))))
```

25. Create a matrix like the one in Figure 6-22 to test your measures.

ProductCategory	Top Sale Day Quantity	Top Sale Day
Audio	2477	11/26/2009
Cameras and camcorders	9251	11/2/2007
Cell phones	27438	8/20/2009
Computers	12898	12/20/2009
Games and Toys	9674	7/15/2009
Home Appliances	12096	11/15/2009
Music, Movies and Audio Books	2344	2/15/2007
TV and Video	4334	12/18/2007
Total	**69757**	**12/27/2009**

Figure 6-22. *Testing the Top Sale Day measure*

26. Change the Product Sales Ratio measure to use variables:

```
Product Sales Ratio =
Var SumSales = Sum(Sales[SalesAmount])
Var AllProductSales = Calculate(Sum(Sales[SalesAmount]),ALL('Product'))
Return
Divide(SumSales,AllProductSales,Blank())
```

27. Verify the Product Sales Ratio is still giving valid results. When done, save the file and close Power BI Desktop.

Summary

This was a long and meaty chapter. You now have a firm grasp of how to create measures in your Power BI model. You should also understand data context and how it affects the measurements. This can be a very confusing concept when you start to develop more complex measures. Don't worry—the more you work with it, the clearer it becomes.

The next chapter extends the concepts of this chapter. One of the most common types of data analysis is comparing values over time. Chapter 7 shows you how to correctly implement time-based analysis in Power BI. It includes setting up a date table and using the various built-in functions for analyzing values to date, comparing values from different periods, and performing semi-additive aggregations.

CHAPTER 7

Incorporating Time Intelligence

One of the most common types of data analysis is comparing values over time. This chapter shows you how to correctly implement time-based analysis in Power BI. It includes setting up a date table and using the various built-in functions for analyzing values to date, comparing values from different periods, and performing semi-additive aggregations.

After completing this chapter, you will be able to

- Create a date table
- Use DAX for time period–based evaluations
- Shift the date context using filter functions
- Create semi-additive measures

Date-Based Analysis

A large percentage of data analysis involves some sort of datetime-based aggregation and comparison. For example, you may need to look at usage or sales for the month-to-date (MTD) or year-to-date (YTD), as shown in Figure 7-1.

D. Clark, *Beginning Microsoft Power BI*, https://doi.org/10.1007/978-1-4842-5620-6_7

Year	Sum of Sales	MTD Sales	YTD Sales
⊟ **Year 2007**			
January	$33,937,209.25	$33,937,209.25	$33,937,209.25
February	$39,627,318.77	$39,627,318.77	$73,564,528.02
March	$43,055,837.85	$43,055,837.85	$116,620,365.87
April	$45,152,238.05	$45,152,238.05	$161,772,603.92
May	$49,678,655.61	$49,678,655.61	$211,451,259.53
June	$49,431,542.53	$49,431,542.53	$260,882,802.06
July	$53,458,586.99	$53,458,586.99	$314,341,389.05
August	$50,771,841.27	$50,771,841.27	$365,113,230.32
September	$49,242,024.87	$49,242,024.87	$414,355,255.19
October	$47,437,924.07	$47,437,924.07	$461,793,179.26
November	$54,262,312.57	$54,262,312.57	$516,055,491.83
December	$55,148,659.27	$55,148,659.27	$571,204,151.10
⊞ **Year 2008**	**$600,175,898.67**	**$56,659,783.37**	**$600,175,898.67**
⊞ **Year 2009**	**$543,817,781.67**	**$46,788,451.58**	**$543,817,781.67**

Channel
- ☐ Catalog
- ☐ Online
- ■ Reseller
- ☐ Store

***Figure 7-1.** Calculating month-to-date and year-to-date sales*

Another common example is looking at performance from one time period to the next. For example, you may want to compare previous months' sales with current sales (see Figure 7-2) or sales for the current month to the same month a year before.

Row Labels	Sum of Sales	Prev Month Sales	Monthly Sales Growth
⊟ **2007**	**$10,697,642.89**		
January	$675,656.58		
February	$615,633.57	$675,656.58	-8.88 %
March	$614,046.66	$615,633.57	-0.26 %
April	$707,348.69	$614,046.66	15.19 %
May	$966,623.91	$707,348.69	36.65 %
June	$987,545.18	$966,623.91	2.16 %
July	$996,175.98	$987,545.18	0.87 %
August	$883,280.80	$996,175.98	-11.33 %
September	$983,234.05	$883,280.80	11.32 %
October	$908,519.60	$983,234.05	-7.60 %
November	$1,172,632.83	$908,519.60	29.07 %
December	$1,186,945.05	$1,172,632.83	1.22 %
⊞ **2008**	**$12,802,000.68**		
⊞ **2009**	**$13,768,349.90**		

RegionCountryNa...
- Armenia
- Australia
- Bhutan
- Canada
- China
- Denmark
- France
- Germany

***Figure 7-2.** Calculating sales growth*

In addition to these common data analytics, there are also times when you need to base your aggregations on measures that are nonadditive, such as account balances or inventory. In these cases, you need to determine the last value entered and use that value to aggregate across the different time periods (see Figure 7-3).

StoreQuantity	Colur										
Row Labels	200901	200902	200903	200904	200905	200906	200907	200908	200909	200910	200911	Grand Total
Contoso 8GB Super-Slim MP3/Video Player M800 Pink	44	61	23	185	74	151	44	121	86	70	100	194
Contoso Catalog Store	22	61	23	29	24	32		53	30	46	42	42
Contoso North America Online Store	22			156	26	45		40	30	24	58	58
Contoso Ottawa No.1 Store					24	24						24
Contoso Ottawa No.2 Store								28	26			26
Contoso Toronto No.1 Store						27	22					22
Contoso Toronto No.2 Store						23	22					22

Figure 7-3. *Aggregating inventory amounts*

DAX contains many functions that help you create the various datetime-based analyses you may need. In the next section, you will see how to create a date table that is required to use many of the datetime-based functions.

Creating a Date Table

To use the built-in time intelligence functions in DAX, you need to have a date table in your model for the functions to reference. The only requirement for the table is that it needs a distinct row for each day in the date range at which you are interested in looking. Each of these rows needs to contain the full date of the day. The date table can, and often does, have more columns, but it doesn't have to.

There are several ways to create the date table. You can either import a date table from the source if one is available or use the DAX Calendar function. To use the Calendar function, on the Modeling tab, click the New Table button. In the formula bar, enter the CALENDAR function with the appropriate date range or the CALENDARAUTO function, which looks at the model and determines the correct date range (see Figure 7-4).

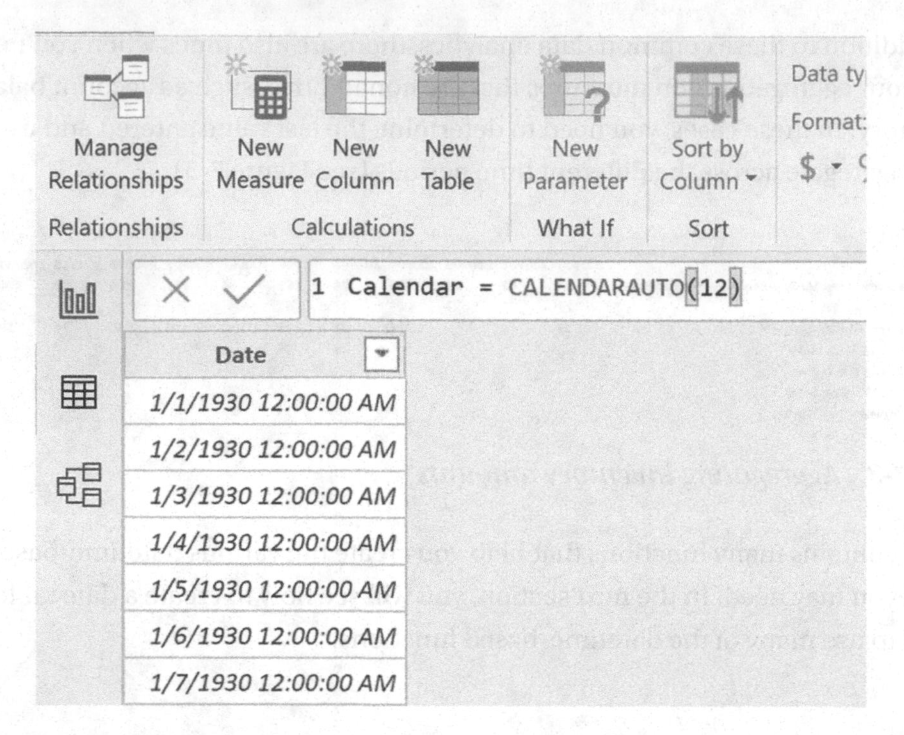

***Figure 7-4.** Creating a date table*

After creating the date table, you can use DAX to create additional calculated columns such as month, year, and weekday. The final step is to create a relationship between the date table and the table that contains the values you want to analyze.

Once you have the table in the model, you need to mark it as the official date table (see Figure 7-5) and indicate which column is the unique key (see Figure 7-6). This tells the DAX query engine to use this table as a reference for constructing the set of dates needed for a calculation. For example, if you want to look at year-to-date sales, the query engine uses this table to get the set of dates it needs.

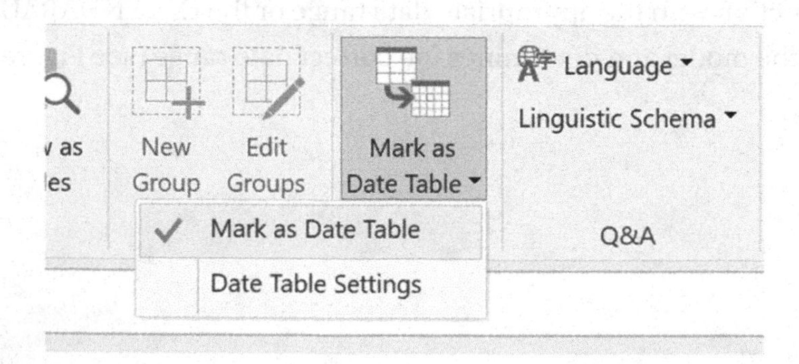

***Figure 7-5.** Identifying the date table*

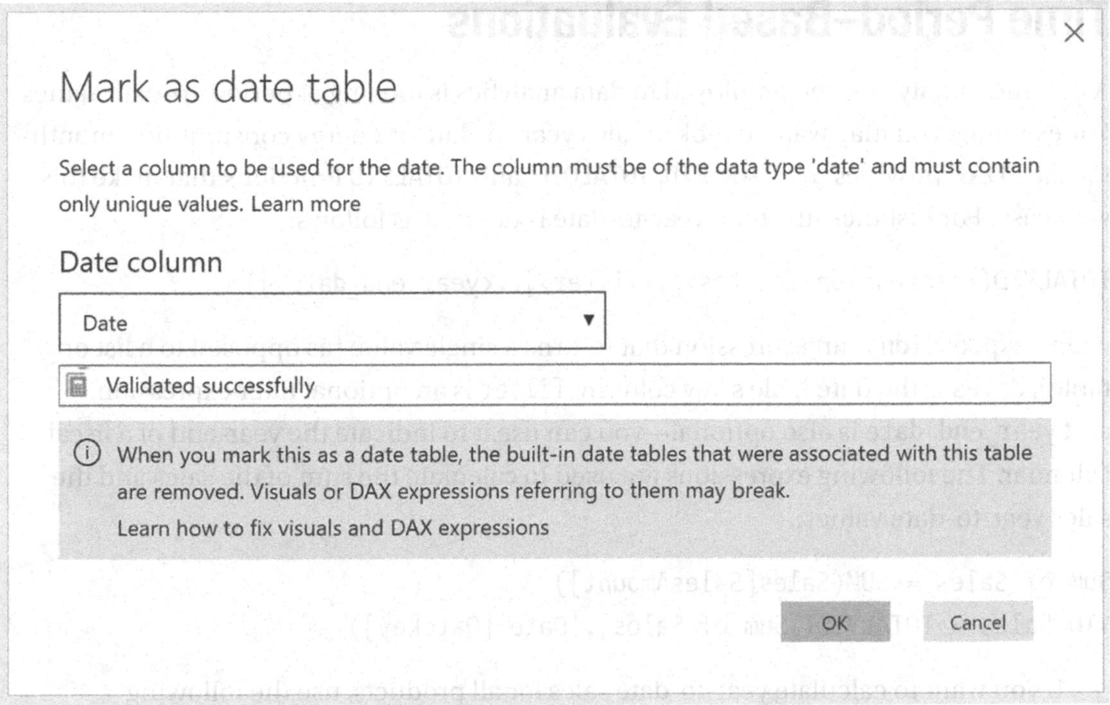

Mark as date table

Select a column to be used for the date. The column must be of the data type 'date' and must contain only unique values. Learn more

Date column

Date

Validated successfully

When you mark this as a date table, the built-in date tables that were associated with this table are removed. Visuals or DAX expressions referring to them may break.

Learn how to fix visuals and DAX expressions

OK Cancel

Figure 7-6. *Setting the date key*

There are many built-in time intelligent functions in DAX. Some of these functions, like `FIRSTNONBLANK`, return a single date. Some return a set of dates, such as `PREVIOUSMONTH`. And still others, like `TOTALYTD`, evaluate expressions over time. At this point, the DAX built-in time intelligence functions support the traditional calendar ending on December 31. They also support a fiscal calendar that has a different year-end date and contains four quarters containing three months each. If you need to use a custom financial calendar, you must create your own custom calculations.

Now that you understand how to create and designate the date table in your model, it is time to look at implementing some of the common time intelligent functions to analyze your data.

Time Period–Based Evaluations

A common analysis often employed in data analytics is looking at period-to-date values. For example, you may want to look at sales year-to-date or energy consumption month-to-date. DAX provides the `TOTALMTD`, `TOTALQTD`, and `TOTALYTD` functions that make this very easy. For instance, the total year-to-date is defined as follows:

```
TOTALYTD(<expression>,<dates>[,<filter>][,<year_end_date>])
```

where `expression` is an expression that returns a single value (as opposed to a list or table), `dates` is the date table's key column, `filter` is an optional filter expression, and `year_end_date` is also optional—you can use it to indicate the year-end of a fiscal calendar. The following expressions are used to calculate the sum of the sales and the sales year-to-date values:

```
Sum of Sales = SUM(Sales[SalesAmount])
YTD Sales = TOTALYTD([Sum of Sales],'Date'[Datekey])
```

If you want to calculate year-to-date sales for all products, use the following expression:

```
YTD Sales ALL Products = TOTALYTD([Sum of Sales],'Date'[Datekey],
ALL('Product'))
```

Figure 7-7 shows the measures in a matrix. You can use these base measures to calculate further measures, such as percent of year-to-date sales and percent of all product sales.

Category
- ☐ Audio
- ☐ Cameras and camcor...
- ☐ Cell phones
- ■ Computers
- ☐ Games and Toys
- ☐ Home Appliances
- ☐ Music, Movies and Au...
- ☐ TV and Video

Year	Sum of Sales	YTD Sales	YTD Sales ALL Products
⊟ **Year 2007**			
January	$76,580,425.17	$76,580,425.17	$269,835,263.23
February	$88,914,528.27	$165,494,953.44	$568,051,231.58
March	$85,714,037.10	$251,208,990.54	$868,538,158.49
April	$108,206,075.80	$359,415,066.34	$1,268,698,490.08
May	$106,963,757.90	$466,378,824.24	$1,692,127,617.87
June	$104,713,790.47	$571,092,614.71	$2,101,925,163.42
July	$97,226,447.91	$668,319,062.62	$2,491,542,535.70
August	$92,298,097.12	$760,617,159.73	$2,879,972,362.81
September	$88,743,351.91	$849,360,511.64	$3,259,116,962.37
October	$95,184,597.09	$944,545,108.73	$3,682,330,203.21
November	$104,034,888.68	$1,048,579,997.41	$4,136,080,412.45
December	$97,889,999.17	$1,146,469,996.57	$4,561,940,955.02
⊞ **Year 2008**	**$990,173,504.69**	**$990,173,504.69**	**$4,111,233,534.68**
⊞ **Year 2009**	**$1,072,783,640.15**	**$1,072,783,640.15**	**$3,740,483,119.18**

***Figure 7-7.** Calculating year-to-date values*

You can also use another set of functions—DATESMTD, DATESQTD, and DATESYTD—to create the same measures. Just as with the previous to-date measures, you need to pass the date key from the date table to the functions. The following expression uses the CALCULATE function with the DATESYTD filter to get the sales year-to-date measure:

```
YTD Sales 2 = CALCULATE([Sum of Sales],DATESYTD('Date'[Datekey]))
```

Using the CALCULATE function and the DATES functions is more versatile than the total-to-date functions because you can use them for any type of aggregation, not just the sum. The following expressions are used to calculate the average sales year-to-date. The results are shown in Figure 7-8.

```
Ave Sales = AVERAGE([SalesAmount])
YTD Ave Sales = CALCULATE([Ave Sales],DATESYTD('Date'[Datekey]))
```

Category
- ☐ Audio
- ☐ Cameras and camcor...
- ☐ Cell phones
- ■ Computers
- ☐ Games and Toys
- ☐ Home Appliances
- ☐ Music, Movies and Au...
- ☐ TV and Video

Year	Sum of Sales	Ave Sales	YTD Ave Sales
⊟ Year 2007			
January	$76,580,425.17	$2,921.80	$2,921.80
February	$88,914,528.27	$3,460.11	$3,188.30
March	$85,714,037.10	$3,282.68	$3,219.89
April	$108,206,075.80	$3,532.34	$3,307.98
May	$106,963,757.90	$3,428.33	$3,334.83
June	$104,713,790.47	$3,356.10	$3,338.71
July	$97,226,447.91	$3,761.61	$3,394.22
August	$92,298,097.12	$3,607.93	$3,418.80
September	$88,743,351.91	$3,546.19	$3,431.68
October	$95,184,597.09	$2,971.18	$3,378.90
November	$104,034,888.68	$4,061.96	$3,436.23
December	$97,889,999.17	$3,657.66	$3,454.09
⊞ Year 2008	**$990,173,504.69**	**$3,969.79**	**$3,969.79**
⊞ Year 2009	**$1,072,783,640.15**	**$4,505.71**	**$4,505.71**

Figure 7-8. *Calculating average year-to-date values*

Now that you know how to create time period–based calculations, you can use this to compare past performance with current performance. But first, you need to know how to shift the date context to calculate past performance.

Shifting the Date Context

If you want to compare performance from one period to the same period in the past, say sales for the current month to sales for the same month a year ago, you need to shift the date context. DAX contains several functions that do this. One of the most versatile functions for shifting the date context is the PARALLELPERIOD function. As with the other time intelligence functions, you need to pass the key column of the date table to the function. You also need to indicate the number of intervals and the interval type of year, quarter, or month:

```
PARALLELPERIOD(<dates>,<number_of_intervals>,<interval>)
```

One thing to remember is that the PARALLELPERIOD function returns a set of dates that corresponds to the interval type. If you use the year, it returns a year of dates; the month interval returns a month's worth of dates. The following expression calculates the sales totals for the month of the previous year. Figure 7-9 shows the results of the calculation.

```
Month Sales Last Year = Calculate([Sum of Sales],
PARALLELPERIOD('Date'[Datekey],-12,Month))
```

Year	Sum of Sales	Month Sales Last Year
Year 2007		
January	$76,580,425.17	
February	$88,914,528.27	
March	$85,714,037.10	
April	$108,206,075.80	
May	$106,963,757.90	
June	$104,713,790.47	
July	$97,226,447.91	
August	$92,298,097.12	
September	$88,743,351.91	
October	$95,184,597.09	
November	$104,034,888.68	
December	$97,889,999.17	
Year 2008		
January	$68,870,582.97	$76,580,425.17
February	$70,478,255.74	$88,914,528.27
March	$67,535,939.12	$85,714,037.10
April	$83,632,429.02	$108,206,075.80
May	$81,902,933.81	$106,963,757.90
June	$79,488,873.16	$104,713,790.47
July	$90,504,056.82	$97,226,447.91
August	$84,679,692.06	$92,298,097.12

Figure 7-9. *Calculating sales for a parallel period*

Notice that if you drill down to the date level (see Figure 7-10), you still see the month totals for the month of the date for a year ago. As mentioned earlier, this is because the `PARALLELPERIOD` in this case always returns the set of dates for the same month as the row date for the previous year.

Year	Sum of Sales	Month Sales Last Year
⊟ **April**		
4/1/2008	$2,532,050.88	$108,206,075.80
4/2/2008	$2,899,750.52	$108,206,075.80
4/3/2008	$2,593,253.80	$108,206,075.80
4/4/2008	$2,924,097.70	$108,206,075.80
4/5/2008	$2,440,024.98	$108,206,075.80
4/6/2008	$3,092,582.83	$108,206,075.80
4/7/2008	$2,674,293.04	$108,206,075.80
4/8/2008	$2,926,885.56	$108,206,075.80
4/9/2008	$2,723,004.34	$108,206,075.80
4/10/2008	$2,695,557.56	$108,206,075.80
4/11/2008	$2,580,420.75	$108,206,075.80
4/12/2008	$2,975,057.74	$108,206,075.80
4/13/2008	$2,344,679.28	$108,206,075.80
4/14/2008	$2,809,861.70	$108,206,075.80

***Figure 7-10.** Drilling to date level still shows month-level aggregation*

Now that you can calculate the month sales of the previous year, you can combine it with current sales to calculate the monthly sales growth from one year to the next. Figure 7-11 shows the results.

```
YOY Monthly Growth = Divide(([Sum of Sales]-[Month Sales Last Year]),[Month
Sales Last Year],BLANK())
```

Category
- ☐ Audio
- ☐ Cameras and camcor...
- ☐ Cell phones
- ☐ Computers
- ☐ Games and Toys
- ☐ Home Appliances
- ☐ Music, Movies and Au...
- ■ TV and Video

Year	Sum of Sales	Month Sales Last Year	YOY Monthly Growth
⊞ Year 2007	$426,671,354.26		
⊟ Year 2008			
January	$27,863,557.20	$27,495,553.66	1.3%
February	$32,811,966.72	$27,092,949.68	21.1%
March	$32,383,562.83	$26,142,119.90	23.9%
April	$38,726,965.30	$37,007,532.00	4.6%
May	$40,016,450.13	$39,161,448.02	2.2%
June	$39,147,523.99	$39,143,563.16	0.0%
July	$47,064,747.67	$34,862,779.01	35.0%
August	$43,003,479.00	$33,887,767.45	26.9%
September	$42,132,550.02	$33,690,892.69	25.1%
October	$38,639,674.35	$38,766,422.87	-0.3%
November	$45,007,341.23	$43,617,702.87	3.2%
December	$46,468,031.17	$45,802,622.95	1.5%
⊟ Year 2009			

Figure 7-11. *Calculating year over year monthly growth*

Note that if there are no previous year sales, you get an error. You can control this by using the third parameter in the Divide function. This controls what is displayed if there is a divide by zero error. In this case, we are replacing the error with a blank value.

Another function commonly used to alter the date context is the DATEADD function. The DATEADD function is used to add a date interval to the current date context. You can add year, quarter, month, or day intervals.

```
DATEADD(<dates>,<number_of_intervals>,<interval>)
```

The following calculation is used to find the sum of the previous day sales using the DATEADD function as a filter:

```
Prev Day Sales=Calculate([Sum of Sales],DATEADD('Date'[Datekey],-1,day))
```

Now that you know how to shift the date context, let's look at functions you can use in your filters that return a single date.

Using Single Date Functions

DAX contains a set of functions that return a single date. These are typically used when filtering the date context. For example, the FIRSTDATE function returns the first date in the column of dates passed to it. As an example, you can use this in combination with the DATESBETWEEN function to get the range of dates from the 1st day to the 15th day of the current date context set of dates:

```
DATESBETWEEN('Date'[FullDateAlternateKey]
 ,FIRSTDATE('Date'[FullDateAlternateKey])
,DATEADD(FIRSTDATE('Date'[FullDateAlternateKey]), 14, DAY))
```

This can then be used as a filter in the CALCULATE function to get the sales during the first 15 days of the period. The resulting pivot table is shown in Figure 7-12.

Category
- ☐ Audio
- ☐ Cameras and camcor...
- ☐ Cell phones
- ☐ Computers
- ☐ Games and Toys
- ☐ Home Appliances
- ☐ Music, Movies and Au...
- ■ TV and Video

Year	Sum of Sales	First 15 Day Sales
⊟ **Year 2007**		
January	$27,495,553.66	$13,084,676.956
February	$27,092,949.68	$14,339,127.0399
March	$26,142,119.90	$12,332,639.0409
April	$37,007,532.00	$18,691,391.1574
May	$39,161,448.02	$18,657,737.963
June	$39,143,563.16	$19,934,461.864
July	$34,862,779.01	$16,293,109.525
August	$33,887,767.45	$15,828,478.558
September	$33,690,892.69	$16,771,575.123
October	$38,766,422.87	$19,009,503.558
November	$43,617,702.87	$21,659,674.509
December	$45,802,622.95	$21,926,793.42

***Figure 7-12.** Calculating sales for the first 15 days of the month*

If you are just looking at the monthly periods, you can use the functions STARTOFMONTH and ENDOFMONTH (there are ones for year and quarter also). The following expression is used to calculate the sum of the sales for the last 15 days of the month:

```
Last 15 Day Sales=CALCULATE(SUM(InternetSales[SalesAmount]),
 DATESBETWEEN('Date'[FullDateAlternateKey]
 , DATEADD(ENDOFMONTH('Date'[FullDateAlternateKey]), -14, DAY)
,ENDOFMONTH('Date'[FullDateAlternateKey])))
```

Although the majority of measures you need to aggregate from a lower level to a higher level (e.g., from days to months) are simple extensions of the base aggregate, at times you need to use special aggregations to roll up the measure. This type of measure is considered semi-additive and is covered in the next section.

Creating Semi-additive Measures

You often encounter semi-additive measures when analyzing data. Some common examples are inventory and account balances. For example, to determine the total amount of inventory at a current point in time, you add the inventory of all stores. But to find the total inventory of a store at the end of the month, you don't add up the inventory for each day.

To deal with these situations, DAX contains the `FIRSTNONBLANK` and `LASTNONBLANK` functions. These functions return the first or last date for a nonblank condition. For example, the following expression determines the last nonblank date for the product inventory entries:

```
LASTNONBLANK ('Date'[DateKey],CALCULATE (SUM(Inventory[UnitsInStock]))
```

This is then combined with the `CALCULATE` function to determine the total units in stock:

```
Product Units In Stock=CALCULATE(SUM(Inventory[UnitsInStock]),
  LASTNONBLANK('Date'[DateKey],
   CALCULATE(SUM(Inventory[UnitsInStock]))))
```

Now if you want to add up the units in stock across products, you can use the following expression:

```
Total Units In Stock=SUMX(VALUES('Inventory'[ProductKey]),[Product Units In
Stock])
```

Figure 7-13 shows the resulting pivot table. Note that the Total Units In Stock measure is additive across the products but nonadditive across the dates.

Category
- ■ Audio
- ☐ Cameras and camcor...
- ☐ Cell phones
- ☐ Computers
- ☐ Games and Toys
- ☐ Home Appliances
- ☐ Music, Movies and Au...
- ■ TV and Video

Year	Audio	TV and Video	**Total**
⊟ **Year 2007**	**16913**	**77246**	**94159**
⊞ **January**	**2948**	**20786**	**23734**
⊟ **February**	**7100**	**50300**	**57400**
2/3/2007	7229	51917	**59146**
2/10/2007	6974	50684	**57658**
2/17/2007	7381	50259	**57640**
2/24/2007	7100	50300	**57400**
⊟ **March**	**11340**	**49565**	**60905**
3/3/2007	8744	52111	**60855**
3/10/2007	9834	51232	**61066**
3/17/2007	9414	51650	**61064**
3/24/2007	8849	53249	**62098**
3/31/2007	11340	49565	**60905**

***Figure 7-13.** Calculating units in stock*

One of the advantages of DAX is that once you learn a pattern, you can extend it to other scenarios. For example, you can employ the same techniques used in this inventory calculation when calculating measures in a cash flow analysis. The following calculation is used to calculate the ending balance:

```
Balance=CALCULATE ( SUM ( Finance[Amount]),
  LASTNONBLANK ('Date'[DateKey],
   CALCULATE(SUM ( Finance[Amount]))))
```

In the previous examples, the inventory and balance were only entered as a row in the table when a change in inventory or balance occurred. Often the balance or inventory is entered every day, and the same value is repeated until there is a change. In these cases, you can use the DAX functions `CLOSINGBALANCEMONTH`, `CLOSINGBALANCEQUARTER`, and `CLOSINGBALANCYEAR`. These functions look at the last date of the time period and use that as the value for the time period. In other words, whatever the value is on the last day of the month is returned by the `CLOSINGBALANCEMONTH` function.

At this point, you should have a good grasp of how the various time functions work. In the following lab, you will gain experience implementing some of these functions.

HANDS-ON LAB: IMPLEMENTING TIME INTELLIGENCE IN POWER PIVOT

In the following lab, you will

- Create a date table
- Use time intelligence functions to analyze data
- Create a month over month growth matrix
- Create an inventory level report

1. In the LabStarterFiles\Chapter7Lab1 folder, open the Chapter7Lab1.pbix file. This file contains inventory and sales data from the Contoso test database.
2. Select the Model view tab to see the tables and relationships (see Figure 7-14). Notice there is no Date table. We could load one from the data source or create one in Power BI Desktop.

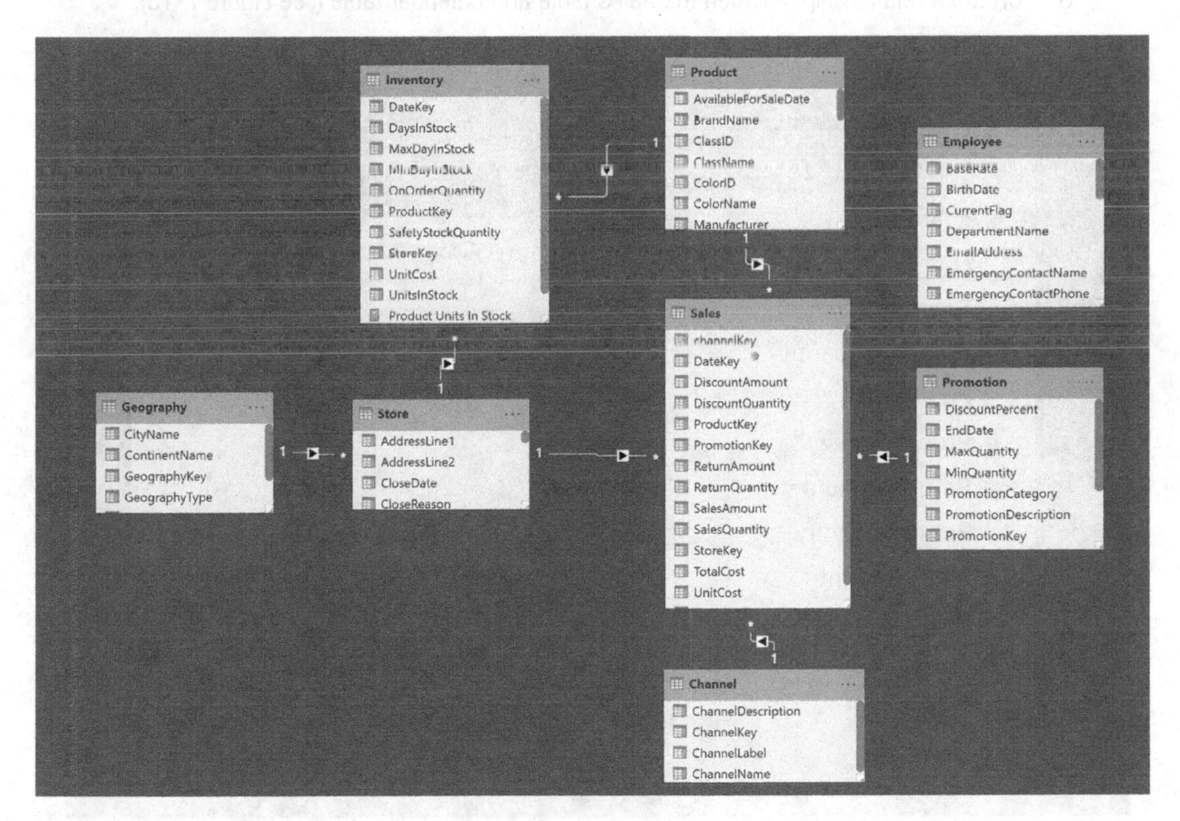

Figure 7-14. *The current Contoso data model*

3. To create the Date table, switch to the Data view. Under the Modeling tab, select New Table.

4. Enter the following DAX to create the set of dates based on the minimum and maximum sales dates:

```
Calendar = CALENDAR(Min('Sales'[DateKey]),MAX(Sales[DateKey]))
```

5. Under the Modeling tab, select Mark as date table. Make sure the Calendar table is marked as the Date table and the Date column is selected as the Date column.

6. Using DAX, add a Year, Month, and Month No column to the Calendar table (this was covered in Chapter 5).

7. Sort the Month column by the Month No column and format the Date column to show the date with no time.

8. Create a relationship between the Sales table and Calendar table (see Figure 7-15).

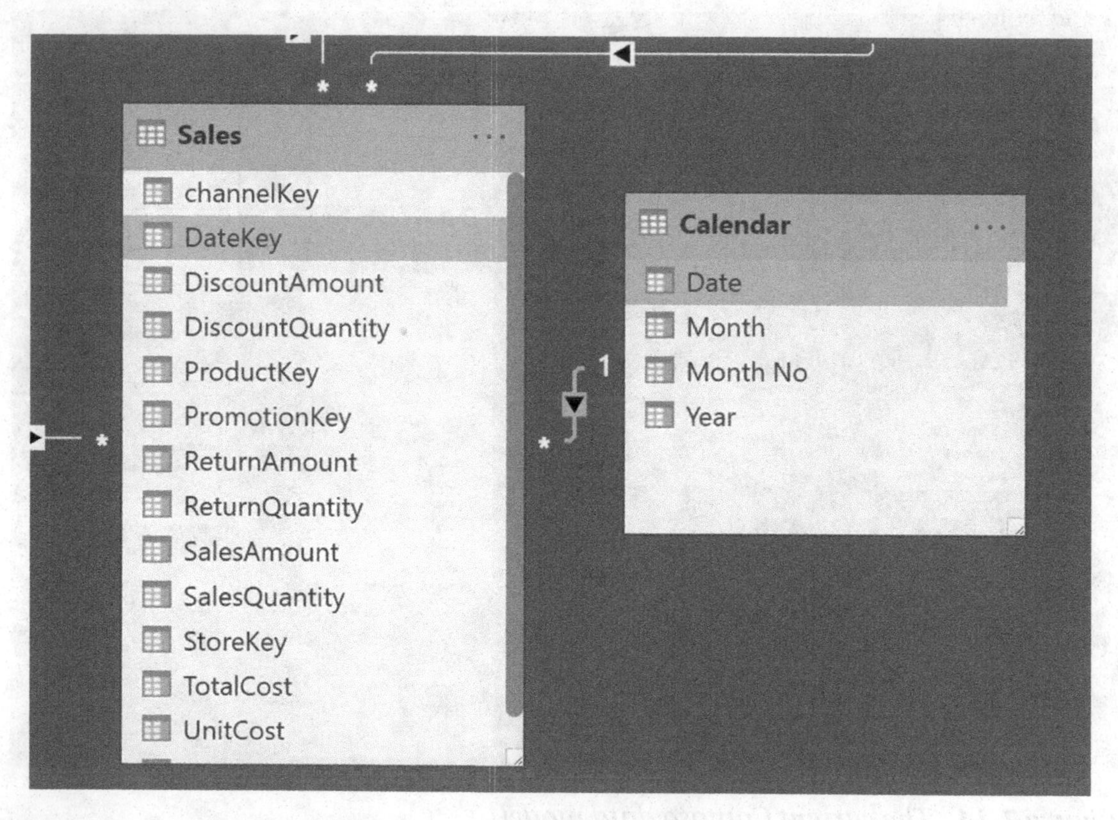

Figure 7-15. *Creating the relationship between the Sales and Date table*

9. Select the Sales table in the Data view window. Add the following measures and format them as currency:

```
Sum of Sales=SUM(Sales[SalesAmount])
YTD Sales=TOTALYTD([Sum of Sales],'Calendar'[Date])
MTD Sales=TOTALMTD([Sum of Sales],'Calendar'[Date])
```

10. To test the measures, switch to the Report view and create a matrix like the one shown in Figure 7-16.

Year	Sum of Sales	MTD Sales	YTD Sales
⊞ 2007	$4,561,940,955.02	$425,860,542.57	$4,561,940,955.02
⊟ 2008	**$4,111,233,534.68**	**$398,786,172.94**	**$4,111,233,534.68**
⊞ January	**$279,460,806.88**	**$279,460,806.88**	**$279,460,806.88**
⊞ February	**$288,852,634.02**	**$288,852,634.02**	**$568,313,440.90**
⊞ March	**$290,060,560.62**	**$290,060,560.62**	**$858,374,001.52**
⊟ April	**$356,139,170.42**	**$356,139,170.42**	**$1,214,513,171.94**
4/1/2008	$11,337,103.99	$11,337,103.99	$869,711,105.51
4/2/2008	$11,930,264.81	$23,267,368.80	$881,641,370.32
4/3/2008	$11,418,575.20	$34,685,943.99	$893,059,945.51
4/4/2008	$12,993,709.97	$47,679,653.96	$906,053,655.48
4/5/2008	$11,354,401.90	$59,034,055.86	$917,408,057.38
4/6/2008	$12,860,403.00	$71,894,458.85	$930,268,460.37
4/7/2008	$12,174,234.71	$84,068,693.56	$942,442,695.08
4/8/2008	$12,158,808.09	$96,227,501.65	$954,601,503.17
4/9/2008	$11,620,966.13	$107,848,467.78	$966,222,469.30
4/10/2008	$11,865,557.43	$119,714,025.21	$978,088,026.73
4/11/2008	$11,983,754.09	$131,697,779.30	$990,071,780.82
4/12/2008	$12,710,318.17	$144,408,097.47	$1,002,782,098.99
4/13/2008	$11,273,925.30	$155,682,022.77	$1,014,056,024.29
4/14/2008	$12,068,443.34	$167,750,466.11	$1,026,124,467.63
4/15/2008	$11,509,504.55	$179,259,970.66	$1,037,633,972.19
4/16/2008	$11,825,456.95	$191,085,427.61	$1,049,459,429.14
4/17/2008	$11,277,835.48	$202,363,263.09	$1,060,737,264.61
4/18/2008	$12,141,840.55	$214,505,103.64	$1,072,879,105.16
4/19/2008	$12,106,695.72	$226,611,799.36	$1,084,985,800.88
4/20/2008	$11,624,443.88	$238,236,243.23	$1,096,610,244.76
4/21/2008	$11,030,369.35	$249,266,612.59	$1,107,640,614.11
Total	$12,413,657,608.89	$330,734,413.46	$3,740,483,119.18

Figure 7-16. *Verifying the year-to-date sales measure*

11. Switch back to the Sales table in the Data tab. Create a rolling three-month sales measure with the following expression. Format the measure as currency:

```
Rolling 3 Month Sales=CALCULATE([Sum of Sales],
DATESINPERIOD('Calendar'[Date],LASTDATE('Calendar'[Date]),-3,MONTH))
```

12. Add the measure to the matrix you created and verify it is working as expected.

13. Next, you want to compare sales growth from one month to the next. First, create a previous month's sales measure in the Sales table:

```
Prev Month Sales=CALCULATE([Sum of Sales],
PARALLELPERIOD('Calendar'[Date],-1,MONTH))
```

14. Next, use the previous sales and current sales to create a sales growth measure. Format the measure as percent.

```
Monthly Sales Growth = Divide
([Sum of Sales] - [Prev Month Sales],[Prev Month Sales],BLANK())
```

15. Create a new page in Report view and add a matrix like the one shown in Figure 7-17 to test your results.

Year	Sum of Sales	Prev Month Sales	Monthly Sales Growth
⊟ 2007	**$4,561,940,955.02**	**$4,136,080,412.45**	**10.30%**
January	$269,835,263.23		
February	$298,215,968.35	$269,835,263.23	10.52%
March	$300,486,926.90	$298,215,968.35	0.76%
April	$400,160,331.60	$300,486,926.90	33.17%
May	$423,429,127.79	$400,160,331.60	5.81%
June	$409,797,545.55	$423,429,127.79	-3.22%
July	$389,617,372.27	$409,797,545.55	-4.92%
August	$388,429,827.11	$389,617,372.27	-0.30%
September	$379,144,599.56	$388,429,827.11	-2.39%
October	$423,213,240.84	$379,144,599.56	11.62%
November	$453,750,209.24	$423,213,240.84	7.22%
December	$425,860,542.57	$453,750,209.24	-6.15%
⊞ 2008	**$4,111,233,534.68**	**$4,138,307,904.31**	**-0.65%**
⊞ 2009	**$3,740,483,119.18**	**$3,808,534,878.67**	**-1.79%**
Total	**$12,413,657,608.89**	**$12,082,923,195.43**	**2.74%**

***Figure 7-17.** Testing the Monthly Sales Growth measure*

16. To investigate semi-additive measures, you are going to create a pivot table that shows inventory counts. First, create a relationship between the Inventory and Calendar tables.

17. Add the following inventory count measure to the Inventory table:

    ```
    Inventory Count = SUM('Inventory'[UnitsInStock])
    ```

18. To test the measures, create a new report page as shown in Figure 7-18. If you look at the monthly totals, they are adding up the inventory for the days in the month. What we want is to carry over the last entry as the inventory for the month.

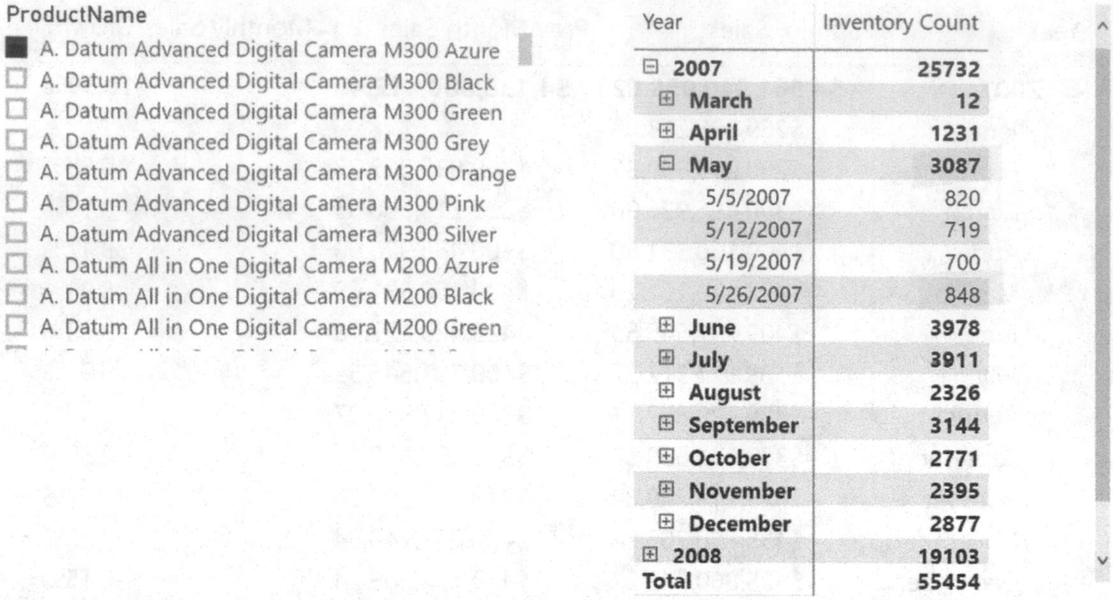

ProductName

- ■ A. Datum Advanced Digital Camera M300 Azure
- ☐ A. Datum Advanced Digital Camera M300 Black
- ☐ A. Datum Advanced Digital Camera M300 Green
- ☐ A. Datum Advanced Digital Camera M300 Grey
- ☐ A. Datum Advanced Digital Camera M300 Orange
- ☐ A. Datum Advanced Digital Camera M300 Pink
- ☐ A. Datum Advanced Digital Camera M300 Silver
- ☐ A. Datum All in One Digital Camera M200 Azure
- ☐ A. Datum All in One Digital Camera M200 Black
- ☐ A. Datum All in One Digital Camera M200 Green

Year	Inventory Count
⊟ **2007**	**25732**
⊞ **March**	**12**
⊞ **April**	**1231**
⊟ **May**	**3087**
5/5/2007	820
5/12/2007	719
5/19/2007	700
5/26/2007	848
⊞ **June**	**3978**
⊞ **July**	**3911**
⊞ **August**	**2326**
⊞ **September**	**3144**
⊞ **October**	**2771**
⊞ **November**	**2395**
⊞ **December**	**2877**
⊞ **2008**	**19103**
Total	**55454**

Figure 7-18. *Testing the inventory count measure*

19. Using the last nonblank filter and the `CALCULATE` function, you can calculate the last nonblank quantity for a product. Add the following measure to the Inventory table:

```
Product Quantity = CALCULATE([Inventory Count],LASTNONBLANK('Calendar'
[Date],[Inventory Count]))
```

20. Verify the results by adding the Product Quantity measure to the matrix.

21. The final step is to add up the inventory across all the stores selected using the following measure:

```
Inventory Level=SUMX(Values(Store[StoreKey]),[Product Quantity])
```

22. To test the measure, create a report page like the one in Figure 7-19.

ProductName

- ■ A. Datum Advanced Digital Camera M300 Azure
- ■ A. Datum Advanced Digital Camera M300 Black
- ☐ A. Datum Advanced Digital Camera M300 Green
- ☐ A. Datum Advanced Digital Camera M300 Grey
- ☐ A. Datum Advanced Digital Camera M300 Orange
- ☐ A. Datum Advanced Digital Camera M300 Pink
- ☐ A. Datum Advanced Digital Camera M300 Silver
- ☐ A. Datum All in One Digital Camera M200 Azure
- ☐ A. Datum All in One Digital Camera M200 Black
- ☐ A. Datum All in One Digital Camera M200 Green

StoreName

- ☐ Contoso Albany Store
- ■ Contoso Alexandria Store
- ■ Contoso Amsterdam Store
- ☐ Contoso Anchorage Store
- ☐ Contoso Annapolis Store
- ☐ Contoso Appleton Store
- ☐ Contoso Arlington Store
- ☐ Contoso Ashgabat No.2 Store
- ☐ Contoso Ashgabat No.1 Store
- ☐ Contoso Asia Online Store

Year	Contoso Alexandria Store	Contoso Amsterdam Store	Total
⊟ 2007	**12**	**18**	**30**
⊟ February	**13**	**14**	**27**
2/3/2007	13	13	**26**
2/24/2007		14	**14**
⊟ March	**9**	**26**	**35**
3/10/2007	9	26	**35**
3/17/2007	9	26	**35**
3/24/2007	9		**9**
⊟ June	**12**	**10**	**22**
6/16/2007		13	**13**
6/30/2007	12	10	**22**
⊟ July	**15**	**14**	**29**
7/7/2007	12	10	**22**
7/14/2007	12		**12**
Total	**3**	**24**	**27**

***Figure 7-19.** Testing the inventory measure*

23. Notice currently blanks are being treated as zero inventory. You can fix this by replacing any blanks with the previous value.

Summary

This chapter provided you with the basics you need to successfully incorporate datetime-based analysis using Power Pivot and DAX. You should now understand how to shift the date context to compare measures based on parallel periods. You should also feel comfortable aggregating measures using the period-to-date DAX formulas. When you combine the concepts you learned in Chapter 6 and this chapter, you should start to recognize patterns in your analysis. Based on these patterns, you can identify the DAX template you need to solve the problem.

Now that you are familiar with how to create a solid data model in Power BI, you are ready to create interactive reports. Power BI has a very rich set of interactive visuals you can use to view and gain insight into the data. These visuals can then be deployed to the Power BI portal, where they can be used to create powerful dashboards and shared with others.

CHAPTER 8

Creating Reports with Power BI Desktop

In the previous chapters, you saw how to import, clean, and shape data using Power BI Desktop. In addition, you created the data model and augmented it with calculated columns and measures. In this chapter, you will investigate some of the common visualizations used to create reports in Power BI Desktop. You will become familiar with how to control visual interactions, along with report filtering. You will build standard visualizations such as column, bar, and pie charts. In addition, you will look at line and scatter charts. Finally, you will investigate how to use maps to analyze data geographically.

After completing this chapter, you will be able to

- Create tables and matrices
- Construct bar, column, and pie charts
- Build line and scatter charts
- Create map-based visualizations

Note For copies of color figures, go to Source Code/Downloads at `https://github.com/Apress/beginning-power-bi-3ed`.

Creating Tables and Matrices

Although it is one of the most basic types of data visualization, a *table* is still one of the most useful ways to look at your data. This is especially true if you need to look up detailed information. To create a visual on the report page, select the Report view.

D. Clark, *Beginning Microsoft Power BI*, https://doi.org/10.1007/978-1-4842-5620-6_8

In this view, you will see a Visualizations toolbox and a Fields list on the right side of the designer (see Figure 8-1).

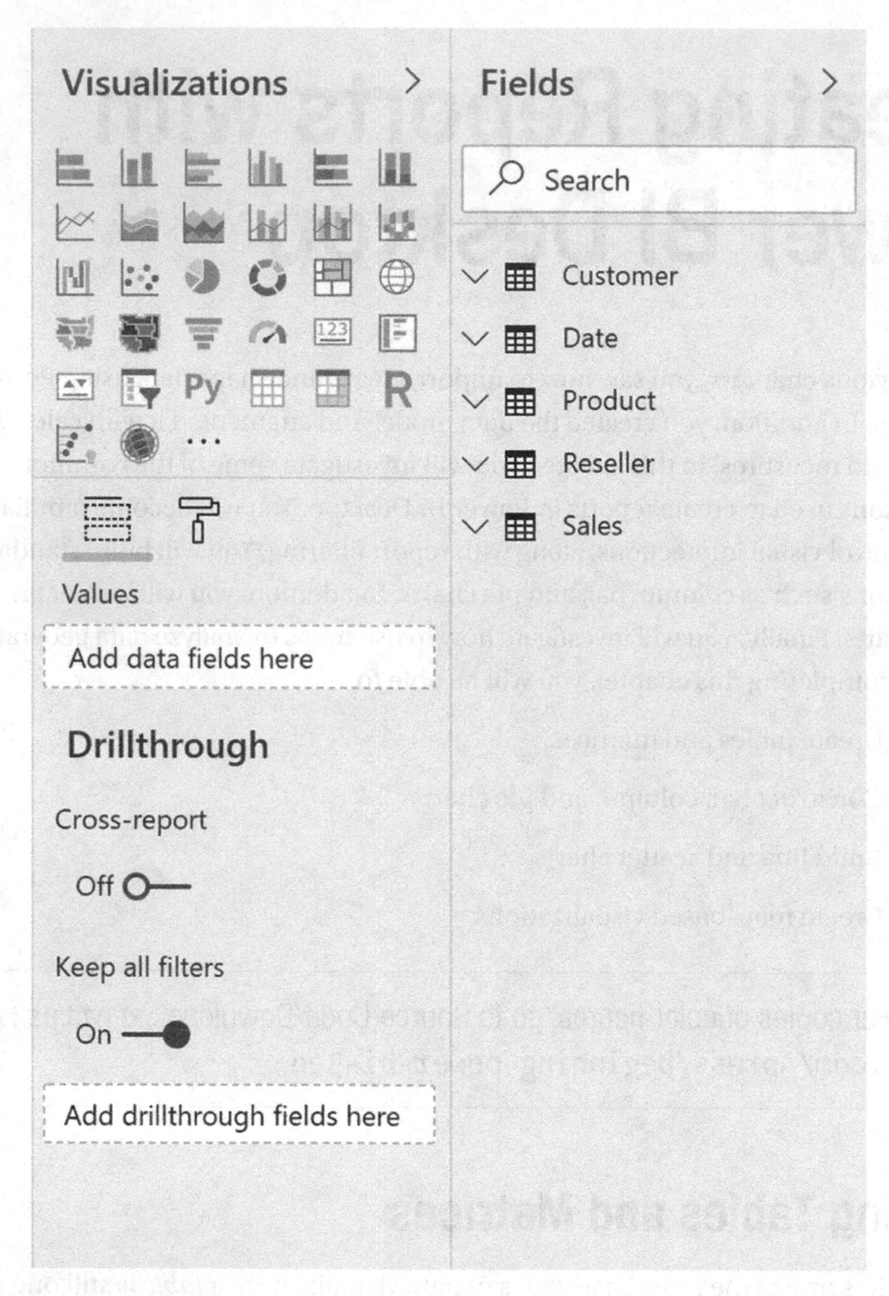

***Figure 8-1.** The Visualizations toolbox*

Selecting the table visualization will reveal a Values drop area where you will drag fields from the Fields list and drop them into the Values area. This will create a table with the fields as columns. Figure 8-2 shows a table displaying customer contact information.

Full Name	AddressLine1	City	State/Province	Country/Region	Phone
Aaron Adams	4116 Stanbridge Ct.	Downey	California	United States	417-555-0154
Aaron Alexander	5021 Rio Grande Drive	Kirkland	Washington	United States	548-555-0129
Aaron Allen	6695 Black Walnut Court	Sooke	British Columb...	Canada	648-555-0141
Aaron Baker	8054 Olivera Rd.	Renton	Washington	United States	488-555-0125
Aaron Bryant	2325 Candywood Ct	Redwood City	California	United States	754-555-0137
Aaron Butler	9761 Darnett Circle	Lebanon	Oregon	United States	466-555-0180
Aaron Campbell	3310 Harvey Way	Bellflower	California	United States	187-555-0177
Aaron Carter	3450 Rio Grande Dr.	Woodland Hills	California	United States	180-555-0167
Aaron Chen	4633 Jefferson Street	Los Angeles	California	United States	969-555-0160
Aaron Coleman	3393 Alpha Way	Santa Monica	California	United States	914-555-0128
Aaron Collins	6767 Stinson	Santa Cruz	California	United States	170-555-0177
Aaron Diaz	9413 Maria Vega Court	Melton	Victoria	Australia	1 (11) 500 555-0130
Aaron Edwards	663 Contra Loma Blvd.	Beverly Hills	California	United States	355-555-0115
Aaron Evans	5623 Detroit Ave.	Daly City	California	United States	150-555-0128
Aaron Flores	8225 Northridge Road	Edmonds	Washington	United States	831-555-0184
Aaron Foster	4461 Centennial Way	Newton	British Columb...	Canada	477-555-0181
Aaron Gonzales	4247 Bellows Ct	Novato	California	United States	647-555-0136
Aaron Gonzalez	1431 Semillon Circle	Grossmont	California	United States	628-555-0119
Aaron Green	8486 Hazelwood Lane	Seattle	Washington	United States	658-555-0113

Figure 8-2. *Customer contact information*

When you combine a table with the automatic filtering you get with Power BI, this becomes a great way to look up customers based on their demographic data. When you select the table in the view, you will see a Filters pane to the left of the Visualizations toolbox (see Figure 8-3). The fields in the table will show up automatically in the table filter list. To add a field to the filter list that you want to filter on but don't want to show up in the table, just drag the field from the Fields list to the Filters area. Along with visual level filters, you can filter all visuals on a page and all visuals on all pages of the report.

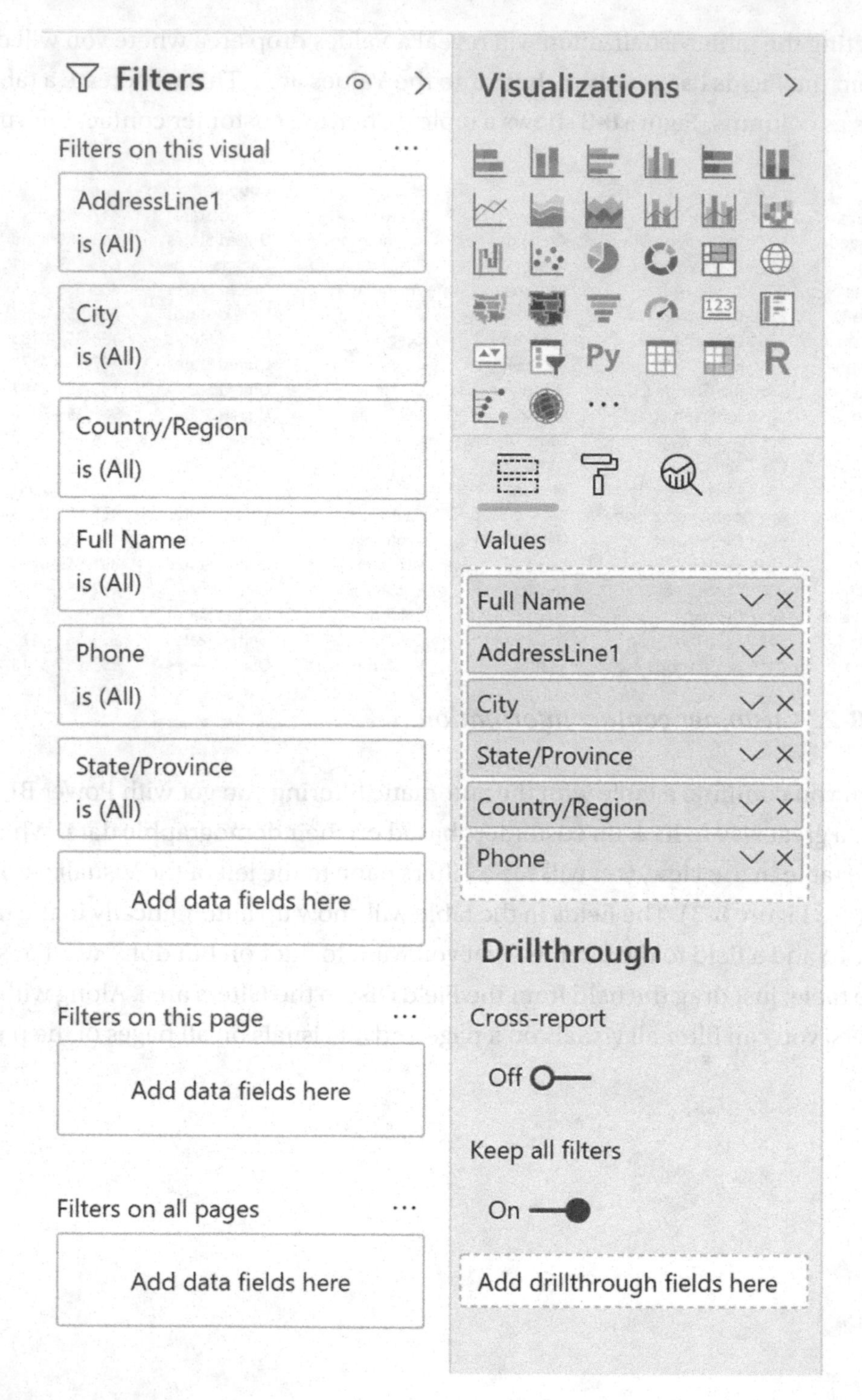

Figure 8-3. *Filtering a table*

If you click the Format tab in the Visualizations toolbox (signified by a paint roller), you will see a pretty extensive set of formatting options available (see Figure 8-4). Figure 8-5 shows a formatted table using one of the built-in styles.

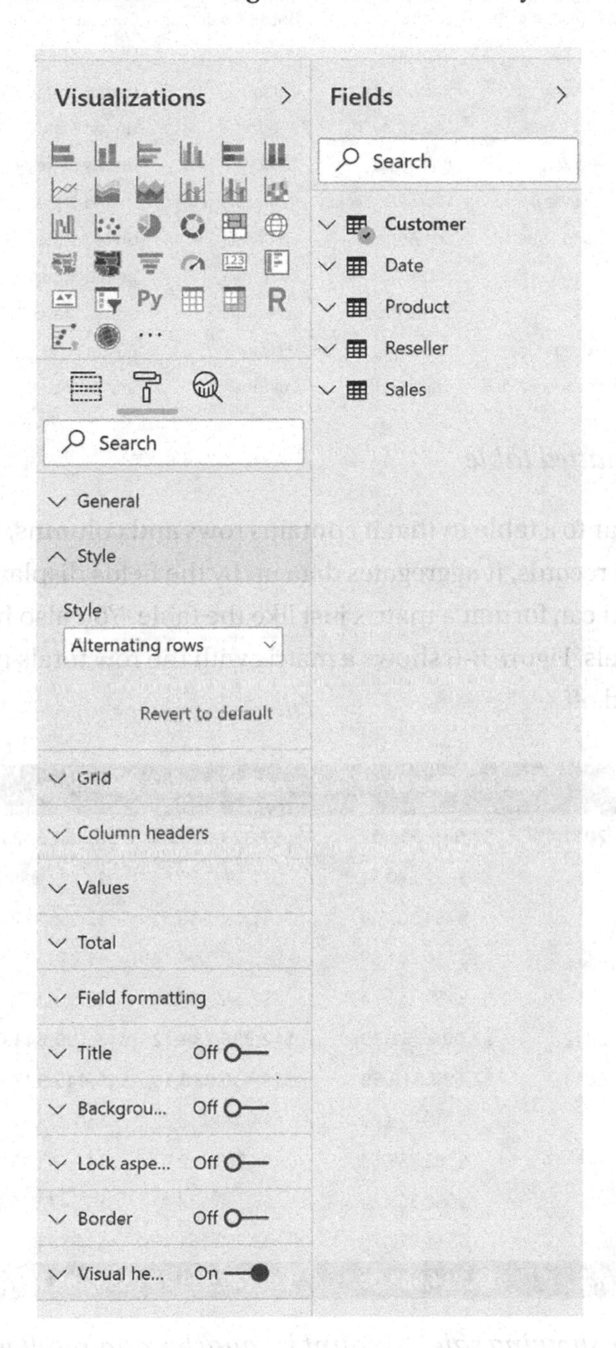

***Figure 8-4.** Table formatting options*

Full Name	AddressLine1	City	State/Province	Country/Region	Phone
Aaron Adams	4116 Stanbridge Ct.	Downey	California	United States	417-555-0154
Aaron Alexander	5021 Rio Grande Drive	Kirkland	Washington	United States	548-555-0129
Aaron Allen	6695 Black Walnut Court	Sooke	British Columb...	Canada	648-555-0141
Aaron Baker	8054 Olivera Rd.	Renton	Washington	United States	488-555-0125
Aaron Bryant	2325 Candywood Ct	Redwood City	California	United States	754-555-0137
Aaron Butler	9761 Darnett Circle	Lebanon	Oregon	United States	466-555-0180
Aaron Campbell	3310 Harvey Way	Bellflower	California	United States	187-555-0177
Aaron Carter	3450 Rio Grande Dr.	Woodland Hills	California	United States	180-555-0167
Aaron Chen	4633 Jefferson Street	Los Angeles	California	United States	969-555-0160
Aaron Coleman	3393 Alpha Way	Santa Monica	California	United States	914-555-0128
Aaron Collins	6767 Stinson	Santa Cruz	California	United States	170-555-0177
Aaron Diaz	9413 Maria Vega Court	Melton	Victoria	Australia	1 (11) 500 555-0130
Aaron Edwards	663 Contra Loma Blvd.	Beverly Hills	California	United States	355-555-0115

***Figure 8-5.** A formatted table*

A matrix is similar to a table in that it contains rows and columns, but instead of showing detail-level records, it aggregates data up by the fields displayed in the row and column headers. You can format a matrix just like the table. You also have the option to show or hide the totals. Figure 8-6 shows a matrix with the row totals turned on and the column totals turned off.

Year	Specialty Bike Shop	Value Added Reseller	Warehouse
⊟ **2011**	**$2,009,000.02**	**$7,323,670.04**	**$8,860,132.65**
1	$357,680.11	$1,199,250.53	$1,992,095.75
2	$484,307.26	$1,489,610.61	$2,053,162.46
3	$716,261.89	$2,285,904.14	$1,949,919.95
4	$450,750.76	$2,348,904.76	$2,864,954.49
⊞ **2012**	**$2,006,751.39**	**$12,996,239.12**	**$13,190,641.03**
⊟ **2013**	**$2,698,816.59**	**$14,430,080.10**	**$16,445,937.47**
1	$740,099.94	$4,074,733.82	$5,727,827.66
2	$782,759.64	$3,738,212.53	$4,135,685.28
3	$660,560.24	$3,636,572.44	$3,347,546.96
4	$515,396.78	$2,980,561.30	$3,234,877.56
Total	**$6,714,568.00**	**$34,749,989.26**	**$38,496,711.15**

***Figure 8-6.** Matrix showing sales amount by quarter and reseller type*

There are many formatting options available such as the ability to expand and collapse by the various row and column headers. These features vary by visual and you should spend some time investigating these.

In addition to wanting to see aggregated values in a matrix, you often want to show these values using a visual representation such as a bar, column, or pie chart. The next section looks at creating those.

Constructing Bar, Column, and Pie Charts

Some of the most common data visualizations used to compare data are the bar, column, and pie charts. A bar chart and a column chart are very similar. The bar chart has horizontal bars where the x-axis is the value of the measure and the y-axis contains the categories you are comparing. For example, Figure 8-7 shows a bar chart comparing sales by country.

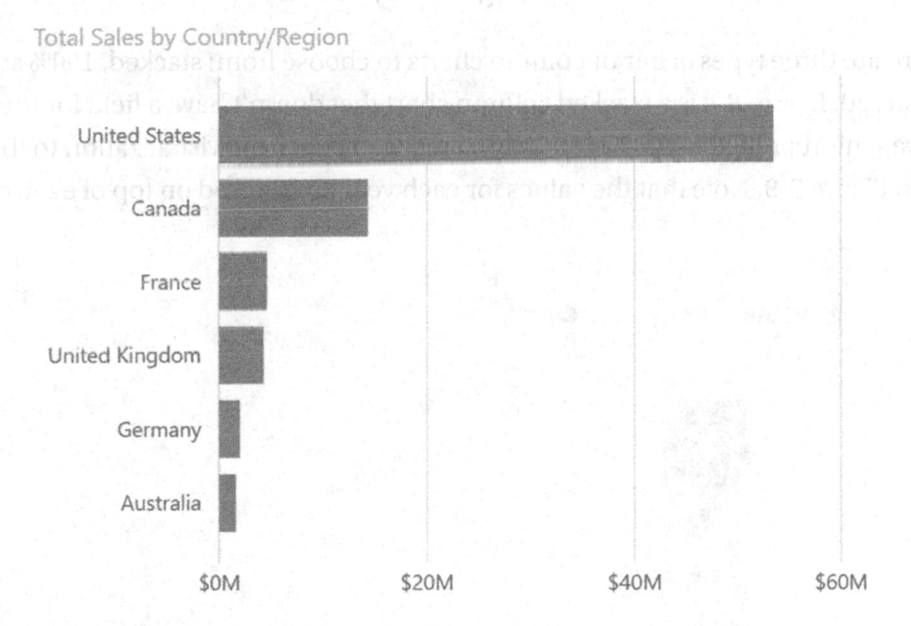

***Figure 8-7.** Bar chart comparing sales by country*

The column chart switches the axes so that the measure amounts are on the y-axis and the categories are on the x-axis. For example, the bar chart in Figure 8-7 can just as easily be displayed as a column chart, as shown in Figure 8-8.

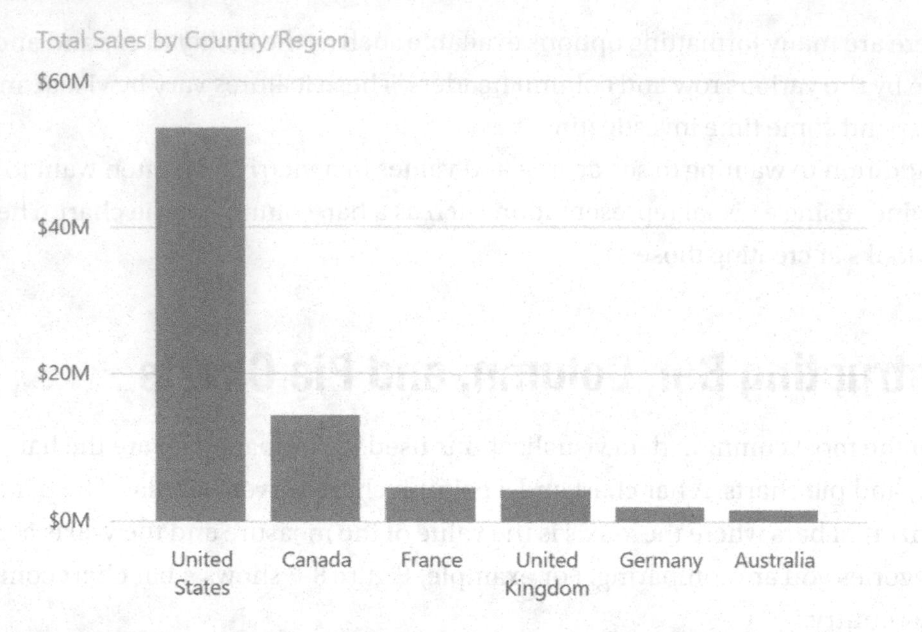

***Figure 8-8.** A column chart comparing sales by country*

There are three types of bar or column charts to choose from: stacked, 100% stacked, and clustered. Figure 8-8 is a stacked column chart that doesn't have a field for the legend. If you drag the Year field to the Legend drop area, it changes the visualization to the one shown in Figure 8-9. Note that the values for each year are stacked on top of each other.

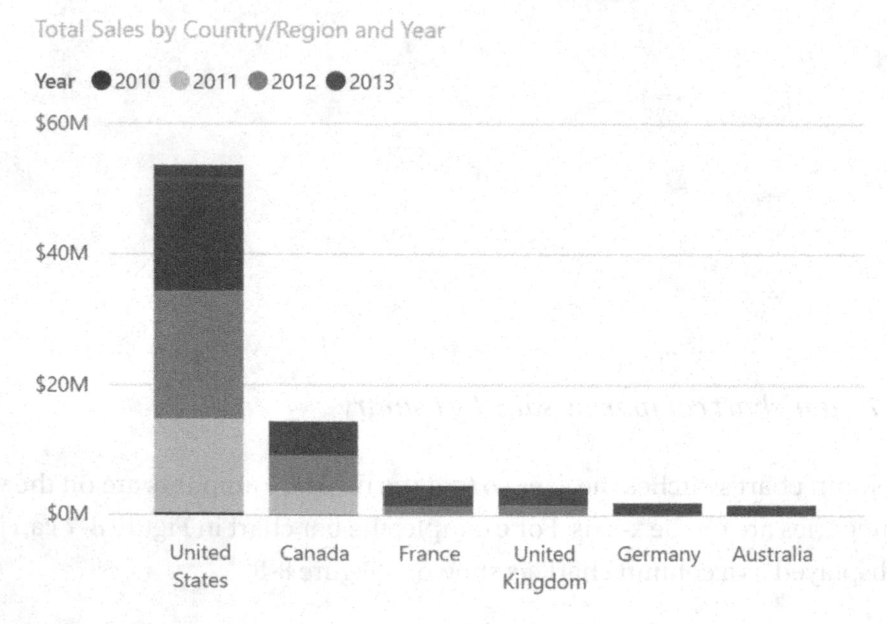

***Figure 8-9.** Creating a stacked column chart*

Although the stacked column chart shows absolute values, you can change it to a 100% stacked chart to show relative values in terms of percentages. In Figure 8-10, the Country field is moved to the Legend and the Year field is placed on the axis. It also has data labels set as Visible for easier comparisons.

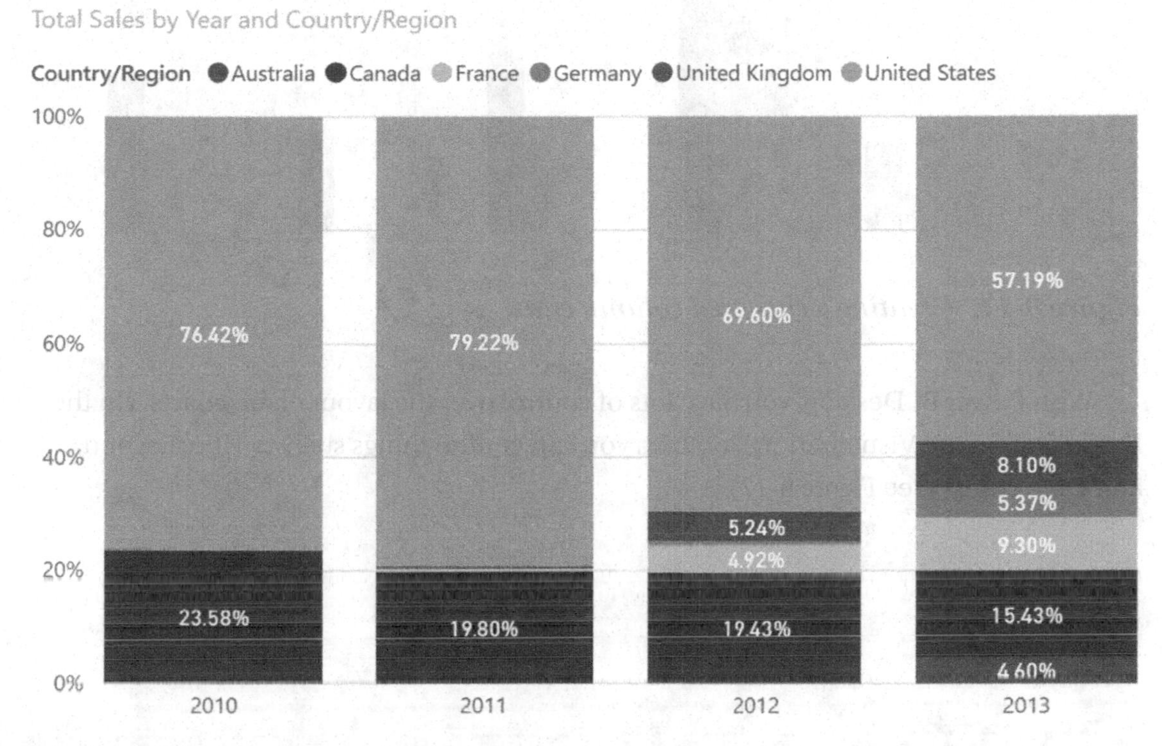

Figure 8-10. *Creating a 100% stacked column chart*

A clustered column chart moves the columns for the various countries side by side instead of stacking them on top of one another. Figure 8-11 shows the same information as Figure 8-9 but as a clustered column chart.

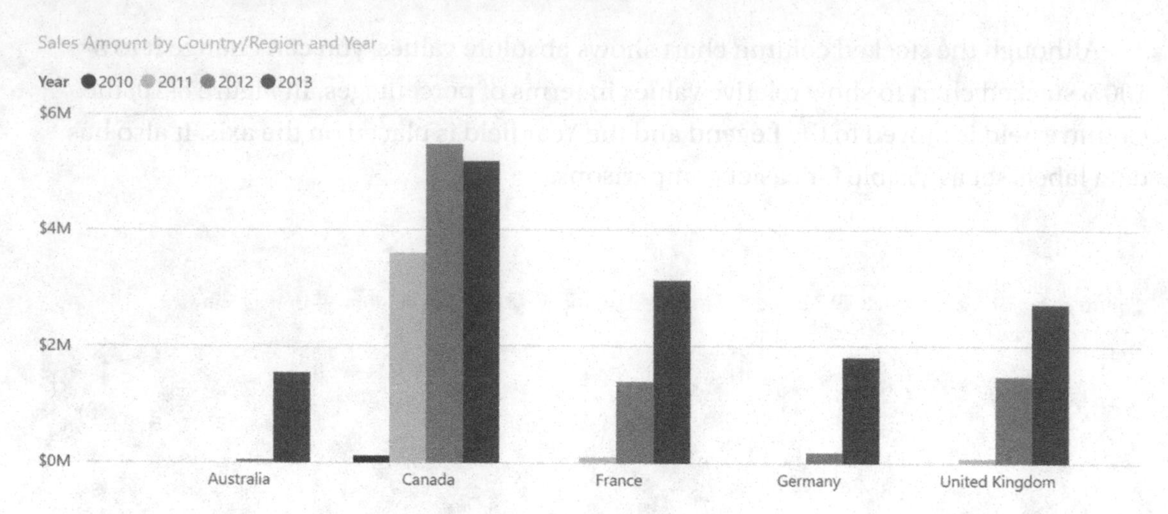

***Figure 8-11.** Creating a clustered column chart*

With Power BI Desktop, you have lots of control over the layout of the charts. On the Format tab in the Visualizations toolbox, you can control things such as Title, Legend, and Data colors (see Figure 8-12).

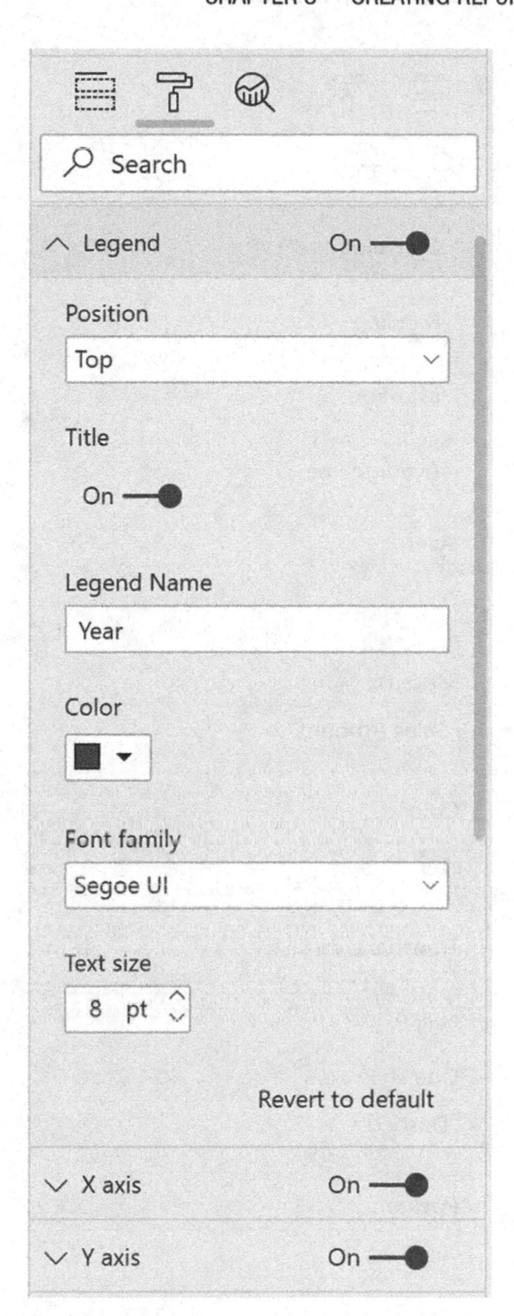

Figure 8-12. *Controlling the Legend layout*

A nice feature associated with the bar and column charts is the ability to add reference lines to aid in the data analysis. If you select the Analytics tab in the Visualizations toolbox (see Figure 8-13), you see the various lines you can add. Figure 8-14 shows a column chart with the average line displayed.

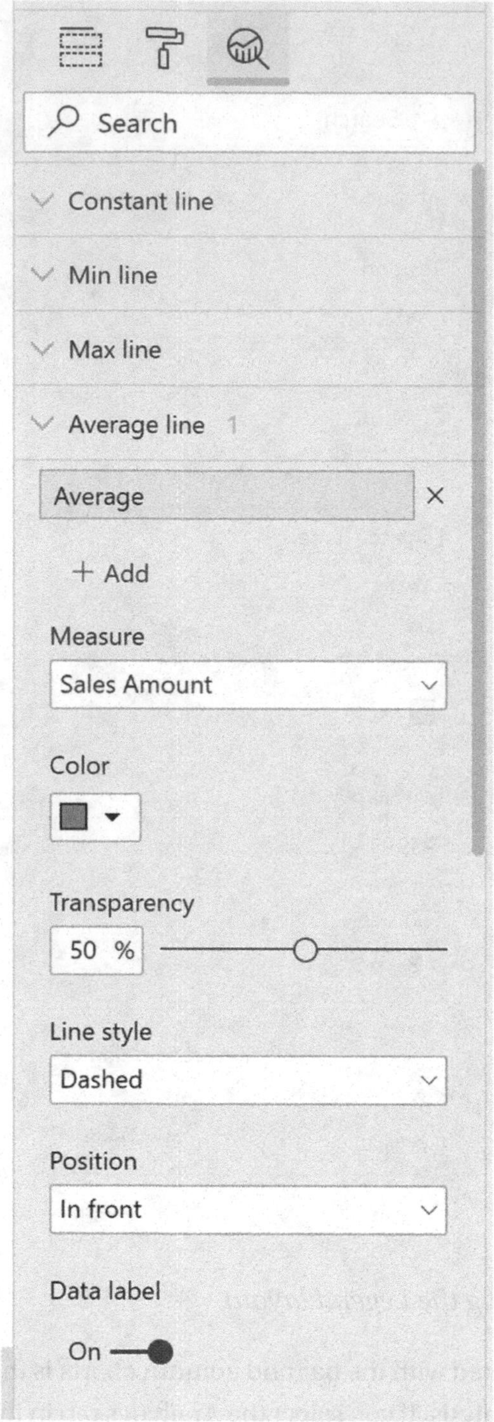

***Figure 8-13.** Adding reference lines to a chart*

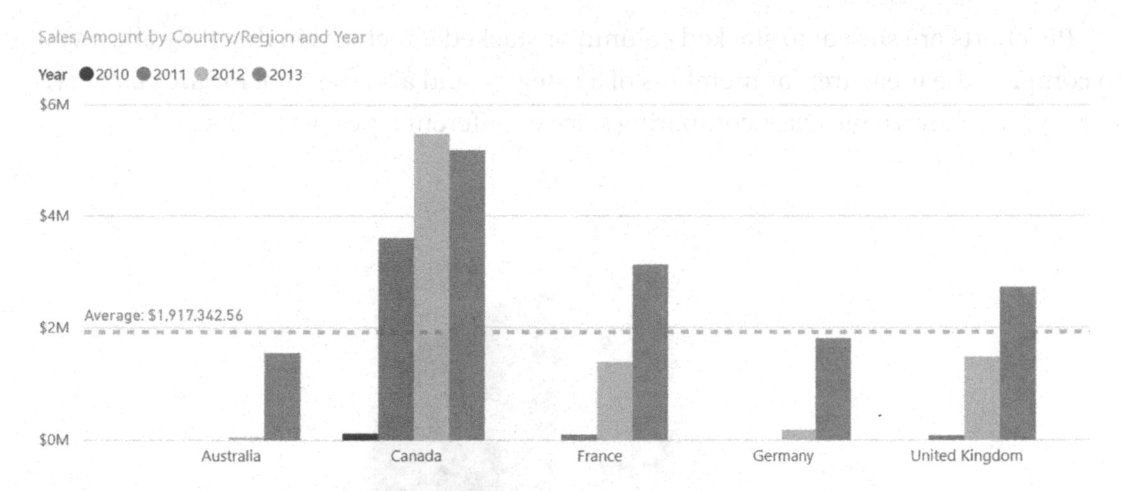

***Figure 8-14.** Displaying the average reference line*

As with most visualizations, in Power BI you get automatic filtering and sorting capabilities. For example, if you select the chart and click the ellipses in the upper right corner of the chart, you get the option to change the sorting by Sales Amount instead of by Country name (see Figure 8-15).

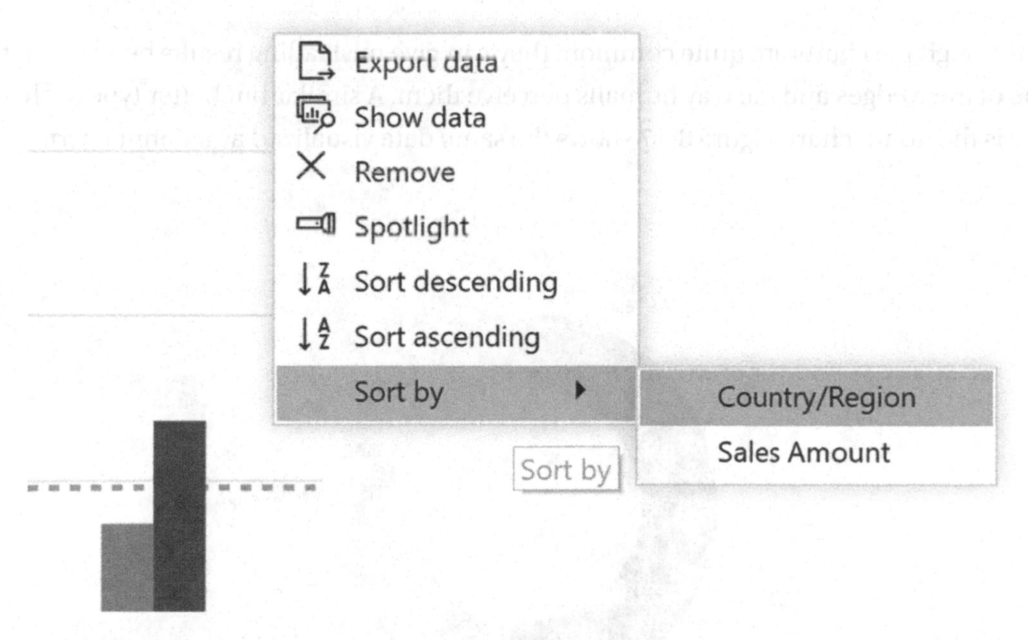

***Figure 8-15.** Changing the sorting of a column chart*

Pie charts are similar to stacked column or stacked bar charts in that they allow you to compare the measures for members of a category and also the total for the category. Figure 8-16 shows a pie chart comparing sales of different types of resellers.

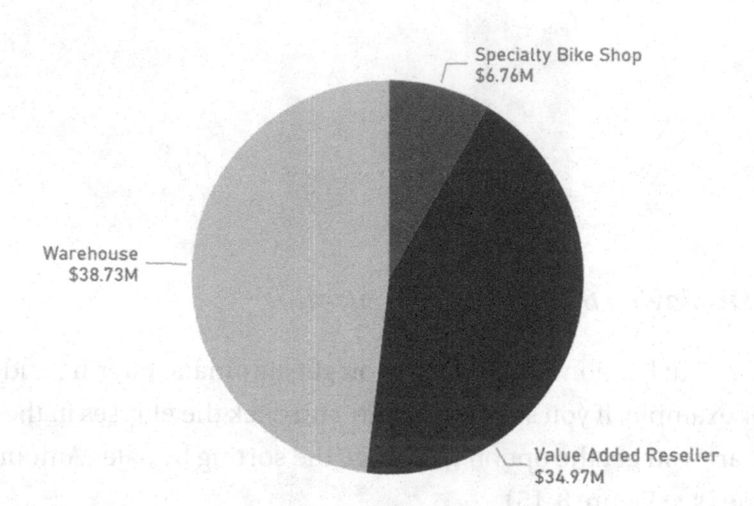

***Figure 8-16.** Comparing data using a pie chart*

Although pie charts are quite common, they can give misleading results because of the shape of the wedges and the way humans perceive them. A similar but better type of chart to use is the donut chart. Figure 8-17 shows the same data visualized as a donut chart.

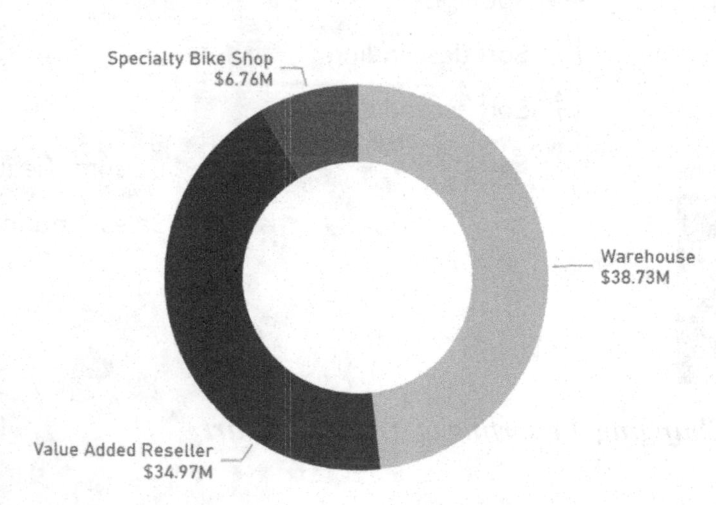

***Figure 8-17.** Comparing data using a donut chart*

Just as with column and bar charts, you have lots of control over the layout of the pie and donut charts, including data labels, data colors, and positioning of the title and the legend.

Although bar, column, pie, and donut charts are great for comparing aggregated data for various categories, if you want to spot trends in the data, line and scatter charts are a better choice. You will investigate these types of charts next.

Building Line and Scatter Charts

A line chart is used to look at trends across equal periods. The periods are often time units consisting hours, days, months, and so on. The time periods are plotted along the x-axis, and the measurement is plotted along the y-axis. Figure 8-18 shows order quantity by month. Each line represents a different year. Using this chart, you can easily spot trends, such as order quantities being up in 2013, while the overall trend is very similar to 2012.

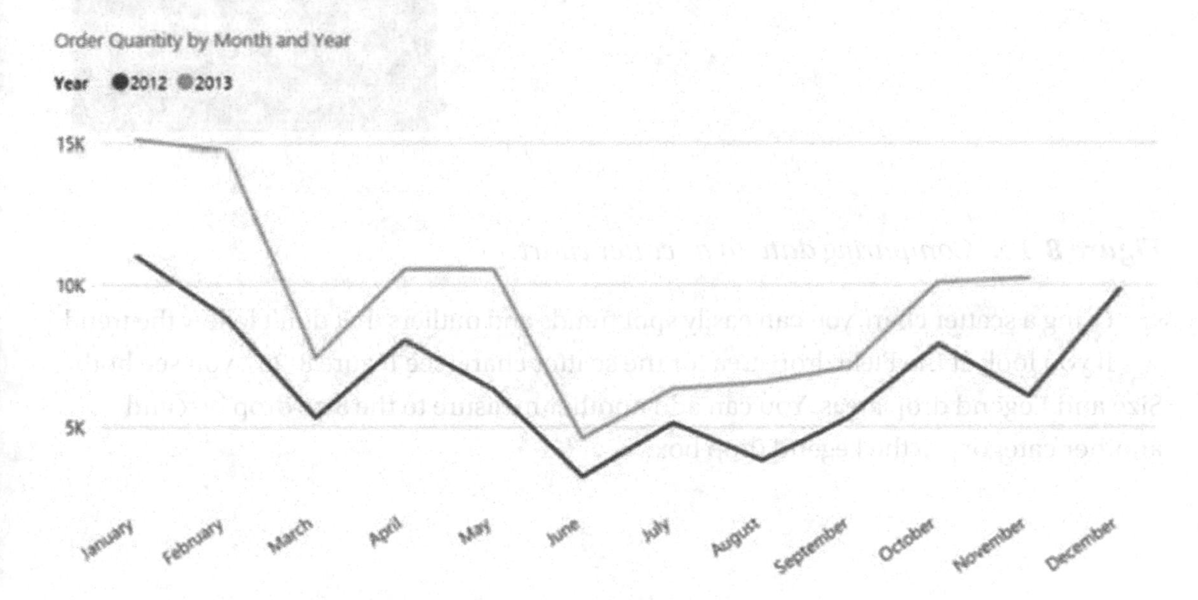

***Figure 8-18.** Spotting trends using a line chart*

If you need to compare trending of two measures at the same time, you can use a scatter chart. Scatter charts plot one measure along the y-axis and the other along the x-axis. To create a scatter chart, select a category field and two numeric fields—for example, reseller name (category), sales amount (numeric), and sales profit (numeric). Figure 8-19 shows the resulting scatter chart.

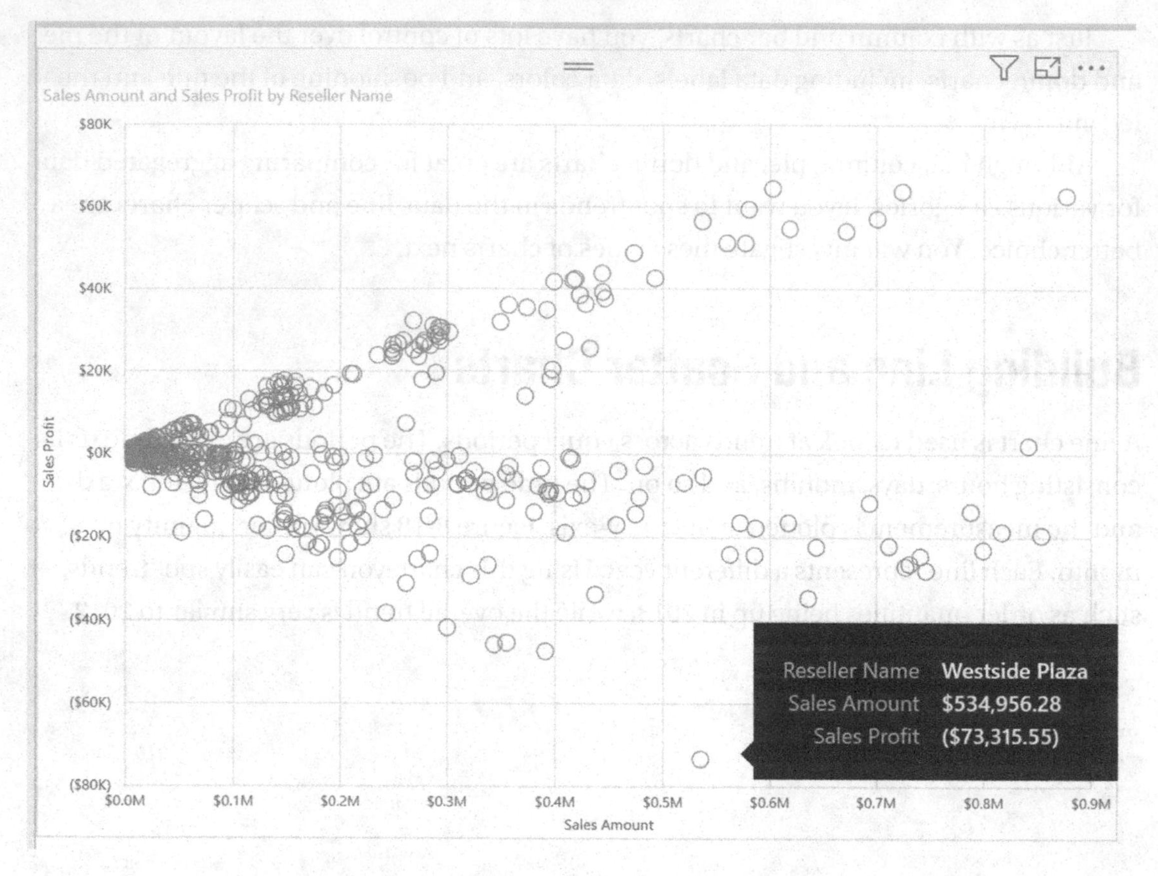

***Figure 8-19.** Comparing data in a scatter chart*

Using a scatter chart, you can easily spot trends and outliers that don't follow the trend.

If you look at the Field drop area for the scatter chart (see Figure 8-20), you see both Size and Legend drop areas. You can add another measure to the Size drop box and another category to the Legend drop box.

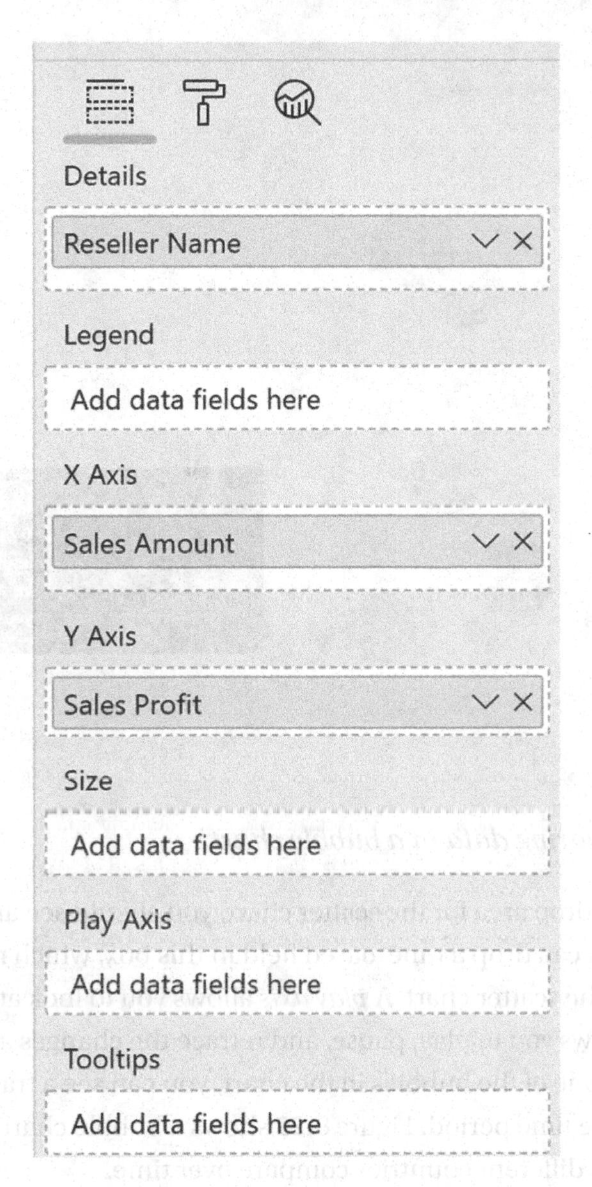

***Figure 8-20.** Adding size and color to a scatter chart*

Once you add the size axis to the scatter chart, it becomes a bubble chart. Figure 8-21 shows a bubble chart where the size of the bubble represents the number of employees of the reseller and the color represents the reseller's business type.

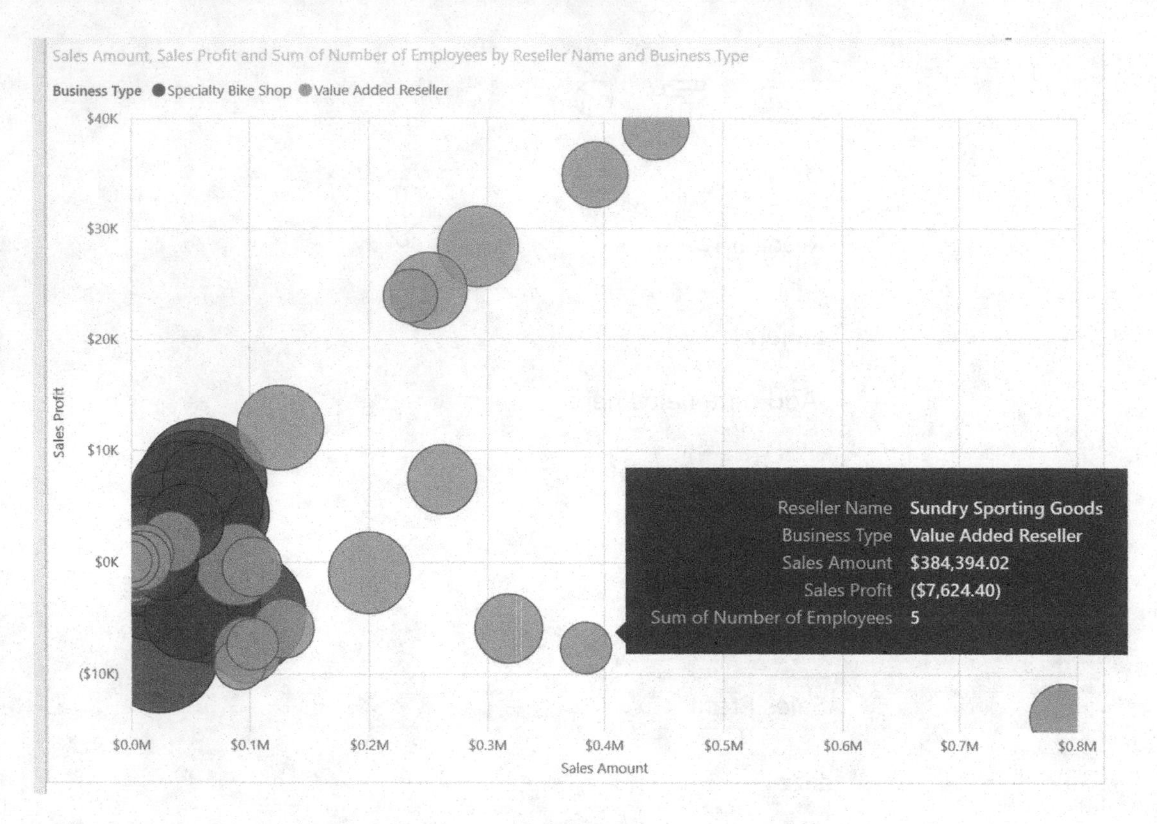

***Figure 8-21.** Comparing data in a bubble chart*

If you look at the drop area for the scatter chart, you should see another drop box labeled play axis. You can drop a time-based field in this box, which results in a play axis being placed below the scatter chart. A *play axis* allows you to look at how the measures vary over time. It allows you to play, pause, and retrace the changes as you perform your analysis. If you click one of the bubbles in the chart, you can see a trace of the changes that occurred over the time period. Figure 8-22 shows a bubble chart that can be used to analyze how sales for different countries compare over time.

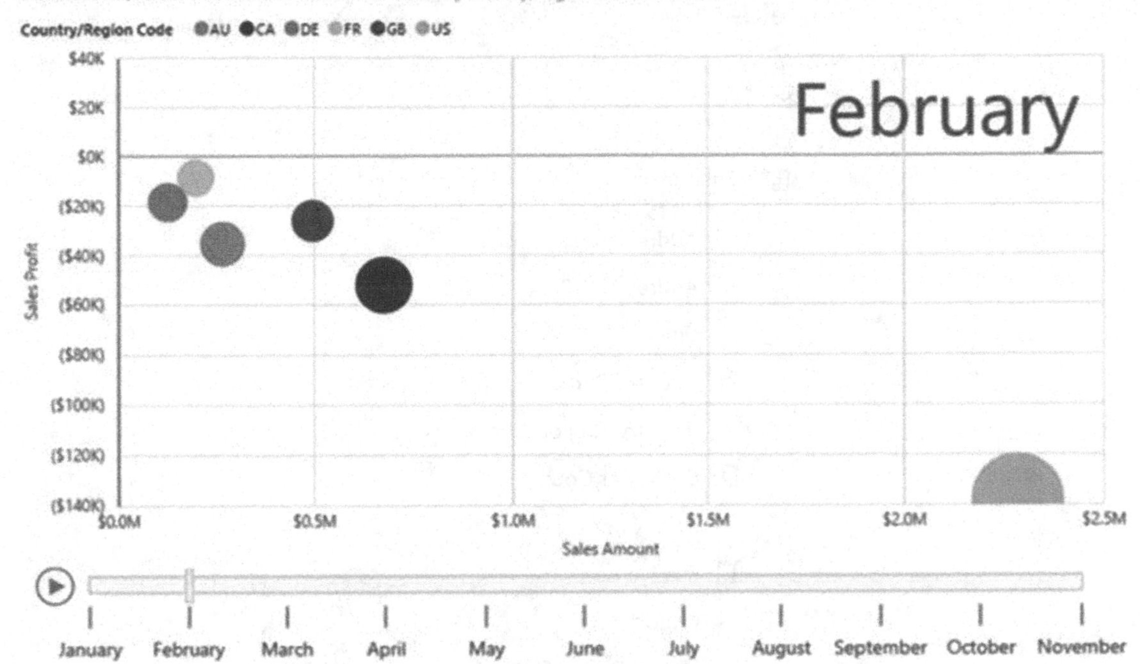

***Figure 8-22.** Adding a play axis to a bubble chart*

In addition to the standard visualizations for comparing data, Power BI Desktop allows you to look at data geographically using maps. You will see how to create map-based visualizations in the next section.

Creating Map-Based Visualizations

One of the nice features of Power BI is that it can use map tiles to create visualizations. If your data contains a geographic field such as city, state/province, and country, it's very easy to incorporate the data into a map. You can tell whether a field can be geolocated by a globe icon in the field list (see Figure 8-23).

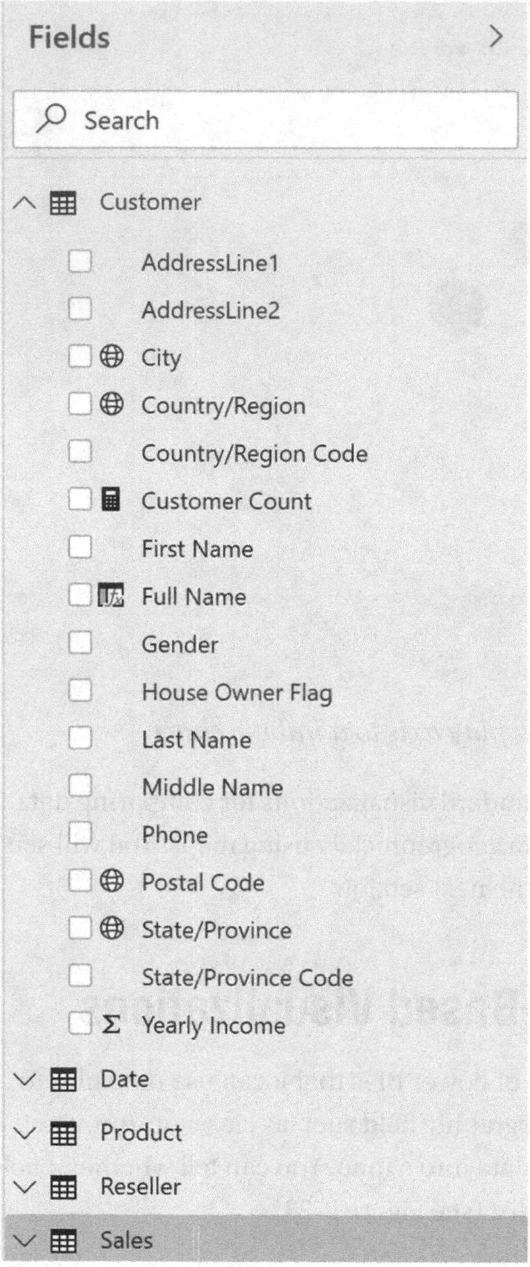

***Figure 8-23.** The globe icon indicates fields that can be mapped*

There are currently two types of maps in Power BI: map and filled map. When you use a map, the locations show up on the map as bubbles with sizes indicating the value of the measure. Figure 8-24 shows number of customers by city, where the map is zoomed in to the Bay Area in Northern California.

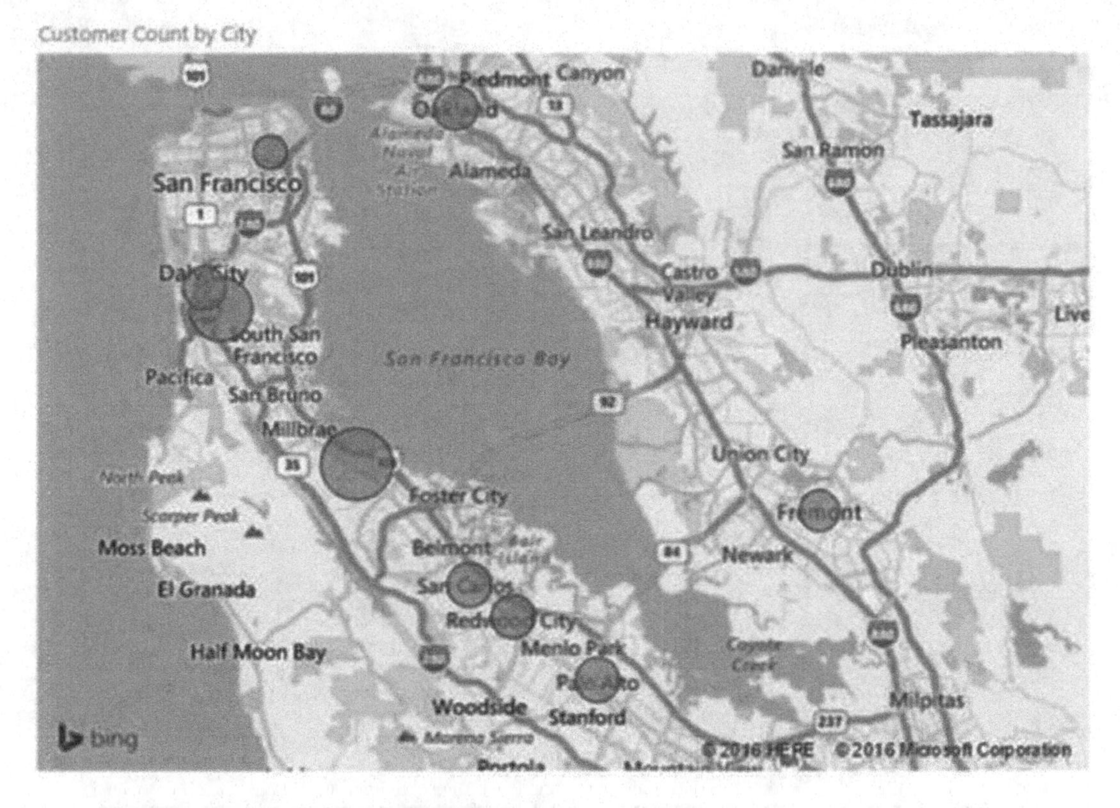

***Figure 8-24.** Viewing data on a map*

If you look at the Field drop areas of the map, you should see Longitude and Latitude drop boxes; this feature allows you to create precise location points on the map. There is also a Legend drop area where you can drop a category field. That will convert the bubbles on the map into pie charts showing each category (see Figure 8-25).

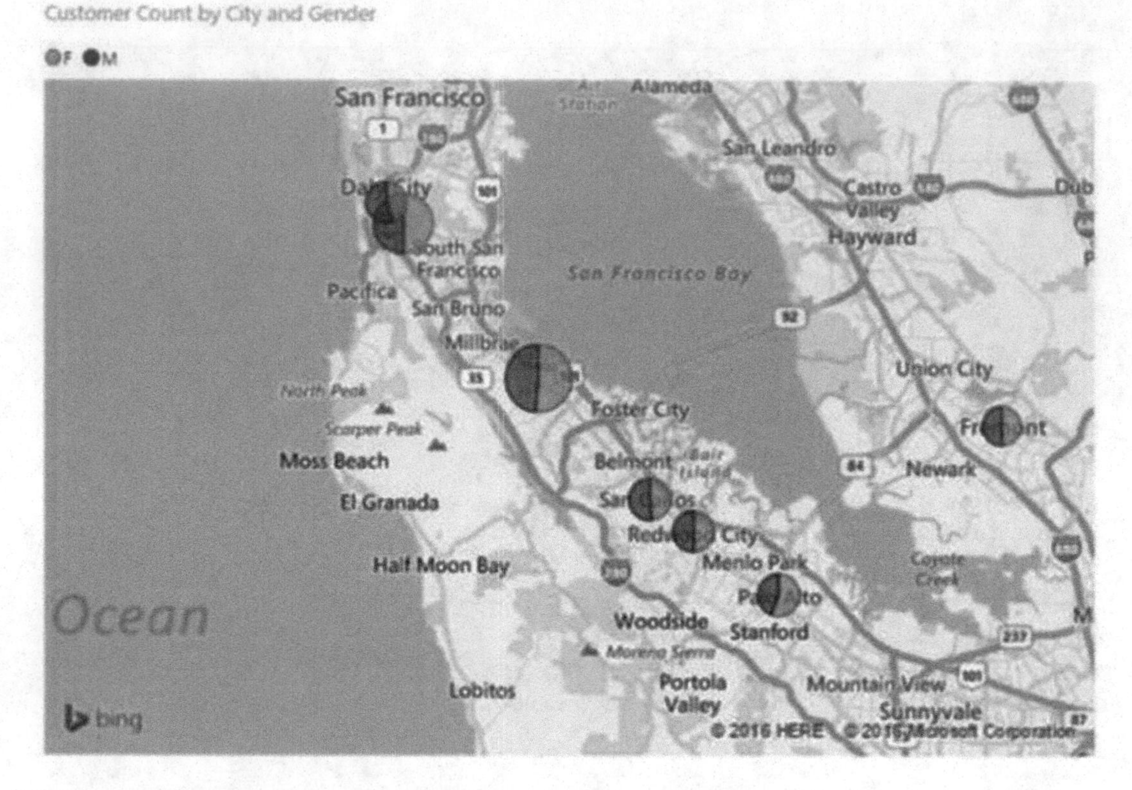

Figure 8-25. *Adding a category to the map*

A filled map uses geospatial areas rather than bubbles on the map. Common areas used are continent, country, region, state, city, or county. Figure 8-26 shows the number of customers in the Western states of the United States. The darker the shading, the larger the number of customers for the state.

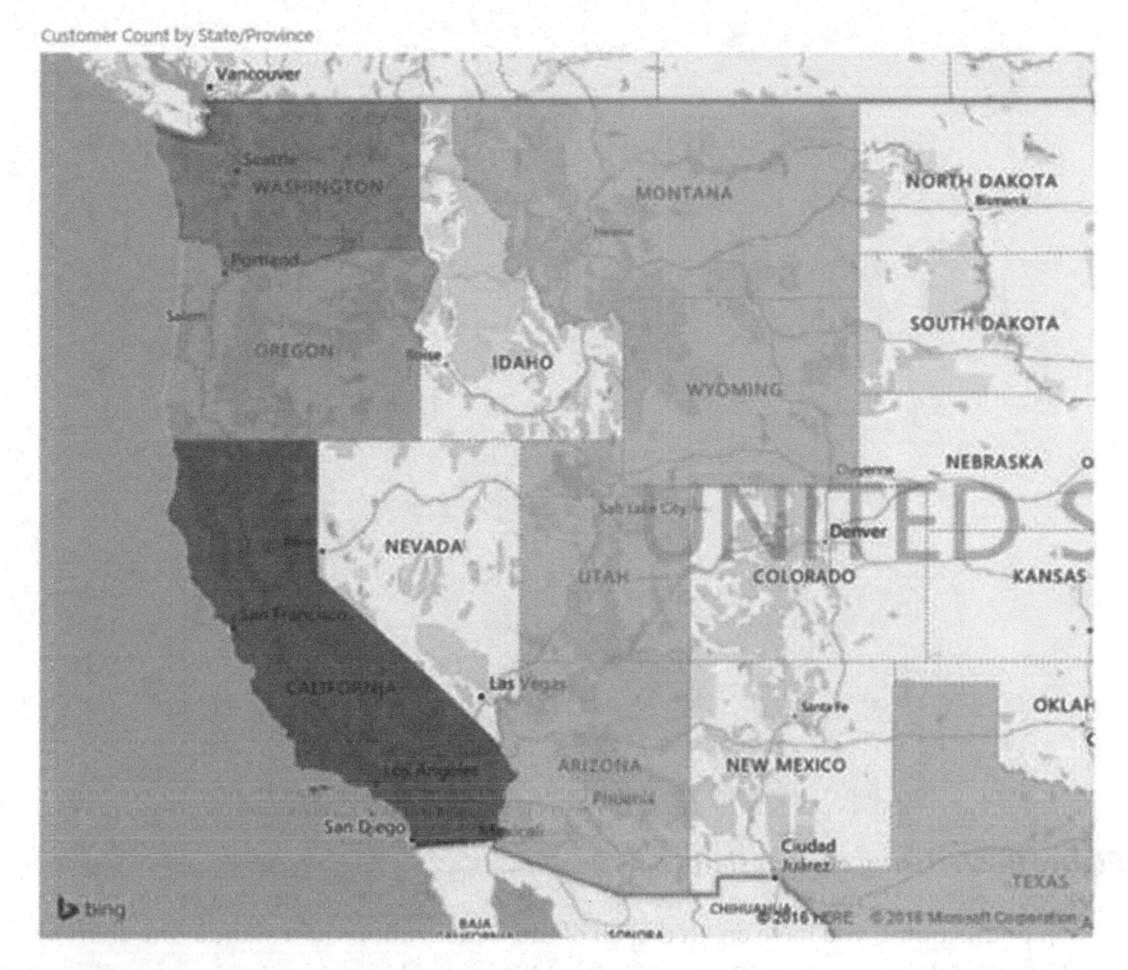

Figure 8-26. *Creating a filled map*

Another type of map (officially, this is in preview at the time of this book's publication) is the shape map. Shape map visuals are based on ESRI/TopoJSON maps. They give you the ability to use custom maps that you can create, such as geographical, seating arrangements, floor plans, and more. Figure 8-27 shows the number of customers in various San Francisco neighborhoods.

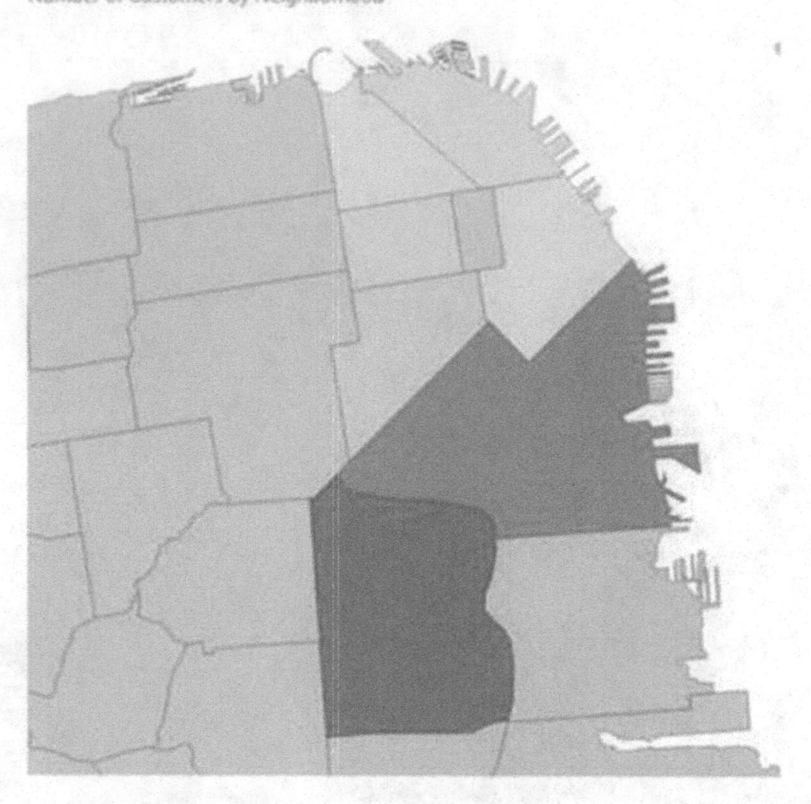

Figure 8-27. *Using custom shapes*

Up to this point, you have created each of the visualizations as a standalone chart or graph. One of the strengths of Power BI is its ability to tie these visualizations together to create interactive reports for data exploration.

Linking Visualizations in Power BI

A great feature of Power BI is that if the data model contains a link between the data used to create the various visualizations, Power BI will use the relationship to implement interactive filtering. *Interactive filtering* is when the process of filtering one visualization automatically filters a related visualization. For example, Figure 8-28 shows a bar chart that displays sales for the different product categories from the Adventure Works sample database and a column chart that shows various countries' sales.

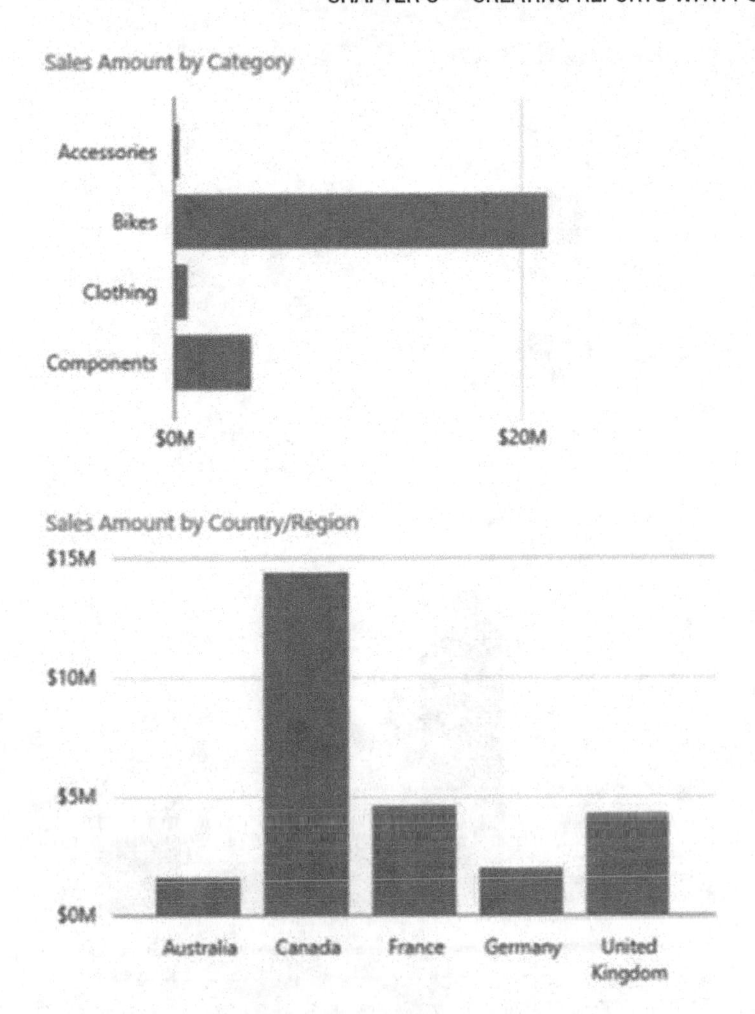

Figure 8-28. *Adding related visualizations to the same page*

Because the Sales table is related to the SalesTerritory table and the Product table, if you select one of the Product category bars, it will highlight the bar chart to show sales for that category (see Figure 8-29).

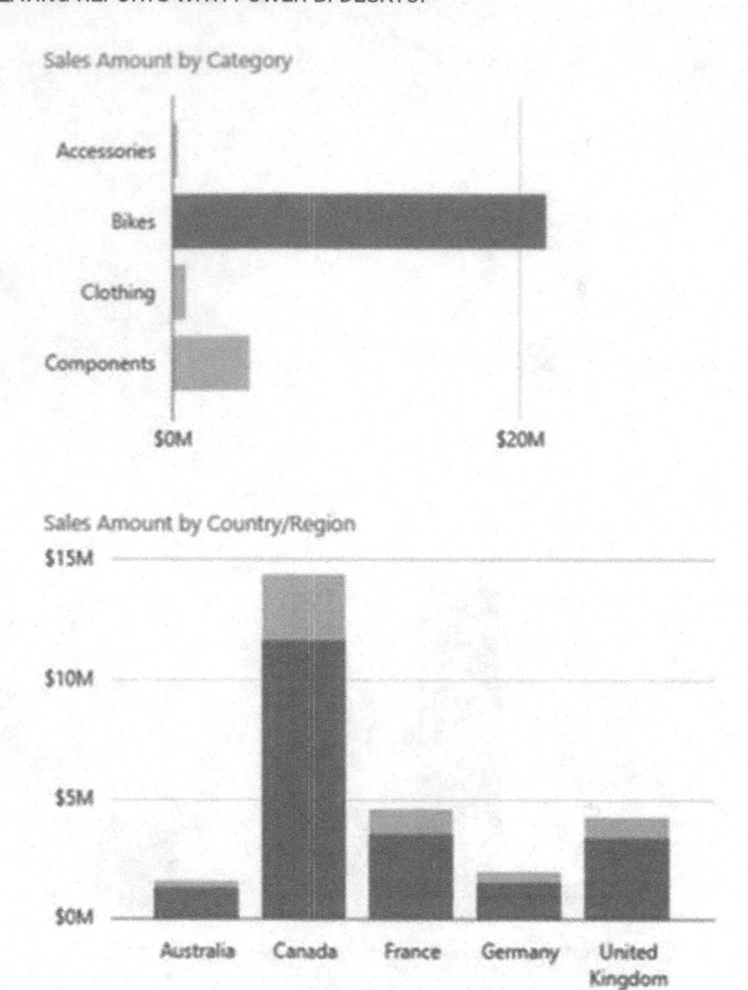

Figure 8-29. *Interactive filtering of related visualizations*

The filtering works both ways, so you can click one of the columns in the lower chart, and it will filter the top chart to show sales for that country. This interactive filtering works for most types of visualizations available in Power BI. Figure 8-30 shows a bubble chart and a table. When you click a bubble, the table is filtered to show the detail records that make up the bubble values.

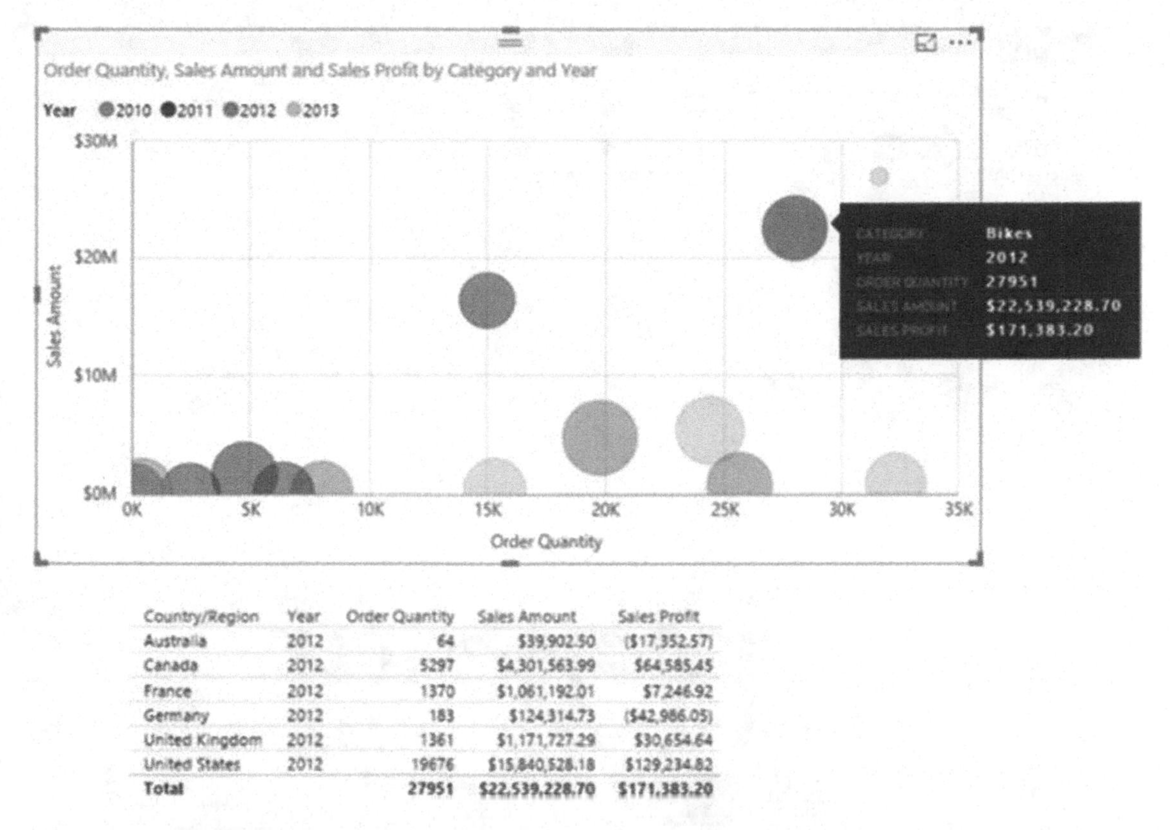

Country/Region	Year	Order Quantity	Sales Amount	Sales Profit
Australia	2012	64	$39,902.50	($17,352.57)
Canada	2012	5297	$4,301,563.99	$64,585.45
France	2012	1370	$1,061,192.01	$7,246.92
Germany	2012	183	$124,314.73	($42,986.05)
United Kingdom	2012	1361	$1,171,727.29	$30,654.64
United States	2012	19676	$15,840,528.18	$129,234.82
Total		**27951**	**$22,539,228.70**	**$171,383.20**

***Figure 8-30.** Filtering to show the details that make up the bubble values*

You can control the visual interactions between visuals by selecting a visual and then clicking the Format tab. The other visuals on the page will then display three icons (see Figure 8-31). Clicking the first one will cause it to filter, clicking the second will cause it to highlight as in Figure 8-29, and clicking the third icon will turn off the interaction.

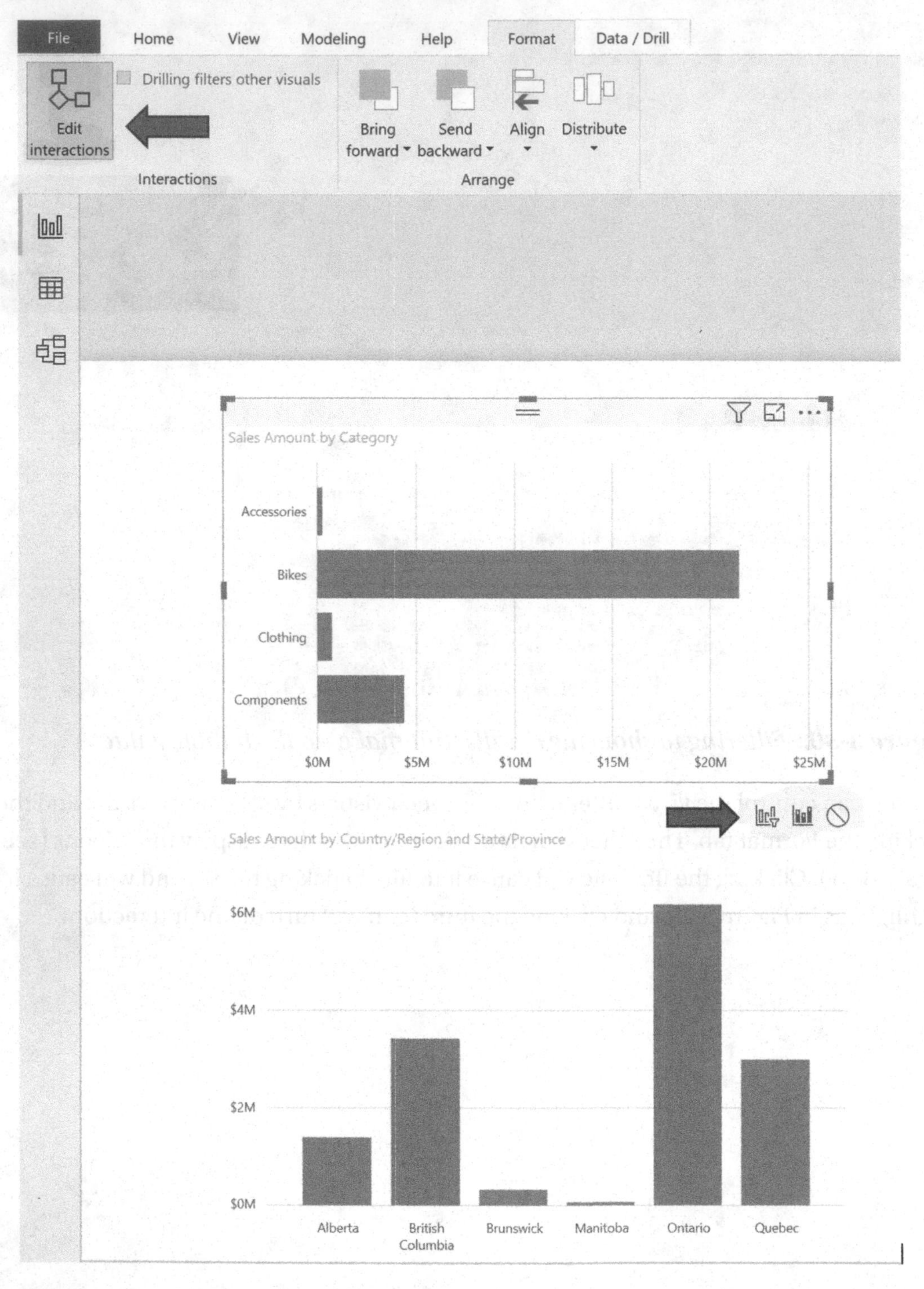

Figure 8-31. *Controlling visual interaction*

Along with built-in visual interaction, another powerful feature of Power BI visuals is the ability to drill down and up through hierarchies, the topic of the next section.

Drilling Through Visualizations

A great feature built-in to most Power BI visualizations is the ability to drill down and up through the various detail levels. This is useful when you have hierarchies such as month, quarter, and year, or products, categories, and subcategories. To enable drilling, just place the different levels of the hierarchy in the appropriate well, depending on the visual. Figure 8-32 shows a hierarchy placed in the Axis well of a columnar chart. You can control the drill through actions using the icons at the top of the chart.

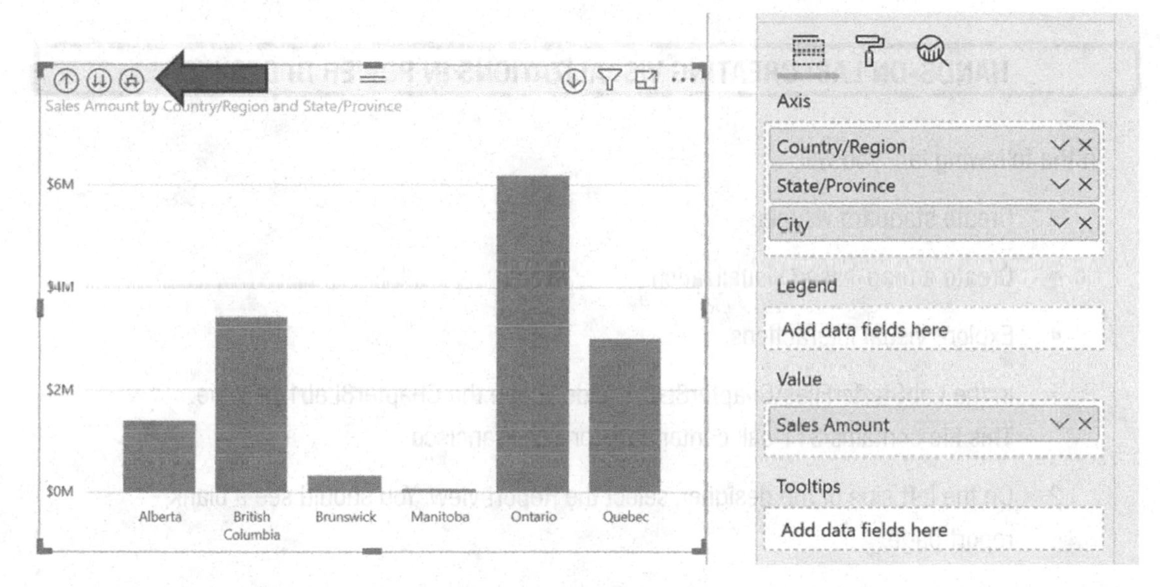

***Figure 8-32.** Enabling drilling on a column chart*

You can also control the drilling action under the Visual tools ➤ Data/Drill tab (see Figure 8-33). You have the option of showing the next level or expanding to the next level. For example, if you select Show next level and select Canada, it will drill through to show provinces in Canada. If you select Expand next level, it will show all states/provinces for all countries.

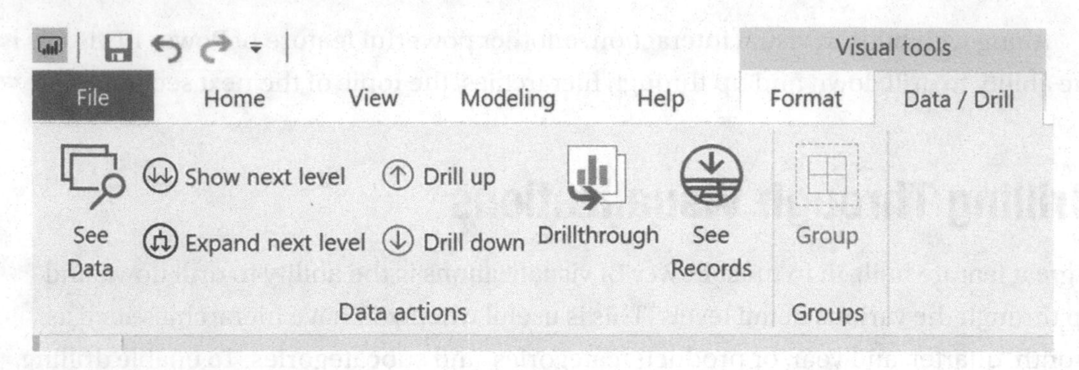

***Figure 8-33.** Controlling drilling actions*

Now that you have seen how to create some of the visualizations available in Power BI, it is time to get some hands-on experience by creating a few.

HANDS-ON LAB: CREATING VISUALIZATIONS IN POWER BI DESKTOP

In the following lab, you will

- Create standard visuals
- Create a map-based visualization
- Explore visual interactions

1. In the LabStarterFiles\Chapter8Lab1 folder, open the Chapter8Lab1.pbix file. This file contains 311 call center data for San Francisco.
2. On the left side of the designer, select the Report view. You should see a blank report page.
3. In the Visualizations toolbox, select the Table visualization. Create a table that lists Type of request, Ave Days Open CY, and Ave Days Open PY.
4. In the Filters pane, filter the table to only show cases where the Status is closed. Format the table with alternating rows and change the font size to 10. Your table should look like Figure 8-34.

Type	Ave Days Open CY	Ave Days Open PY
City Maintenance	5.38	7.16
Enforcement Activities	8.48	11.48
Other	4.40	9.09
Property Damage Reports	12.22	31.58
Service Requests	7.34	24.96
Total	7.32	15.15

Figure 8-34. *Creating a table*

5. Click an empty area of the report page and select the 100% stacked bar chart from the Visualizations toolbox.
6. Add the Number of Cases to the Values well, the Type to the Axis well, and the Year to the Legend well.
7. Change the data colors to gold and orange and turn on the data labels. Change the data labels to black. Your chart should be similar to Figure 8-35.

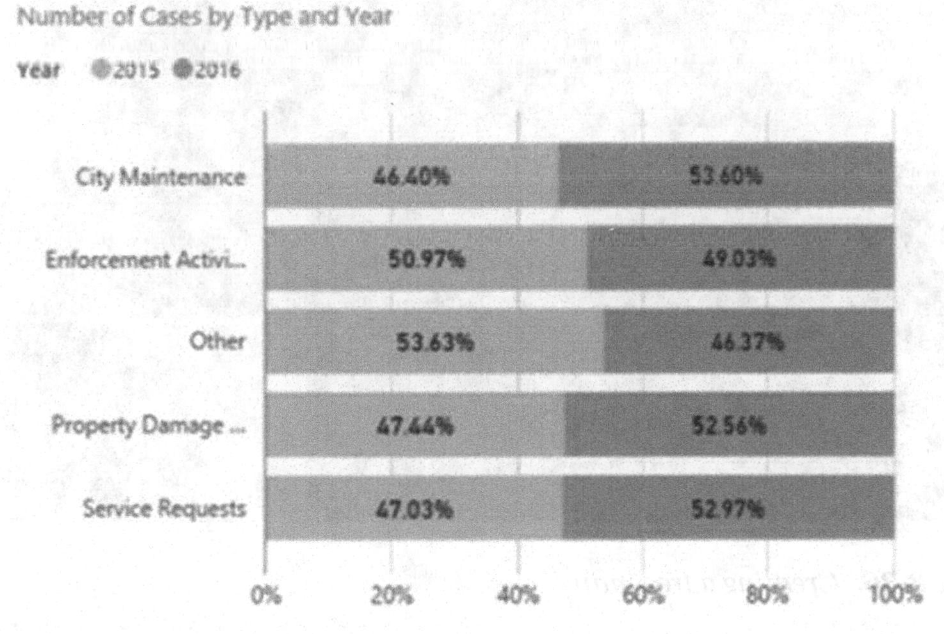

Figure 8-35. *Creating a 100% stacked bar chart*

8. Select a blank area of the page and click the Treemap visual in the toolbox. Add the Number of Cases to the Values well and the Neighborhood to the Group well.

9. Note that when you hover over the neighborhood areas, the tooltip shows the neighborhood and the number of cases. You can also add additional information to the tooltip. Add the Ave Days Open to the Tooltips well. Your final chart should look similar to Figure 8-36.

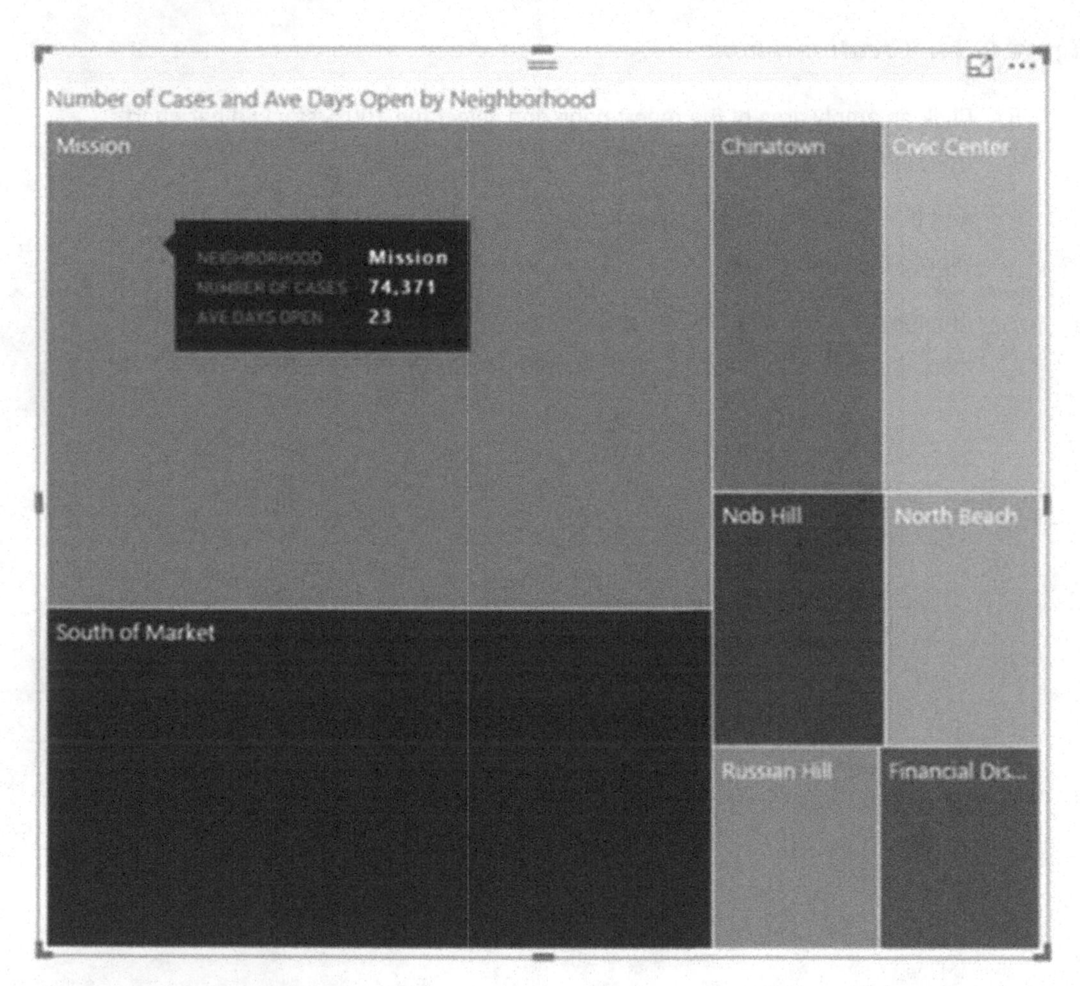

Figure 8-36. *Creating a tree map*

10. If you click one of the neighborhood blocks, you should see that the table and bar chart are filtered by the neighborhood. In this case, we want to turn off the filtering for the table.

11. With the tree map selected, note that a Visual tools ribbon set appears at the top of the program. Go to the Format tab under the Visual tools tab and click the Edit Interactions button. You should see the icons above the table to turn on and off filtering (see Figure 8-37). Turn off filtering for the table.

Type	Ave Days Open CY	Ave Days Open PY
City Maintenance	5.38	7.16
Enforcement Activities	8.48	11.48
Other	4.40	9.09
Property Damage Reports	12.22	31.58
Service Requests	7.34	24.96
Total	**7.32**	**15.15**

Figure 8-37. *Turning off filtering*

12. Click the Edit Interactions button again on the Format tab to turn off the edit mode. Test the filtering by clicking the areas in the tree map. The bar chart should filter, but the table should not.
13. Click the plus sign (+) at the bottom of the designer to add a new report page.
14. On the new page, add a line chart that shows the Week Ending Date on the axis and Number of Cases as the values. By default, the date becomes a date hierarchy when it is dropped in the Axis well. Change this to show just the Week Ending Date by selecting the drop-down next to the Axis field (see Figure 8-38).

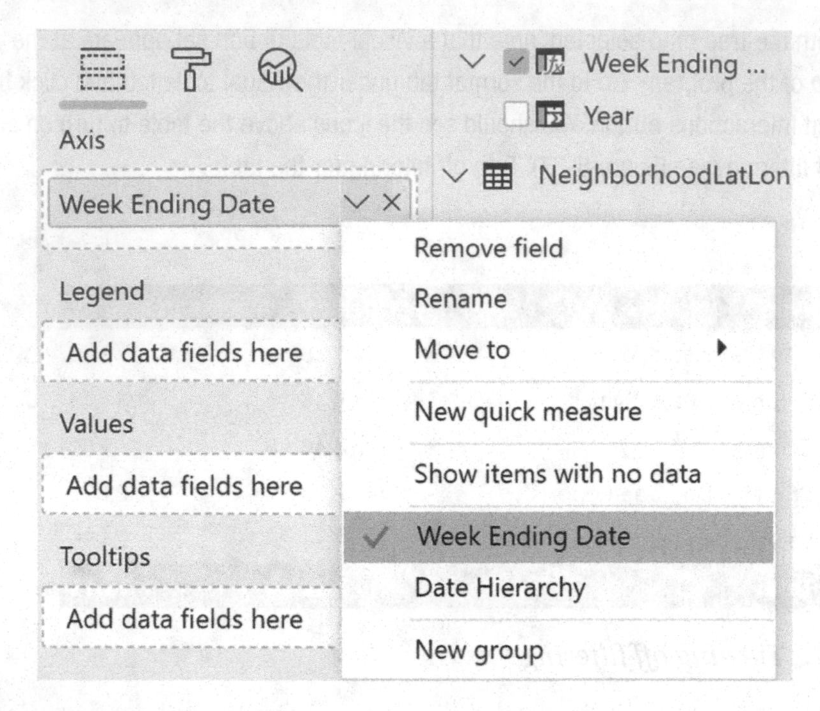

***Figure 8-38.** Showing just the date and not the hierarchy*

15. Click the Analytics icon (the magnifying glass in Figure 8-38) and add a trend line to the graph.
16. Click a blank area of the page and select a slicer from the Visualizations toolbox. Add the Week Ending Date to the Fields well. Use the slicer to adjust the date range in the graph (see Figure 8-39).

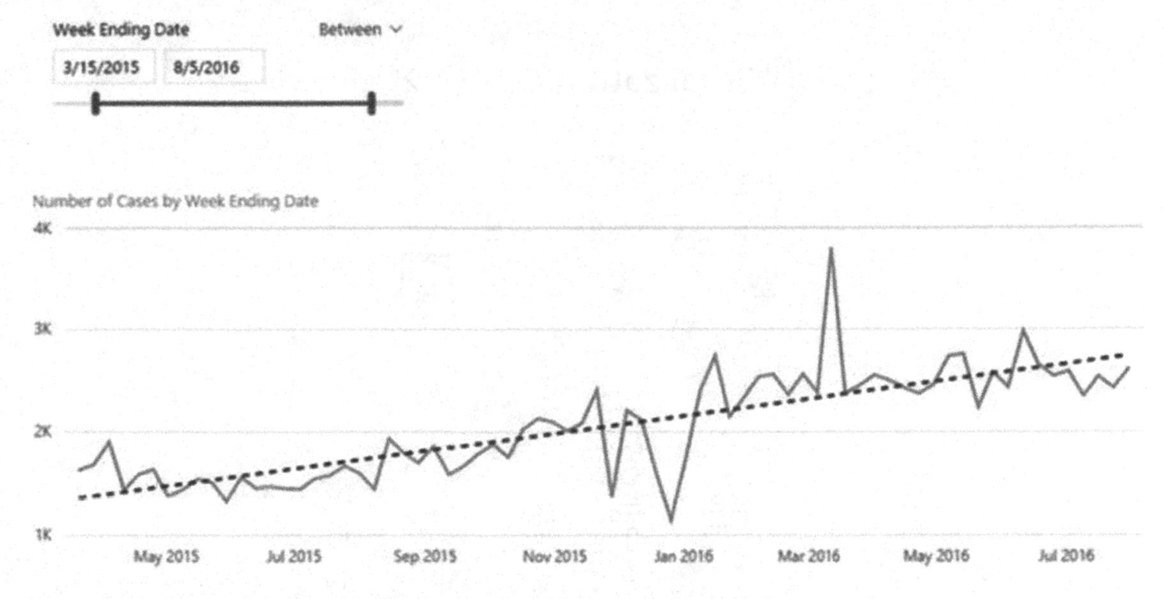

***Figure 8-39.** Adjusting the date range of the graph*

17. To create a map-based visualization, add a new report page. Select the Map visual in the Visualizations toolbox (the globe icon). From the NeighborhoodLatLon table, drag the Latitude and Longitude fields to the Latitude and Longitude field wells (see Figure 8-40).

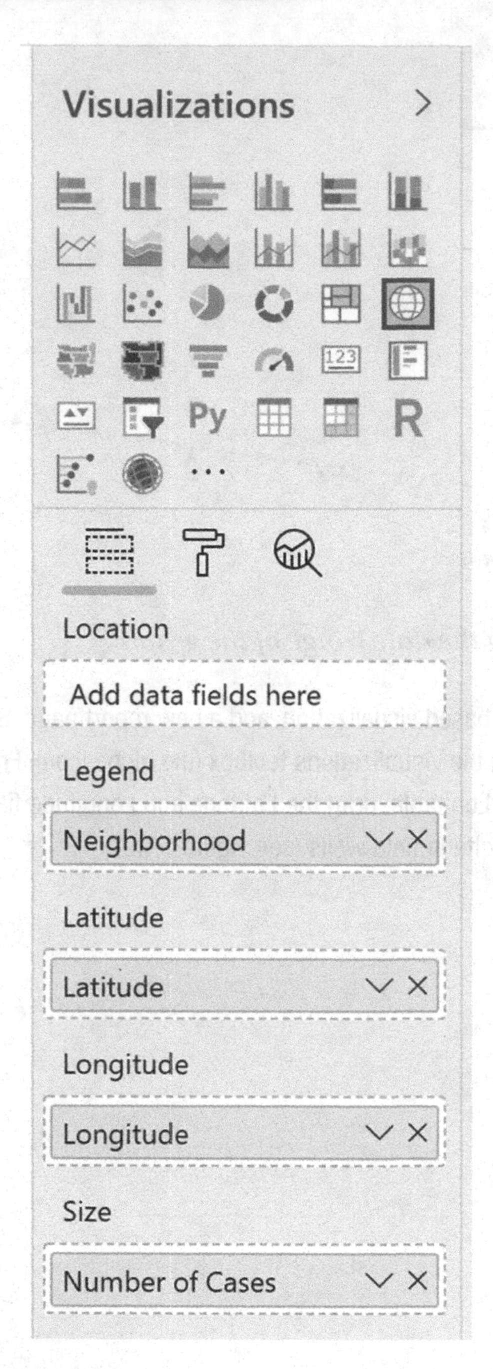

Figure 8-40. *Adding fields to the map*

18. Drag the Neighborhood field to the Legend well. From the SF311Calls table, drag the Number of Cases field to the Size well. Your map should look like Figure 8-41.

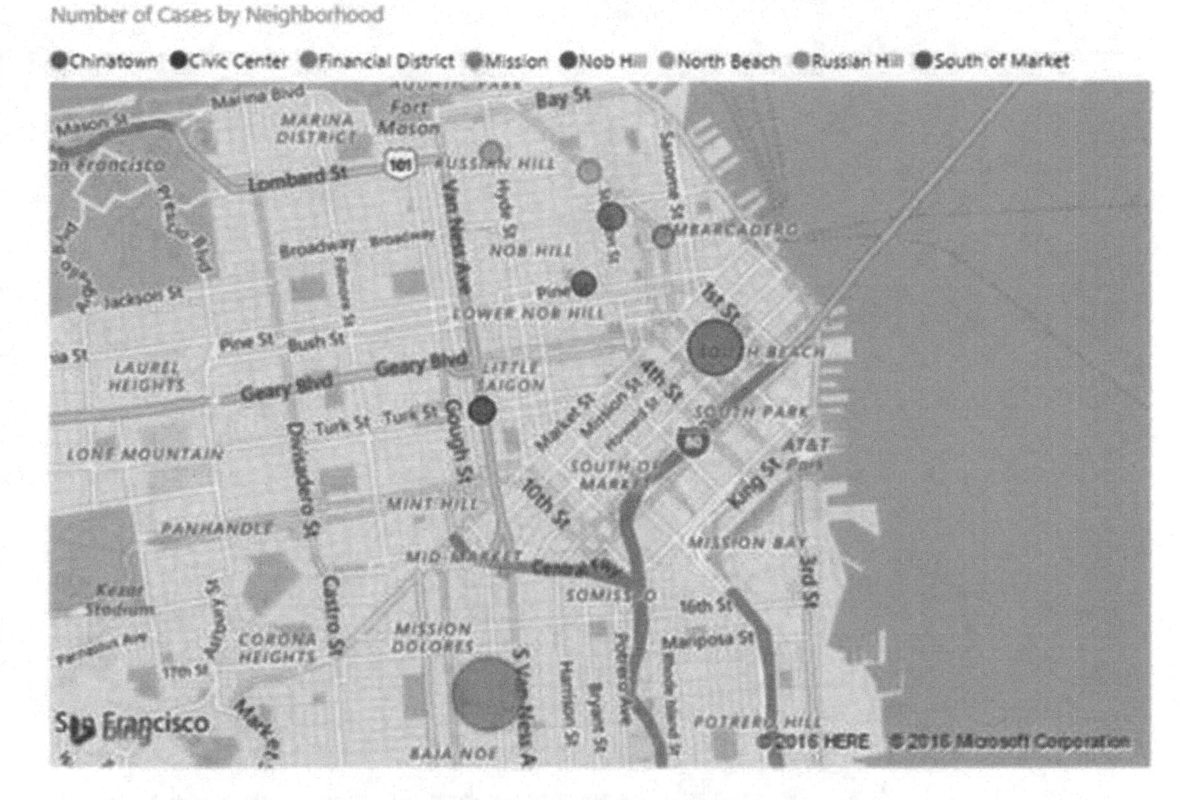

***Figure 8-41.** Mapping the number of cases for each neighborhood*

19. Experiment with mapping other fields and zooming to different areas of the map. Review options available under the Format option for the map. When done, save your changes and exit Power BI Desktop.

Summary

In this chapter, you have seen how to create various visualizations in Power BI Desktop. You created basic bar, line, and tree map charts. In addition, you saw how to present measures tied to a geospatial field on a map. You also saw how to control visual interactions and drilling functionality in your reports. You are now ready to deploy these reports to the Power BI portal. Once deployed, you will create interactive dashboards based on the reports and expose these to others.

CHAPTER 9

Publishing Reports and Creating Dashboards in the Power BI Portal

Now that you know how to create reports in Power BI Desktop, it's time to publish your reports for others to use. In this chapter, you will see how to publish reports created in Power BI Desktop to the Power BI Service (portal). Once the reports are published, you will create dashboards and share them with colleagues. In addition, you will set up an automated data refresh schedule.

After completing this chapter, you will be able to

- Create a user-friendly model
- Publish Power BI Desktop files to the Power BI Service
- Add tiles to a dashboard
- Share dashboards
- Refresh data in published reports

Creating a User-Friendly Model

Before publishing your models and reports for others to use, it is very important that users of your Power BI models have a pleasant experience as they build and explore the various visualizations in Power BI. One of the most useful things you can do is rename the tables and fields so they make sense to business users. You may often find that the names of the fields in the original source are abbreviated or have cryptic names that only make sense to the database developers.

D. Clark, *Beginning Microsoft Power BI*, https://doi.org/10.1007/978-1-4842-5620-6_9

Another good idea is to only expose fields that users find meaningful. It is always wise to hide any nonbusiness key values that are used to relate the tables in the model. To hide a column from clients of the model, right-click the column and select Hide in Report view (this will turn the column gray in the model designer). Figure 9-1 shows hiding an EmployeeKey field that has no business relevance.

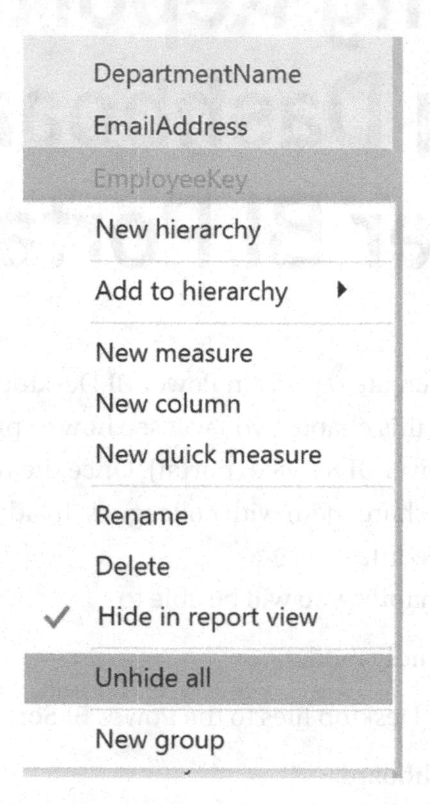

***Figure 9-1.** Hiding fields in Report view*

It's also a good idea to check the data type and format of the fields in the model. Depending on the data source, fields may be exported as text fields and need to be changed in the model. This is particularly true when it comes to date fields. You can use the Model tab in the Data view of Power BI Desktop to set the data type and format of the fields.

As you look at the field list in the Data view, notice the icons in front of the fields. The calculator icon indicates that the field is a measure created in the Power BI model. The summation symbol indicates that the field is numeric and will be aggregated when dragged to the Fields drop area. By default, the aggregate is a summation, which may not be the aggregate needed. In some cases, you don't want to aggregate a number;

for example, in the date table, the Year field is not meant to be aggregated. You can control this behavior in the Power BI model by setting the Summarize By drop-down on the Modeling tab (see Figure 9-2).

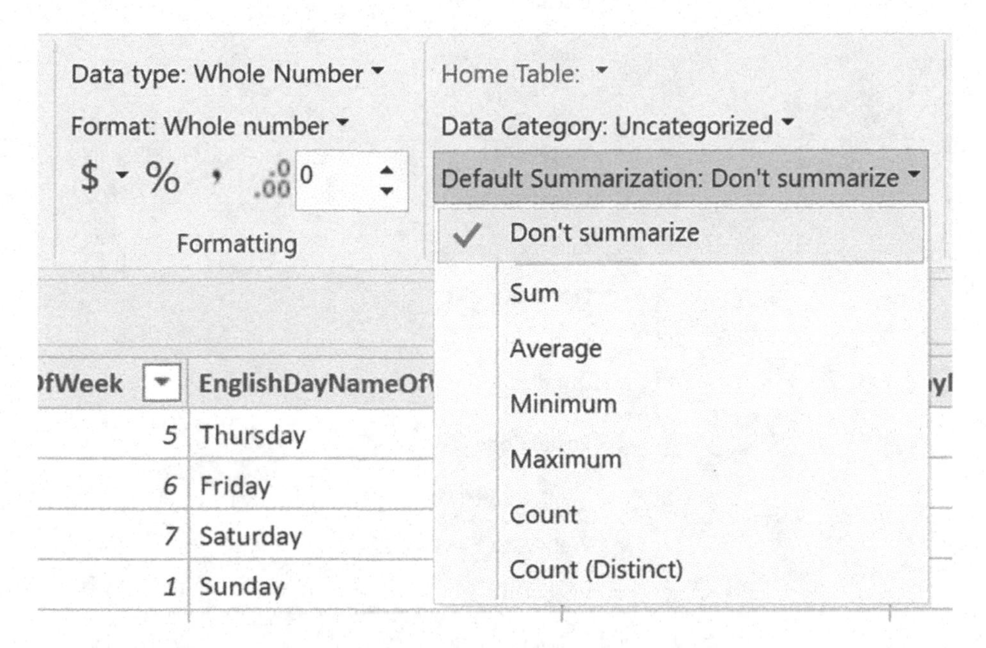

Figure 9-2. *Setting default summarizations for fields*

Another setting available on the Modeling tab of Power BI Desktop is the data category setting for the fields. This comes into play for location types like city, zip codes, and countries. When Power BI knows the column is a location type, it can use the field to implement a visualization using Bing mapping layers. The other data categories that are useful to set are the image and web URLs. When you set these, Power BI will know these are hyperlinks and format them appropriately (see Figure 9-3).

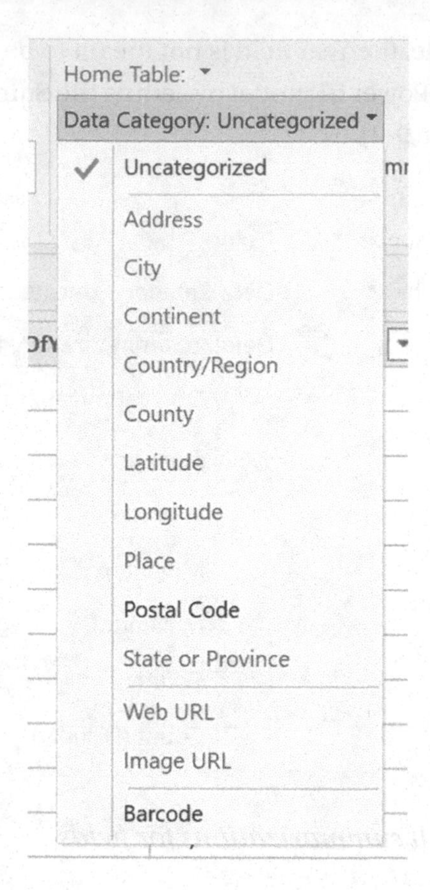

***Figure 9-3.** Setting the data category for a field*

Once you have the model and reports developed in Power BI Desktop, you are ready to deploy them to the Power BI portal.

Publishing Power BI Desktop Files to the Power BI Service

Now that you are ready to deploy the model and reports created in Power BI Desktop to the Power BI Service, make sure you are signed up for the service. If you are not signed up for the service through your organization, you can sign up for a free account at `https://powerbi.microsoft.com/en-us/`. If you are using a nonwork or school account, you have to sign up for an Office 365 trial (see `https://docs.microsoft.com/en-us/power-bi/service-admin-signing-up-for-power-bi-with-a-new-office-365-trial` for instructions). Once you have signed up for the account, you can log into the portal through `https://app.powerbi.com`. Once logged in, you should be on the Home page by default (see Figure 9-4). (Keep in mind that Microsoft is continuously updating the Power BI interface. Your view may not match the screenshots exactly.)

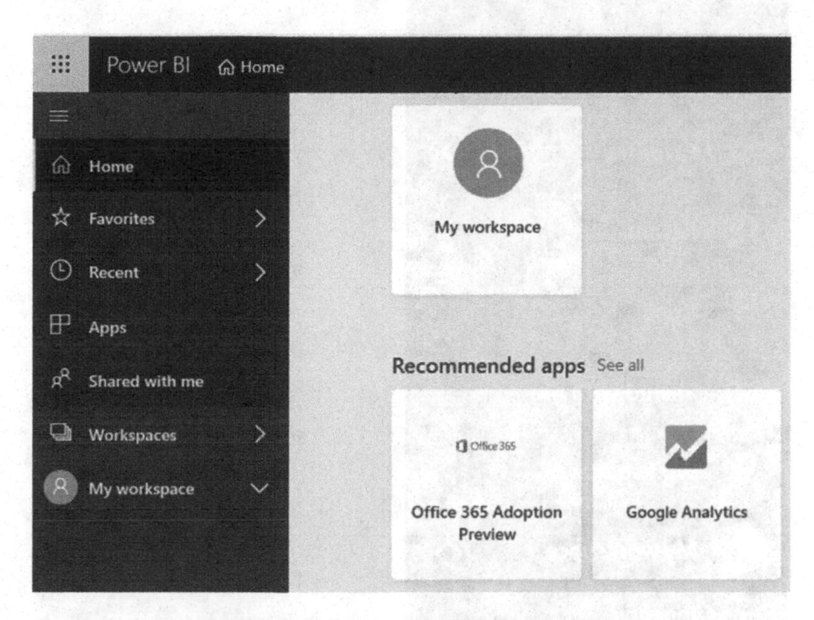

***Figure 9-4.** The Power BI portal Home page*

Click the My workspace icon on the dashboard. On the left side is a navigation pane. Expand the My workspace header. You should see headings for dashboards, reports, and data sets. There is also a button for getting data (see Figure 9-5).

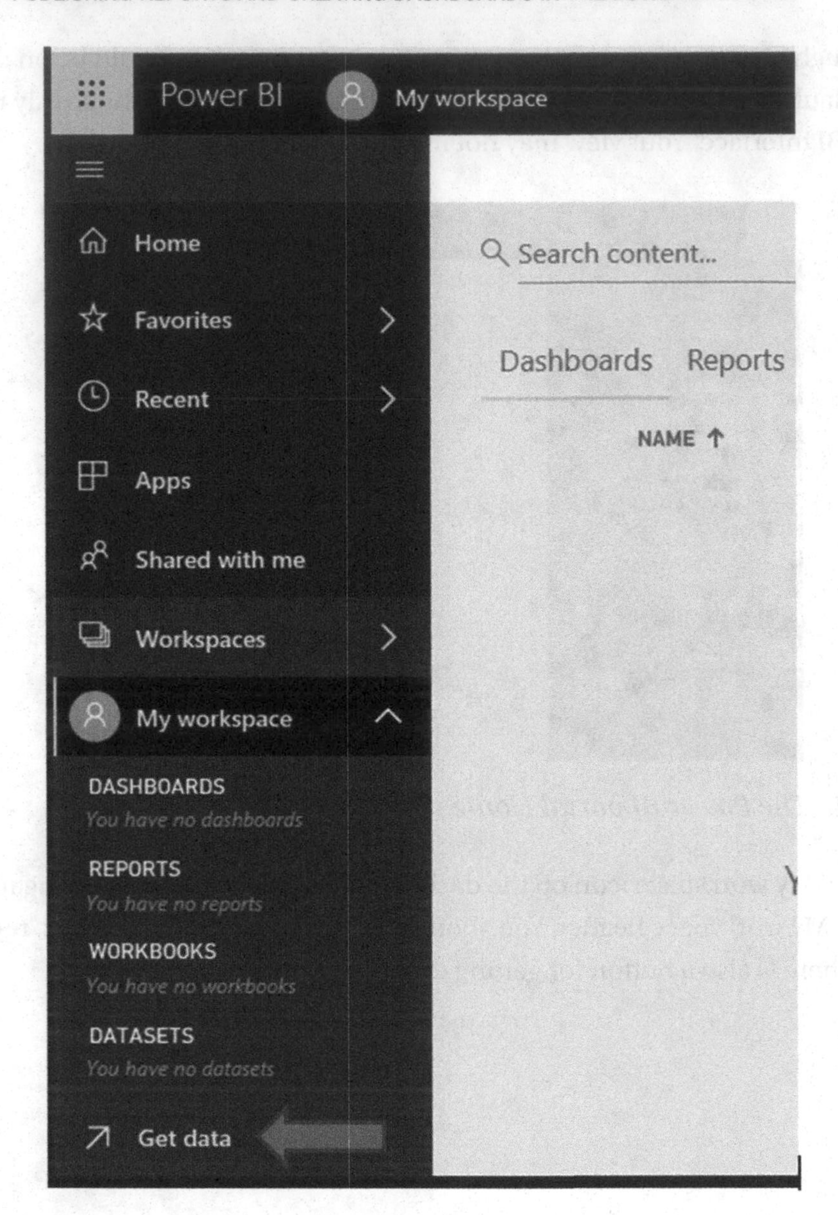

***Figure 9-5.** The navigation pane in Power BI*

After clicking the Get Data button, you can either discover published content or create new content (see Figure 9-6). Because you are publishing a Power BI Desktop file, choose the Files option.

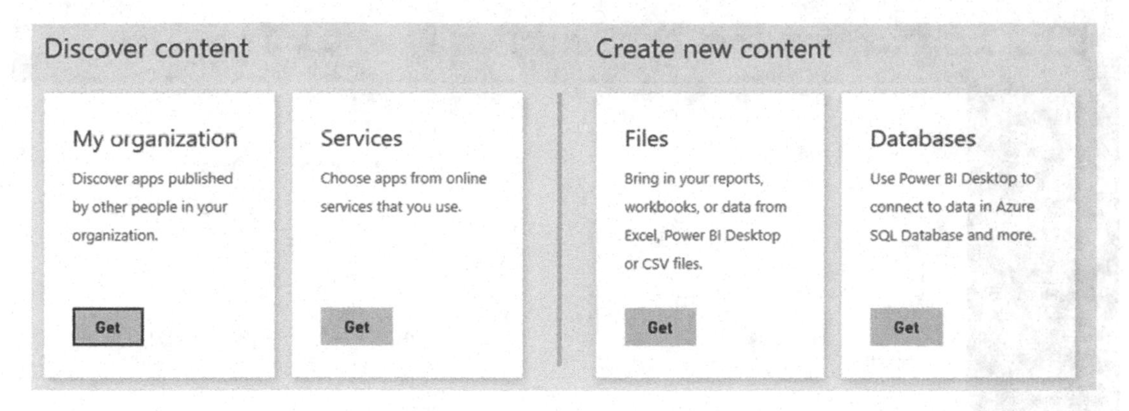

Figure 9-6. *Selecting content*

After selecting Files, the next step is selecting the location of the file. It can be a local file, a file stored in OneDrive, or a file stored in SharePoint (see Figure 9-7).

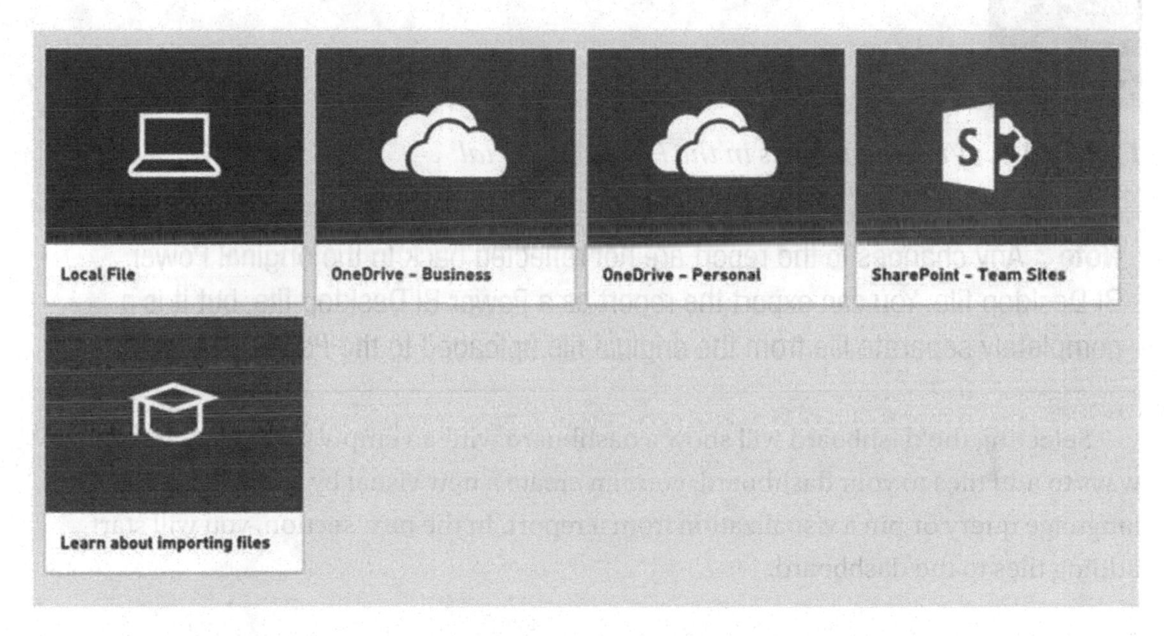

Figure 9-7. *Selecting the file location*

Once you select the file, it is imported into the Power BI Service, and you will see a new data set, report, and dashboard listed in the navigation pane. Selecting the data set will present report designer, which is essentially the same as the Report view in Power BI Desktop. You or a colleague can use this designer to create new reports. Selecting the report will show the report pages you developed in Power BI Desktop (see Figure 9-8). The report is in view mode and is fully interactive. You can switch to edit mode and update the report.

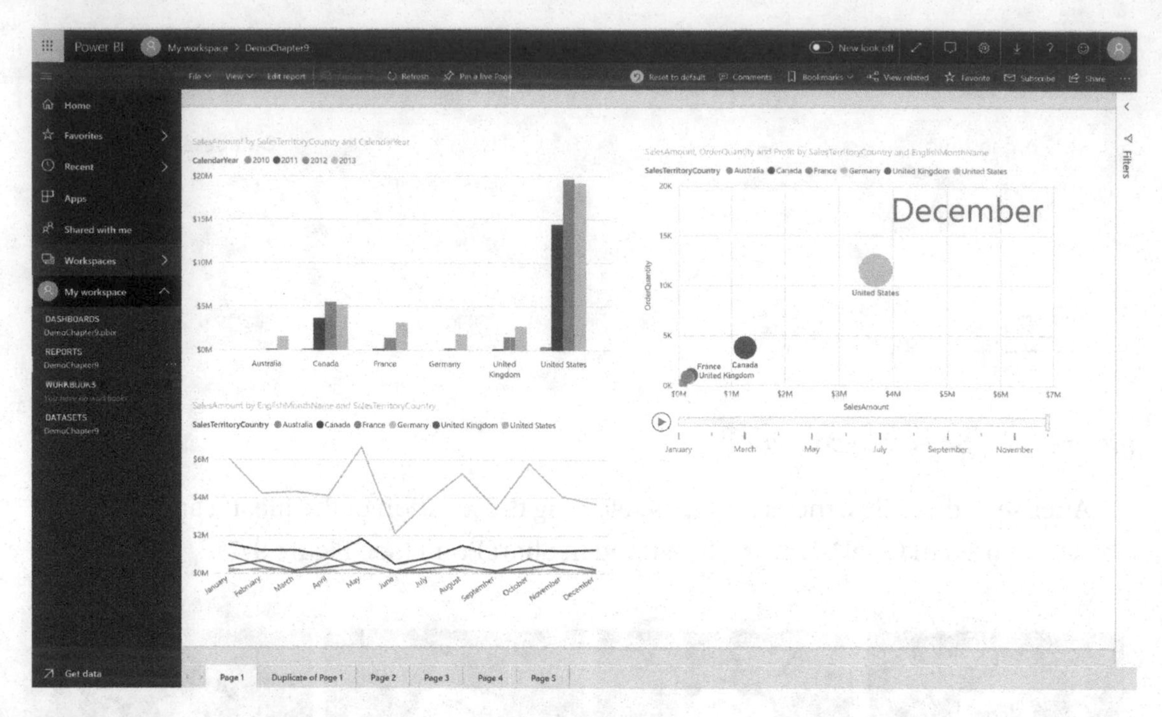

Figure 9-8. Viewing reports in the Power BI portal

Note Any changes to the report are not reflected back to the original Power BI Desktop file. You can export the report as a Power BI Desktop file, but it is a completely separate file from the original file uploaded to the Power BI portal.

Selecting the dashboard will show a dashboard with an empty tile. There are two ways to add tiles to your dashboard: you can create a new visual by using a natural language query or pin a visualization from a report. In the next section, you will start adding tiles to the dashboard.

Adding Tiles to a Dashboard

After deploying your reports to the Power BI portal, you can use the report visuals to create your dashboards. When you select a visual in the report, you will see a pin icon in the upper right corner of the visual (see Figure 9-9).

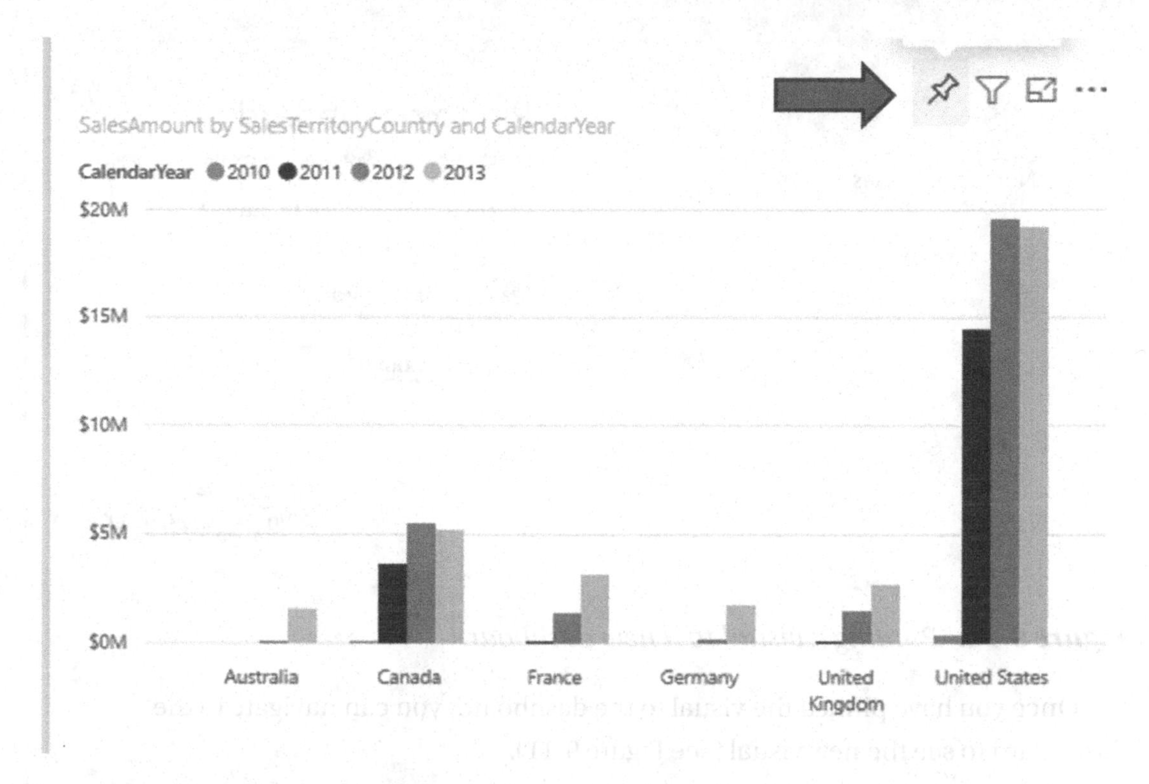

***Figure 9-9.** Selecting a visual to add to a dashboard*

When you click the pin, you are presented with a Pin to dashboard window (see Figure 9-10). You have the option of pinning the visual to an existing dashboard or creating a new one.

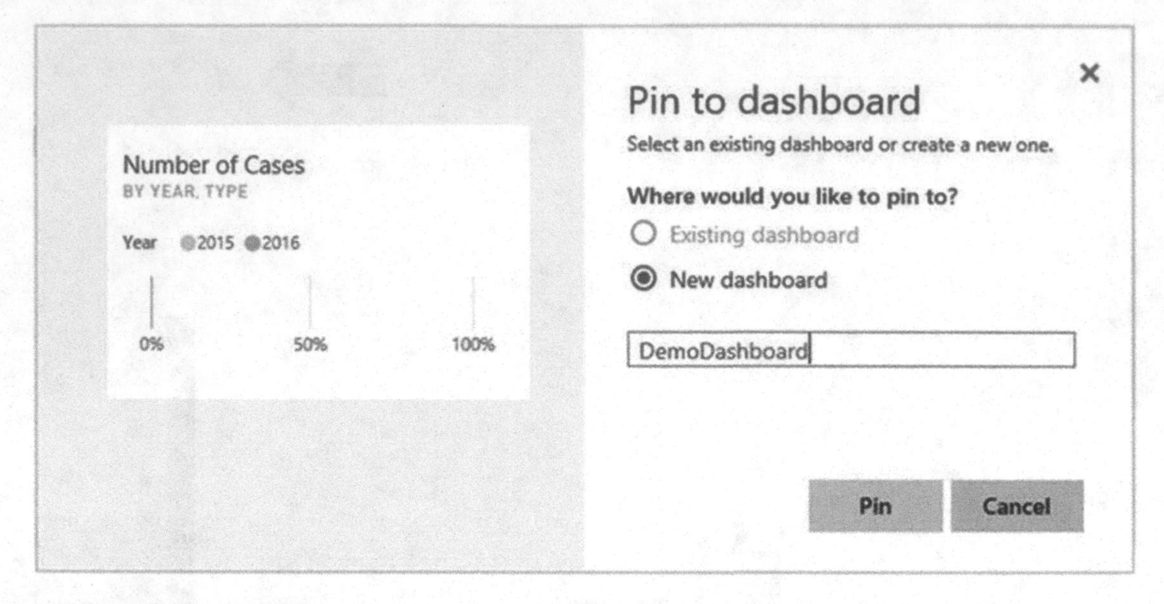

Figure 9-10. Pinning a visual to a new dashboard

Once you have pinned the visual to the dashboard, you can navigate to the dashboard to see the new visual (see Figure 9-11).

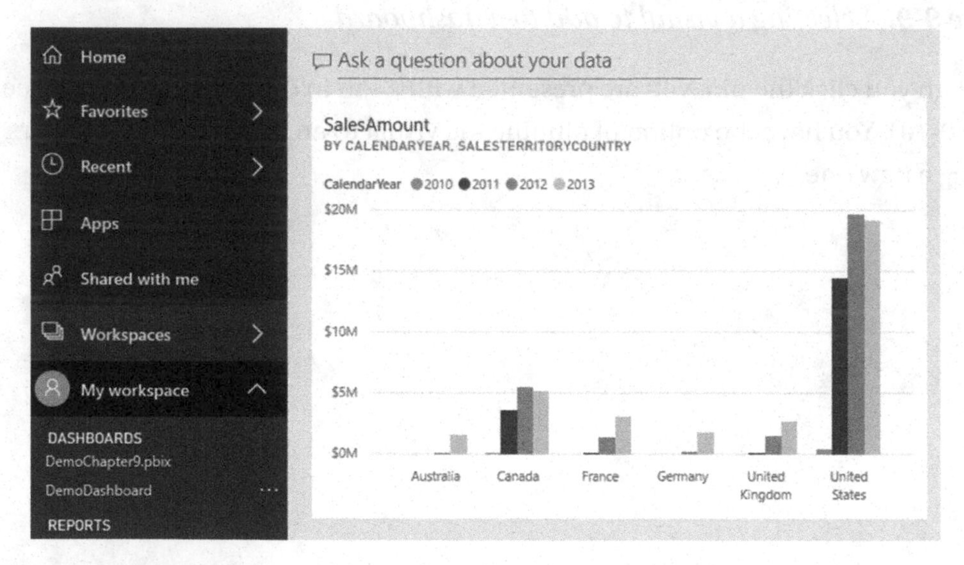

Figure 9-11. Viewing the dashboard

If you click the visual in the dashboard, it will automatically take you to the underlying report where the visual was pinned from.

In addition to pinning visuals to dashboards from reports, you can ask a question to create a visual in the dashboard. Clicking Ask a question about your data at the top of the dashboard will take you to the Q&A entry screen (see Figure 9-12).

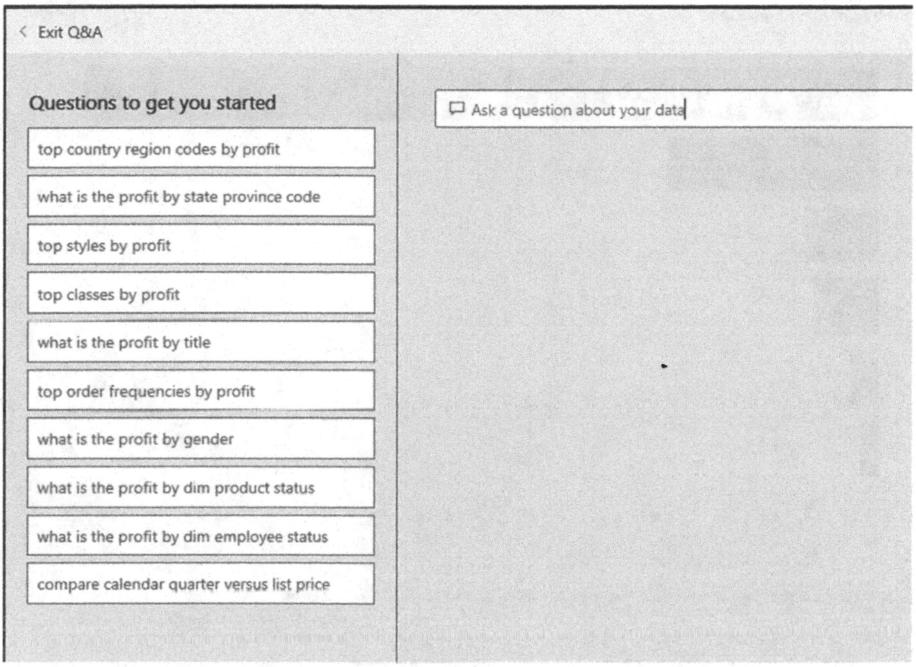

< Exit Q&A

Questions to get you started

top country region codes by profit

what is the profit by state province code

top styles by profit

top classes by profit

what is the profit by title

top order frequencies by profit

what is the profit by gender

what is the profit by dim product status

what is the profit by dim employee status

compare calendar quarter versus list price

Ask a question about your data

***Figure 9-12.** Asking questions about the data*

The Q&A screen shows some suggested questions to get you started. Power BI scans your model and analyzes the tables, fields, and relationships to create these suggestions. Figure 9-13 shows using the Q&A window used to create a bar chart showing profit by country. Once you create the visual, you can pin it to the dashboard.

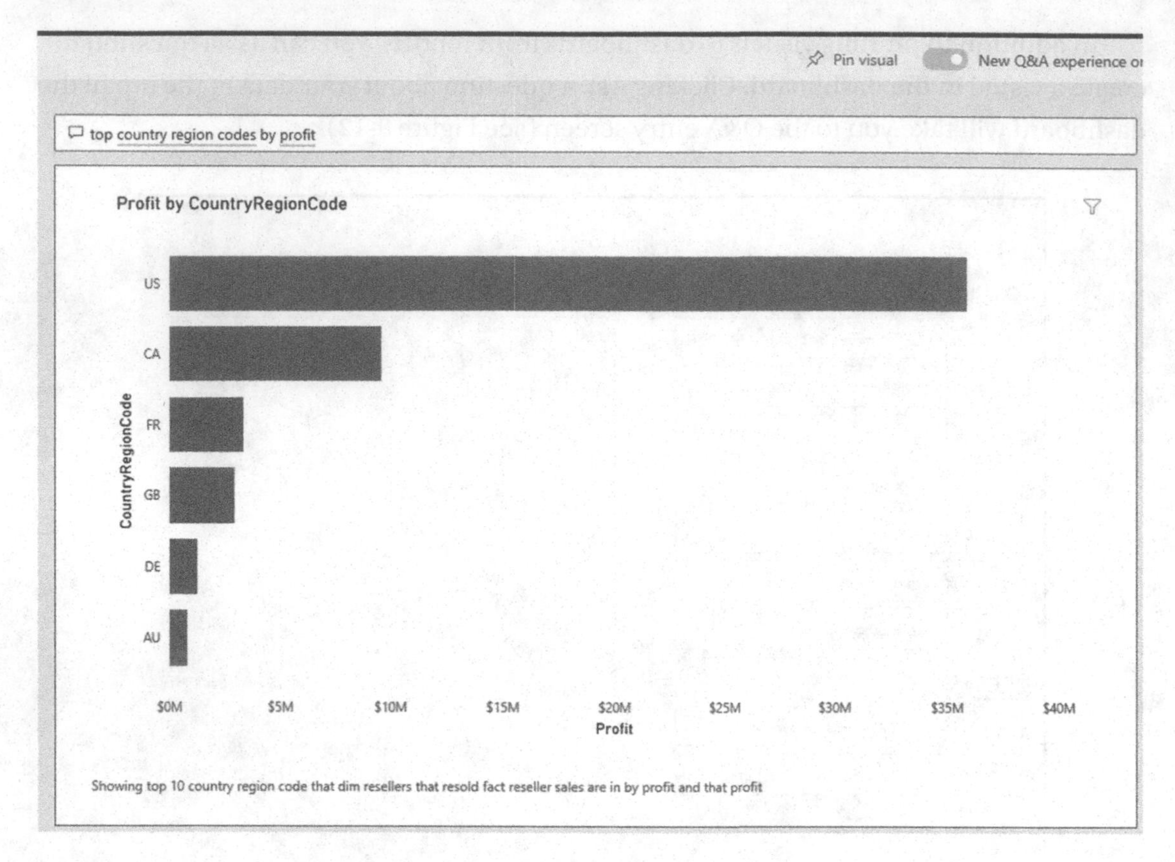

***Figure 9-13.** Creating a bar chart using Q&A*

If you don't want to expose the Q&A window to users of your dashboard, you can turn this feature off in the Dashboard Settings (see Figure 9-14).

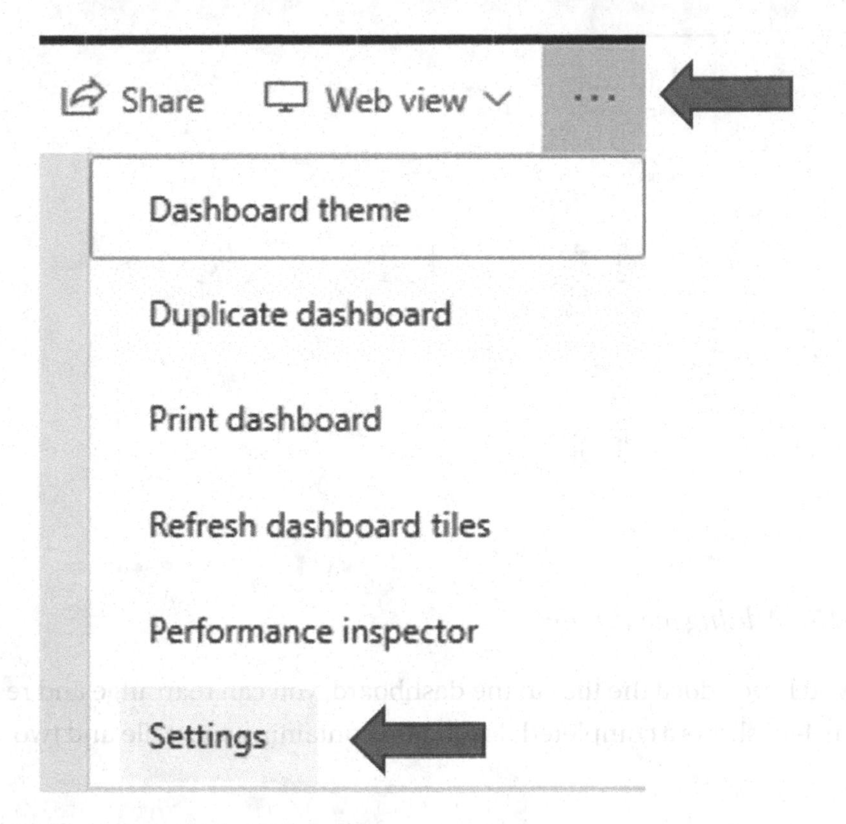

***Figure 9-14.** Turning off the Q&A option*

In addition to adding visuals to the dashboard, you can add media tiles to the dashboard, including text, images, video, and web content. Clicking the Add tile link at the top of the dashboard launches the Add tile window where you can select the media type you want to add (see Figure 9-15).

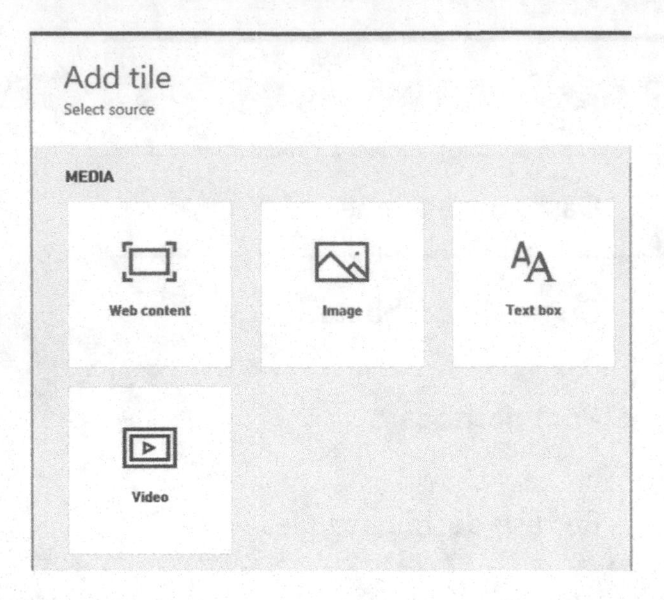

Figure 9-15. *Adding media tiles*

Once you have added the tiles to the dashboard, you can rearrange and resize the tiles. Figure 9-16 shows a completed dashboard containing a text tile and two visual tiles.

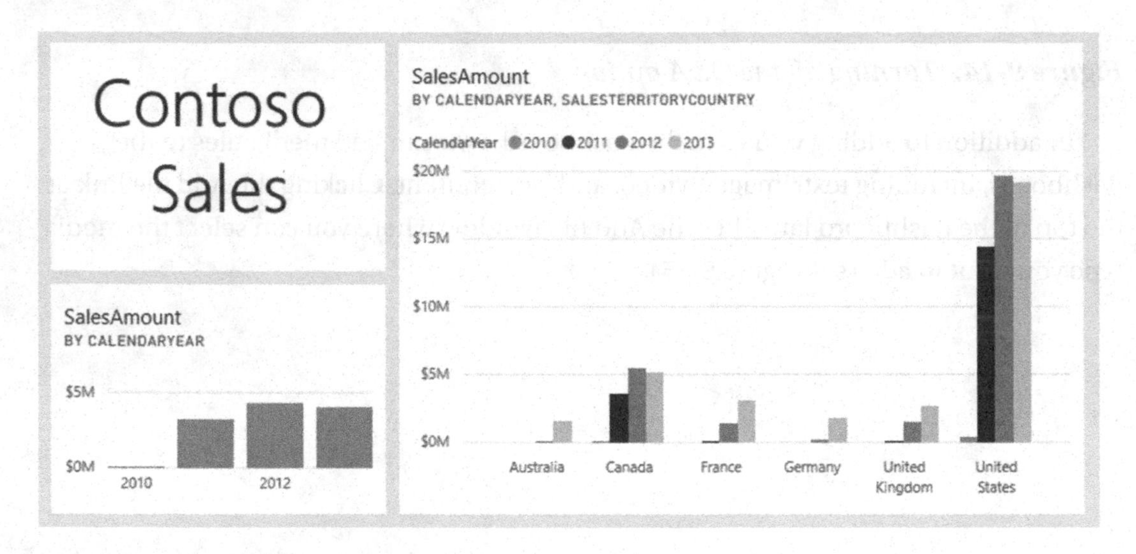

Figure 9-16. *A completed dashboard*

Once the dashboard has been created and designed the way you want it, you are ready to share it with your colleagues.

Sharing Dashboards

You can share dashboards with colleagues both in and outside your organization if they have signed up for the Power BI Service. When you share a dashboard, users can view and interact with the dashboard, but they cannot edit it or create their own copies. To share the dashboard, click the Share link in the upper right corner of the dashboard (refer to Figure 9-14). Using the Share dashboard window, enter the email address of the users you want to share it with. You can also control whether they can share the dashboard, build new content with the underlying data sets, and whether to send them an email indicating that the dashboard has been shared with them (see Figure 9-17).

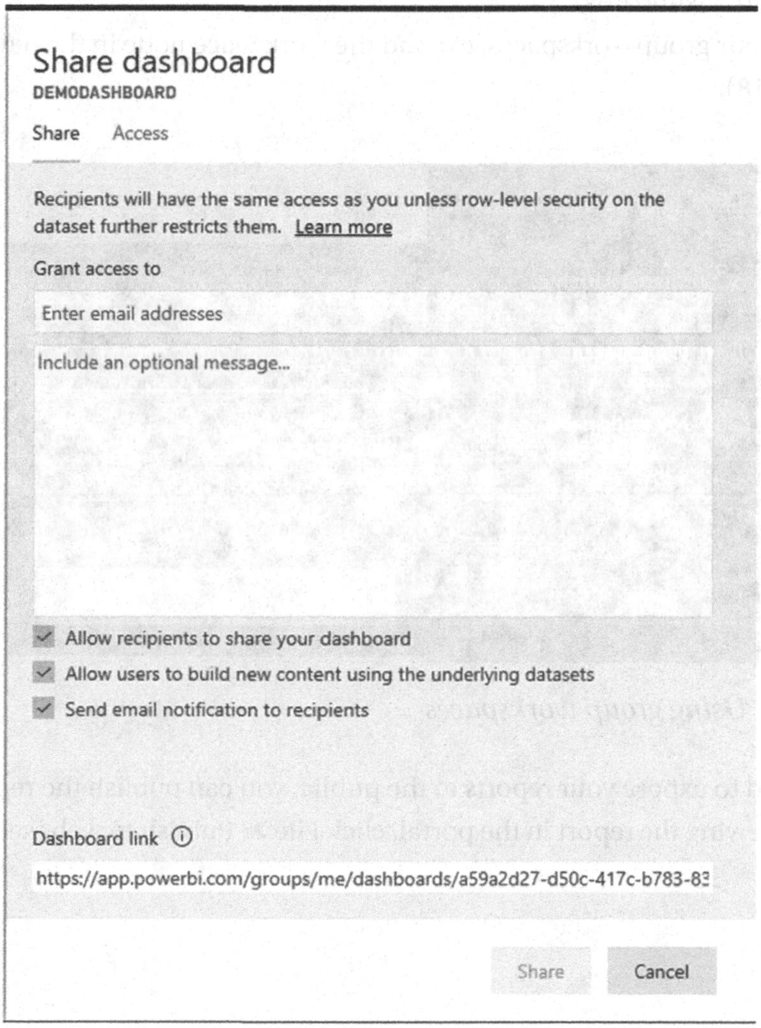

Figure 9-17. *Sharing a dashboard*

After the dashboard is shared, recipients will see the dashboard by selecting the Shared with me header in the portal explorer pane (see Figure 9-11).

One thing to be aware of is that although the underlying reports don't show up in the navigation pane, users can view them by clicking the visuals in the dashboard.

Although sharing dashboards is useful for exposing dashboards to end users, many times you'll want to collaborate with colleagues and allow them to make changes to the reports and dashboards. This is where Power BI *groups* are very useful. Because Power BI groups are based on Office 365 groups, group members must be in the same Office 365 tenant. Group members must also have a Power BI Pro license (currently $10/month). When you create a group, a group workspace is created where members can create and edit reports and dashboards.

To get to your group workspaces, expand the workspace node in the navigation pane (see Figure 9-18).

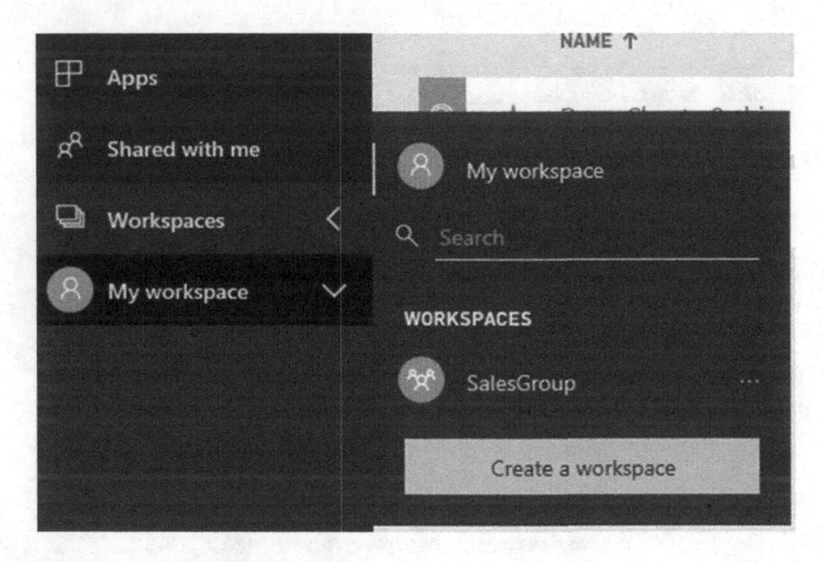

***Figure 9-18.** Using group workspaces*

If you need to expose your reports to the public, you can publish the report to the Web. When viewing the report in the portal, click File ➤ Publish to web (see Figure 9-19).

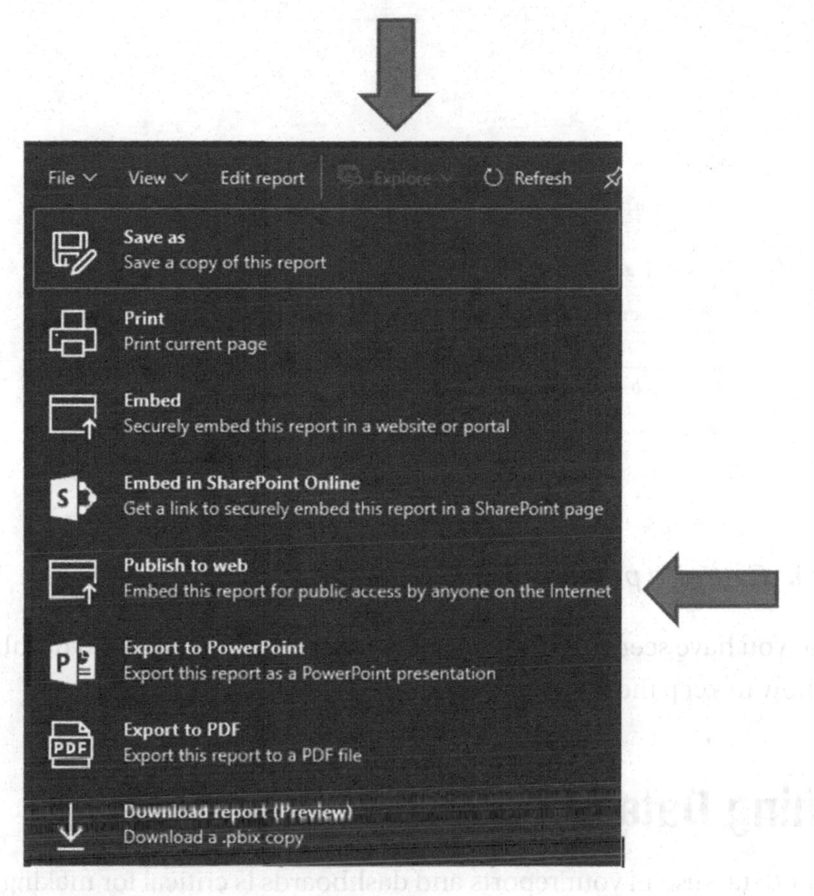

Figure 9-19. *Publishing a report to the Web*

When publishing to the Web, you can either provide a link where it will be open in its own web page or host it in an iframe as part of another web page (see Figure 9-20).

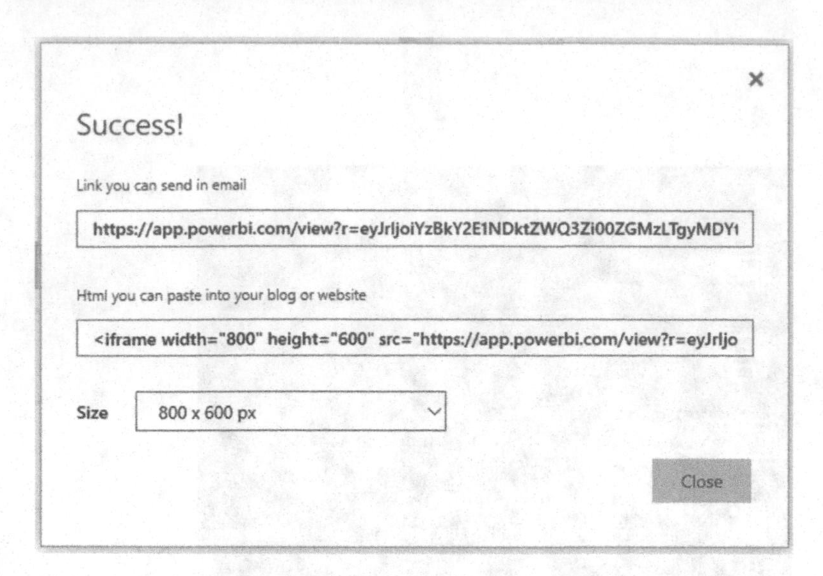

Figure 9-20. *Getting a public link to a report*

Now that you have seen how to publish and share your reports and dashboards, it's time to see how to keep the data up to date.

Refreshing Data in Published Reports

Keeping your data fresh in your reports and dashboards is critical for making informed decisions and investigating trends. How you refresh the data depends on the source and its volatility. Some data doesn't change much and can be refreshed weekly or monthly. Some sources can be updated daily or hourly, whereas others may need to be as close to live as possible. You need to carefully consider the business requirement for refreshing the data. Often a business user will claim they need live data without weighing the complexity and overhead this requires.

Although we're not going to cover every data source scenario, you should be aware of a few common ones. If your data source is a file on OneDrive, by default the data is refreshed hourly. You can turn this off and opt for a manual refresh if you want. If your data source is a web service like Salesforce or Microsoft Project Online, the data is automatically refreshed at a rate that depends on the provider. If your data is from a database in the cloud such as SQL Azure, you can schedule a data refresh. If you are connecting to an on-premises database, you will need to install a data gateway before you can schedule a data refresh.

To schedule a data refresh, in the navigation pane in the Power BI portal, select a data set and click the ellipses to the right of the data set node. In the pop-up menu (see Figure 9-21), you can choose to do a manual refresh or set up a scheduled refresh.

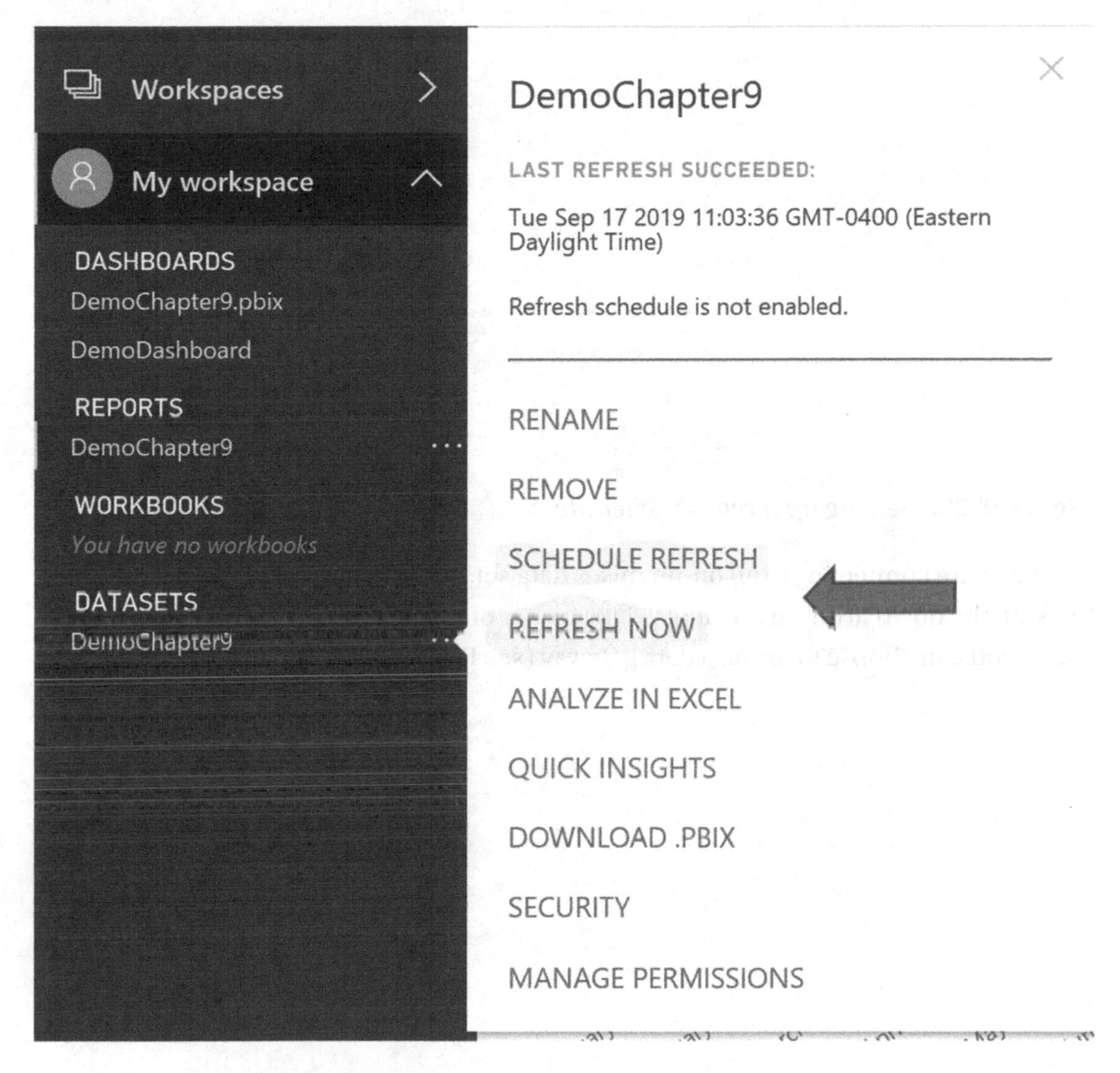

Figure 9-21. *Refreshing a data set*

When you choose SCHEDULE REFRESH, you will see a window where you can set up the refresh frequency. You can set up a daily refresh and indicate what times to refresh the data (see Figure 9-22). You can also set up a weekly refresh and indicate what days of the week to refresh the data.

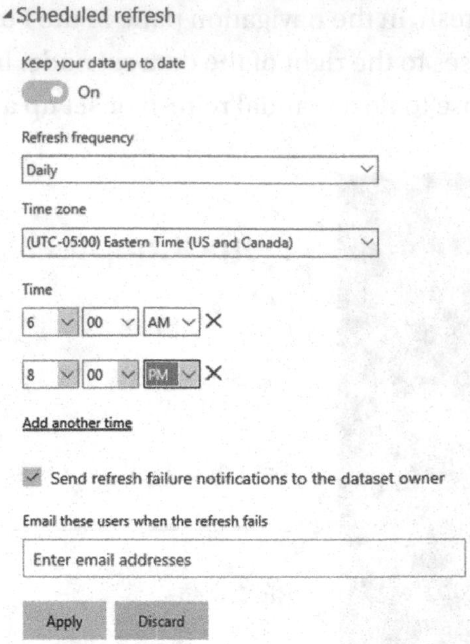

***Figure 9-22.** Setting up a refresh schedule*

If you are connecting to an on-premises data set, you will need to install a gateway. Clicking the down arrow in the upper right corner of the Power BI portal displays a menu where you can choose to install a data gateway (see Figure 9-23).

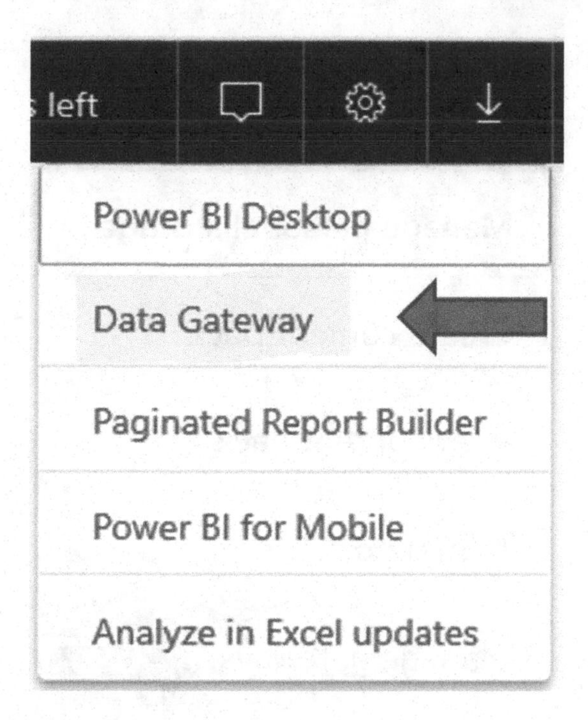

Figure 9-23. *Installing a data gateway*

Once the gateway is installed, click the sprocket icon and select the Manage gateways link (see Figure 9-24).

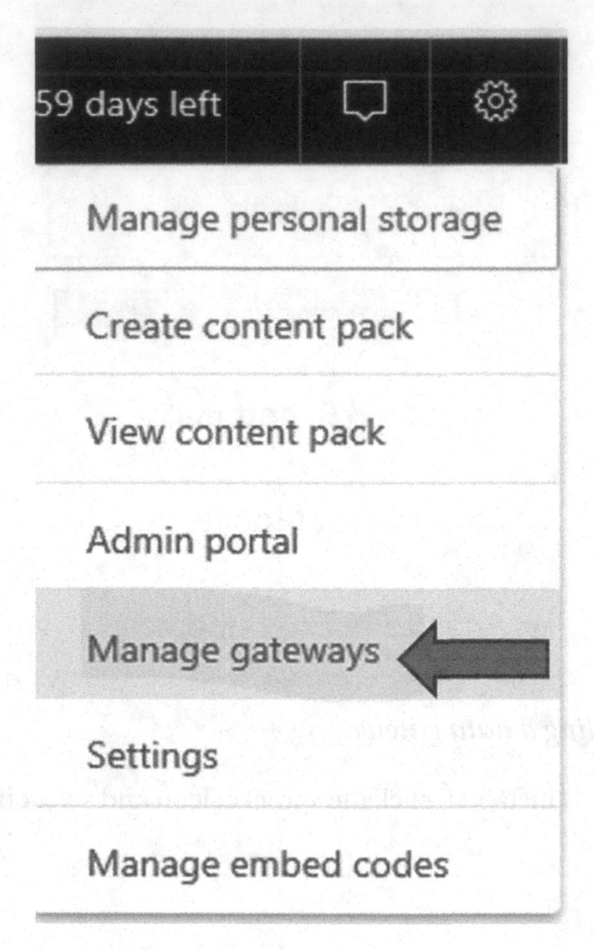

Figure 9-24. *Managing gateways*

In the Gateway Settings window, select Add data sources to use the gateway. In the Data Source Settings window, name the data source and select the type of data source (see Figure 9-25).

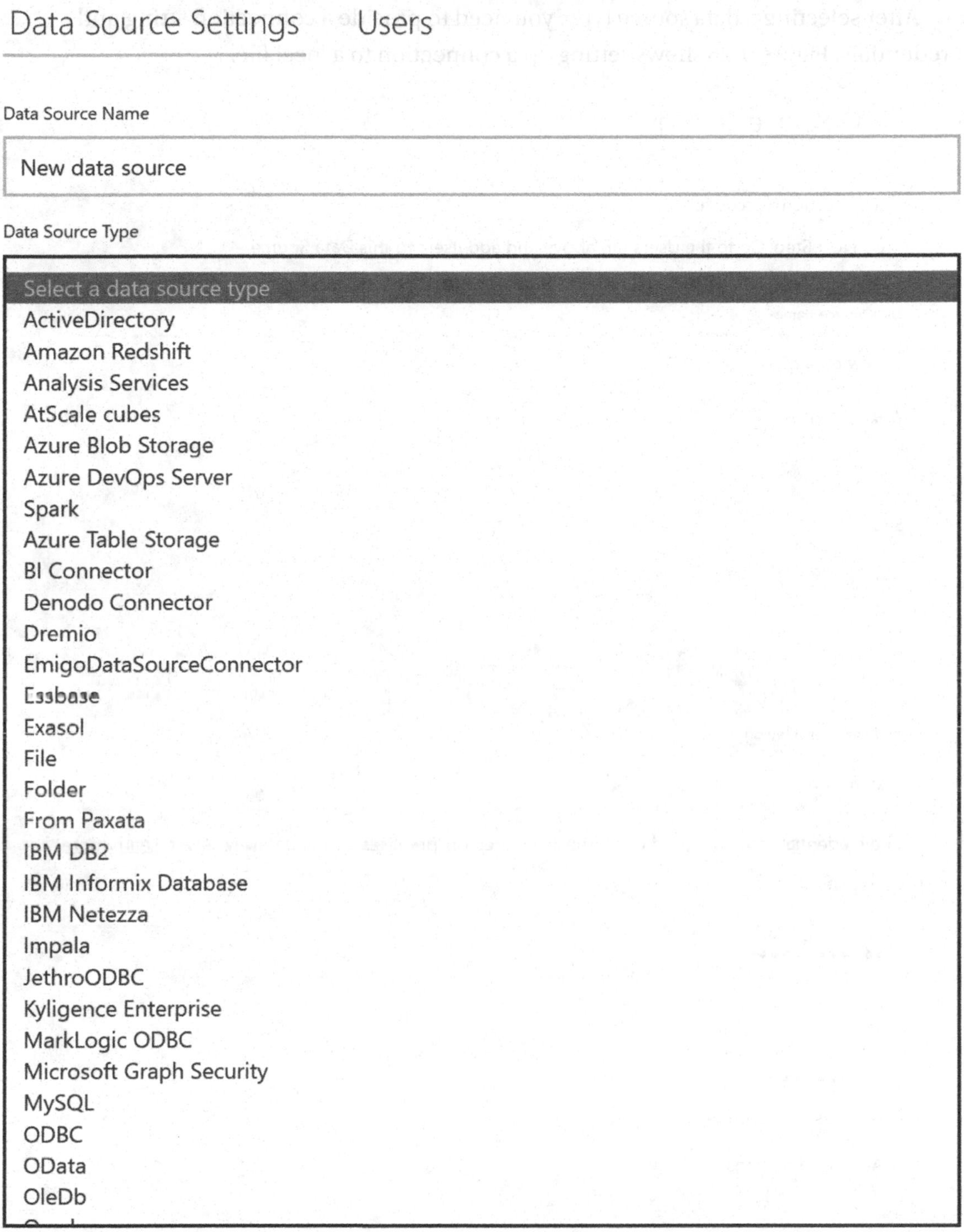

Data Source Settings Users

Data Source Name

New data source

Data Source Type

Select a data source type
ActiveDirectory
Amazon Redshift
Analysis Services
AtScale cubes
Azure Blob Storage
Azure DevOps Server
Spark
Azure Table Storage
BI Connector
Denodo Connector
Dremio
EmigoDataSourceConnector
Essbase
Exasol
File
Folder
From Paxata
IBM DB2
IBM Informix Database
IBM Netezza
Impala
JethroODBC
Kyligence Enterprise
MarkLogic ODBC
Microsoft Graph Security
MySQL
ODBC
OData
OleDb

Figure 9-25. *Adding a data source to the gateway*

After selecting a data source type, you need to provide a connection string and credentials. Figure 9-26 shows setting up a connection to a local file.

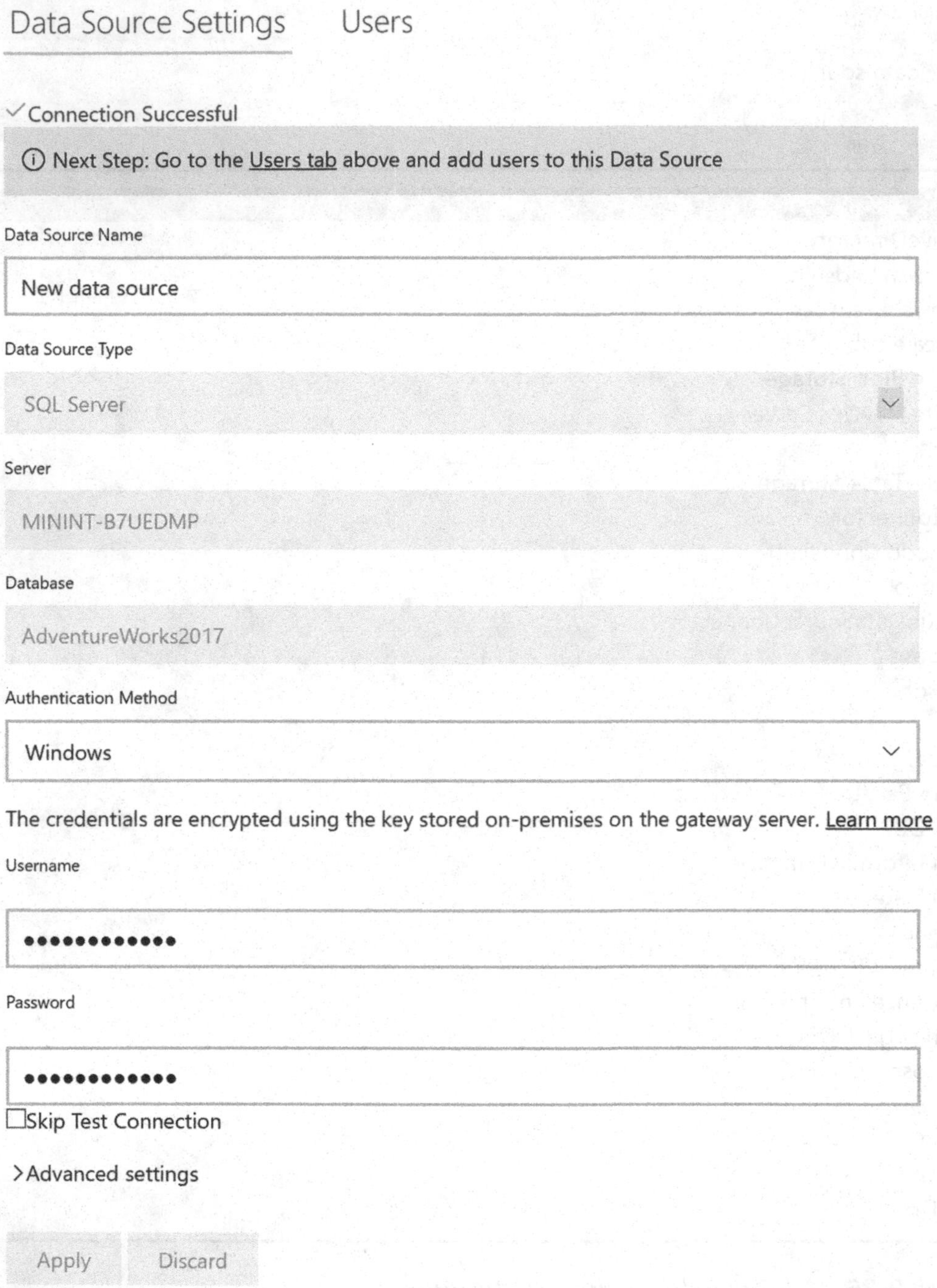

Data Source Settings Users

✓Connection Successful

ⓘ Next Step: Go to the Users tab above and add users to this Data Source

Data Source Name

New data source

Data Source Type

SQL Server

Server

MININT-B7UEDMP

Database

AdventureWorks2017

Authentication Method

Windows

The credentials are encrypted using the key stored on-premises on the gateway server. Learn more

Username

●●●●●●●●●●●●

Password

●●●●●●●●●●●●

☐Skip Test Connection

>Advanced settings

Apply Discard

Figure 9-26. *Connecting to a local file*

This chapter has showed you how to create, publish, and share your reports on the Power BI portal. It's time to get some hands-on experience publishing reports and creating dashboards.

HANDS-ON LAB: CREATING DASHBOARDS ON THE POWER BI PORTAL

In the following lab, you will

- Publish a report to the portal
- Create a dashboard on the portal
- Set up a data refresh schedule

1. In the LabStarterFiles\Chapter9Lab1 folder, open the Chapter9Lab1.pbix file. This file contains sales data from Northwind Traders. You should see a basic report, as shown in Figure 9-27.

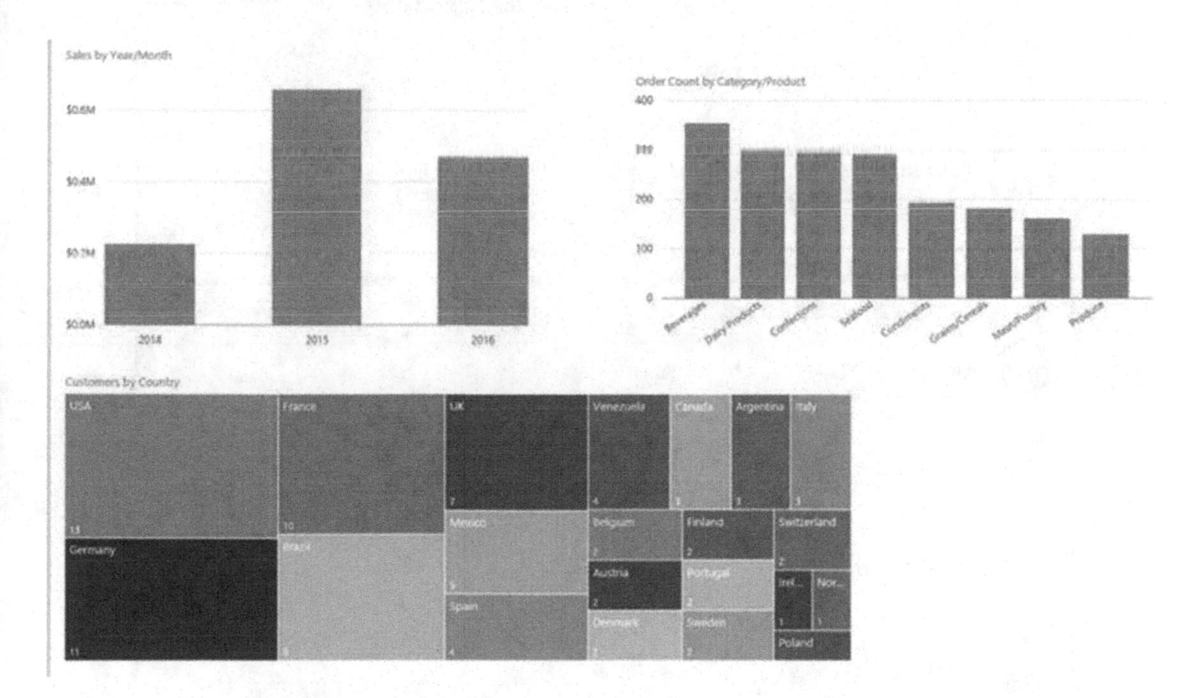

Figure 9-27. Sample sales report

2. If you are not signed into the Power BI Service, click File ➤ Sign In (see Figure 9-28).

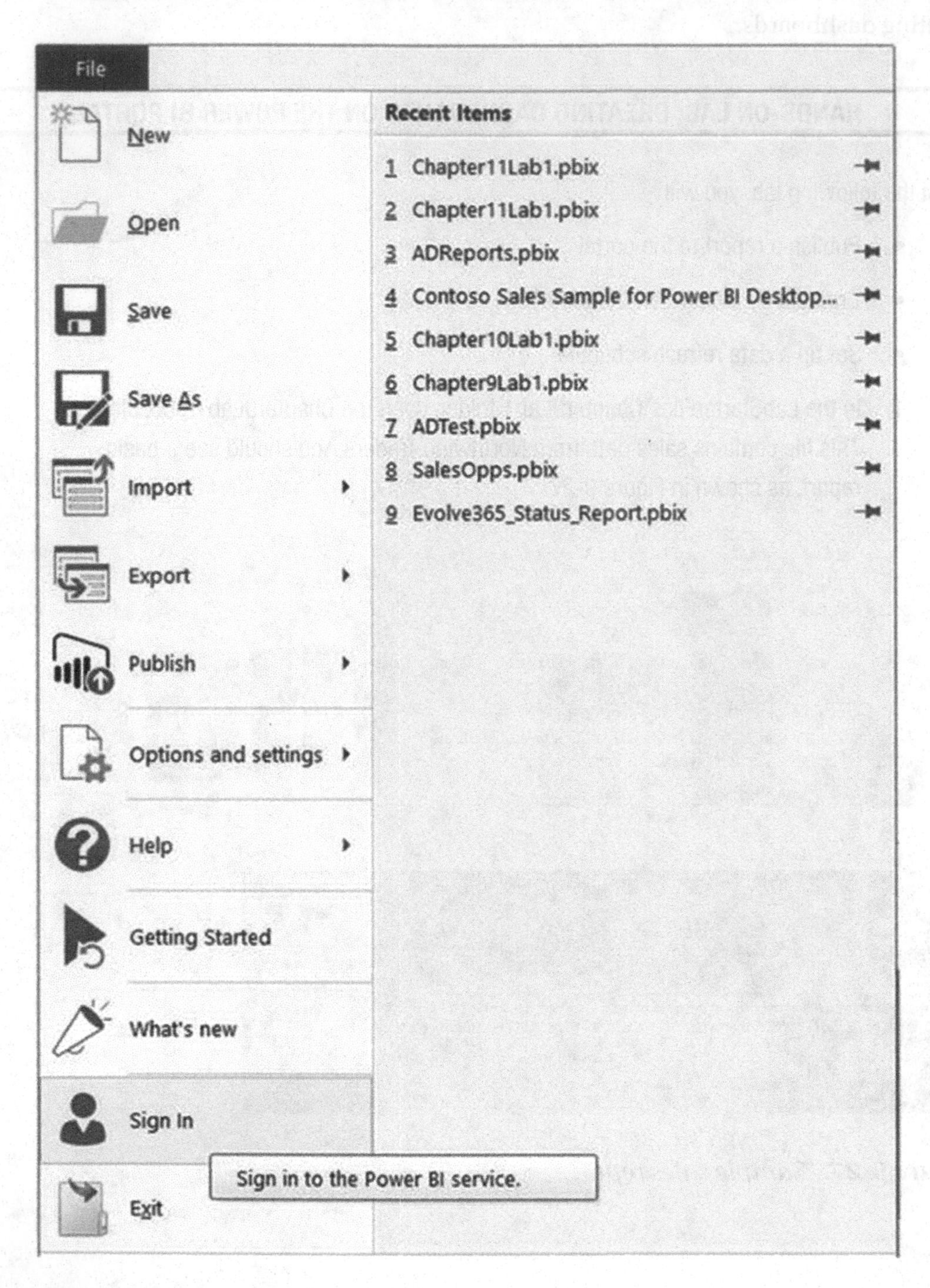

Figure 9-28. *Sign in to the Power BI Service*

3. After signing in, click the Publish button on the Home tab. Select My workspace as the destination.

4. Once published, click the link to open the report in Power BI (see Figure 9-29).

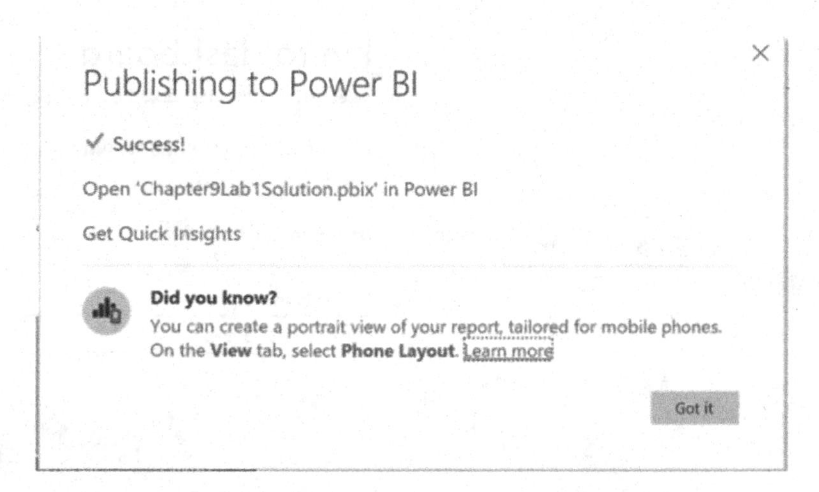

Figure 9-29. *Opening the report in Power BI*

5. In the Report view, hover over the Sales by Year/Month chart. You should see a pin icon in the upper right corner (see Figure 9-30).

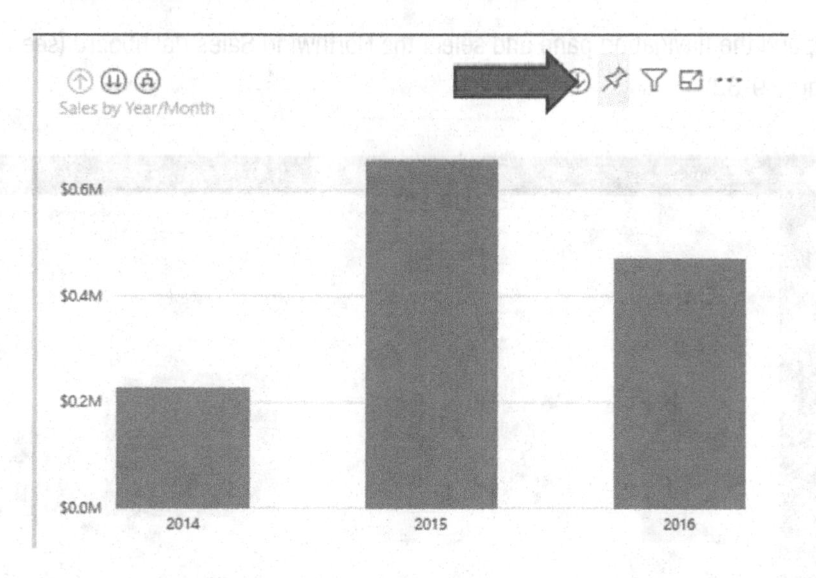

Figure 9-30. *Pinning a tile to a dashboard*

6. In the Pin to dashboard window, select a new dashboard and name it Northwind Sales (see Figure 9-31).

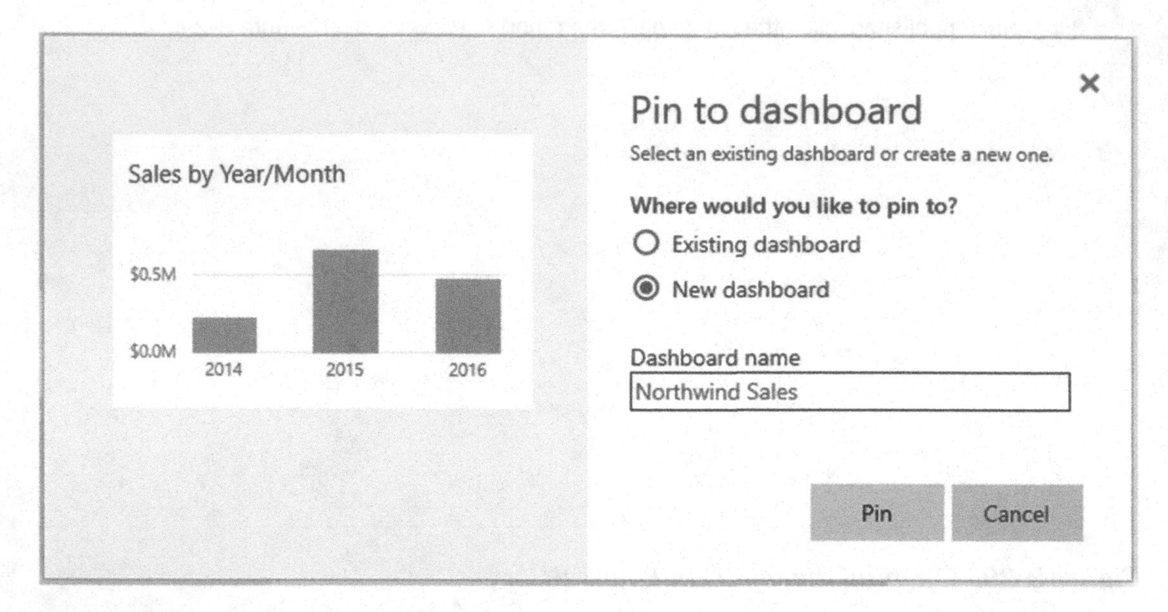

Figure 9-31. *Creating a new dashboard*

7. Add the Order Count by Category/Product to the Northwind Sales dashboard.
8. Expand the navigation pane and select the Northwind Sales dashboard (see Figure 9-32).

Figure 9-32. *Viewing the dashboard*

9. Click the Ask a question about your data link. Enter Product that is discontinued by unit in stock descending. You should see a table as shown in Figure 9-33. Pin this table to the dashboard.

Product that is discontinued by unit in stock descending

ProductID	ProductName	UnitsOnOrder	ReorderLevel	Discontinued	Category	UnitsInStock
9	Mishi Kobe Niku	0	0	True	Meat/Poultry	29
28	Rössle Sauerkraut	0	0	True	Produce	26
42	Singaporean Hokkien Fried Mee	0	0	True	Grains/Cereals	26
24	Guaraná Fantástica	0	0	True	Beverages	20
5	Chef Anton's Gumbo Mix	0	0	True	Condiments	0
17	Alice Mutton	0	0	True	Meat/Poultry	0
29	Thüringer Rostbratwurst	0	0	True	Meat/Poultry	0
53	Perth Pasties	0	0	True	Meat/Poultry	0

***Figure 9-33.** Using Q&A to create a table*

10. Exit the Q&A window to go back to the dashboard. Hover the mouse pointer over the new tile and click the ellipses in the upper right corner. A menu is displayed, as shown in Figure 9-34.

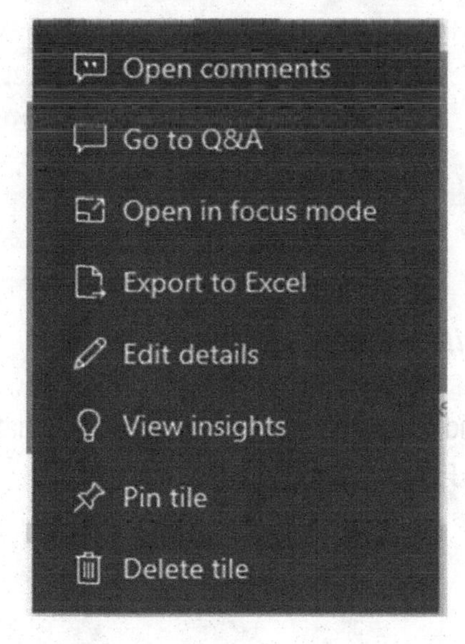

***Figure 9-34.** Displaying a tile menu*

11. Select Edit details. Change the title and put a check mark in Display last refresh time, as shown in Figure 9-35.

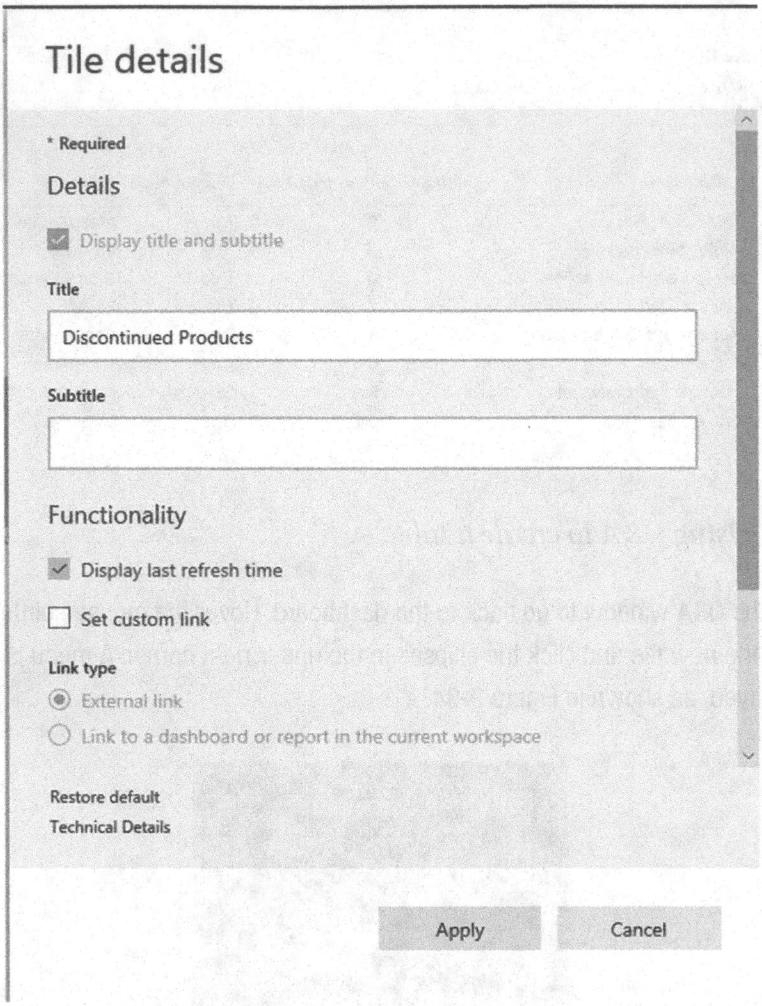

Figure 9-35. *Editing the tile details*

12. Add a text box tile and rearrange and resize the tiles so that your final dashboard looks like Figure 9-36.

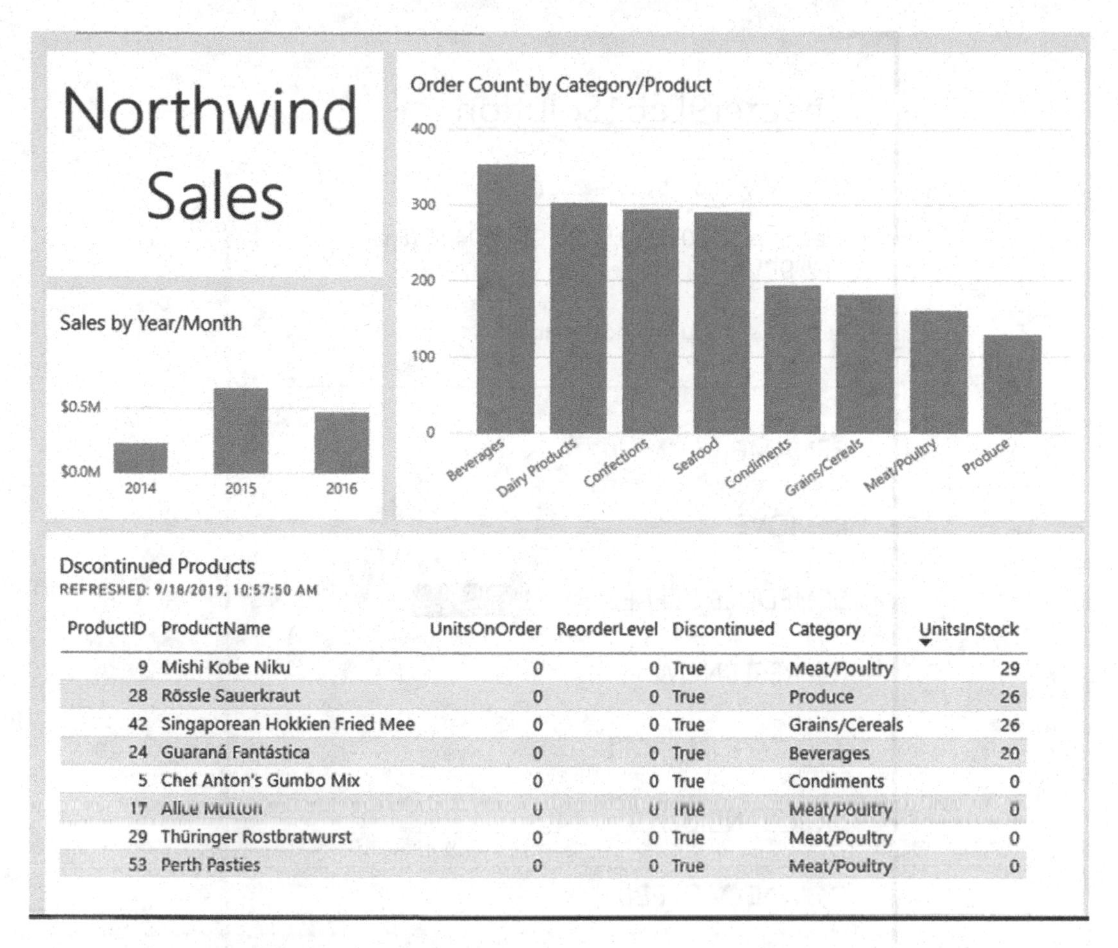

Figure 9-36. *The final dashboard*

13. To set up a refresh schedule, in the navigation pane, click the ellipses to the right of the data set and click SCHEDULE REFRESH in the menu (see Figure 9-37).

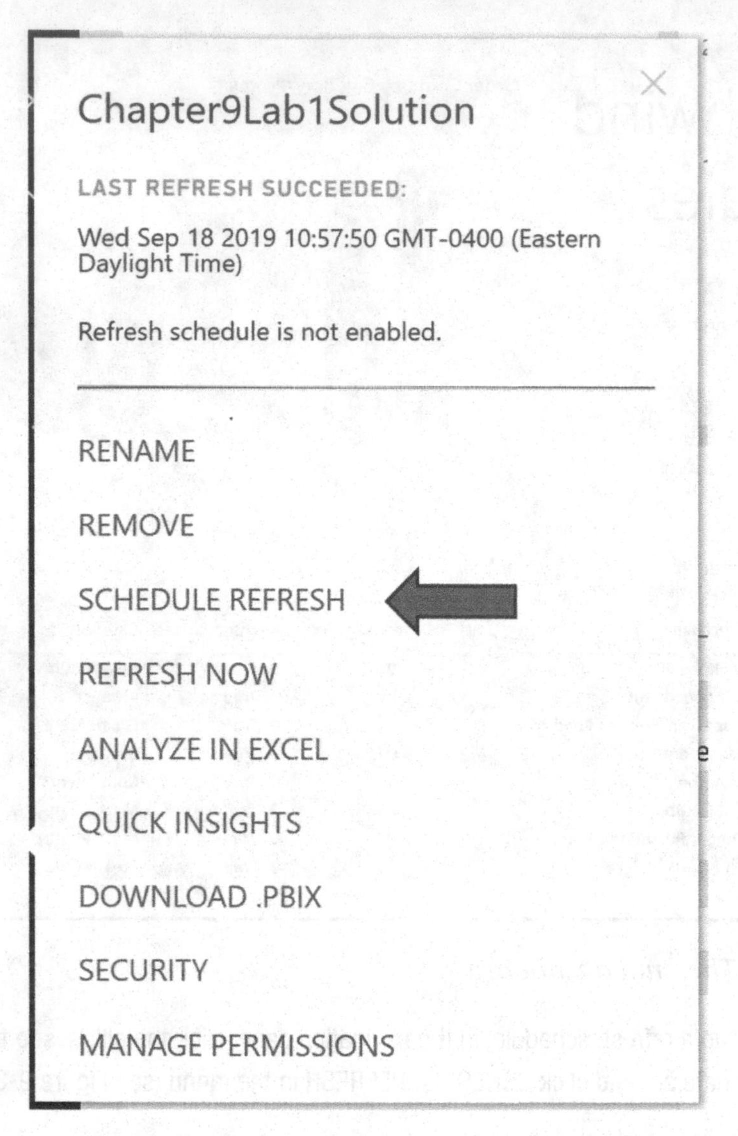

Figure 9-37. *Displaying the data set menu*

14. In the resulting window, expand the data source credentials node and click the Edit credentials link. Make sure Authentication method is set to Anonymous and the Privacy level is set to Public before signing in (see Figure 9-38).

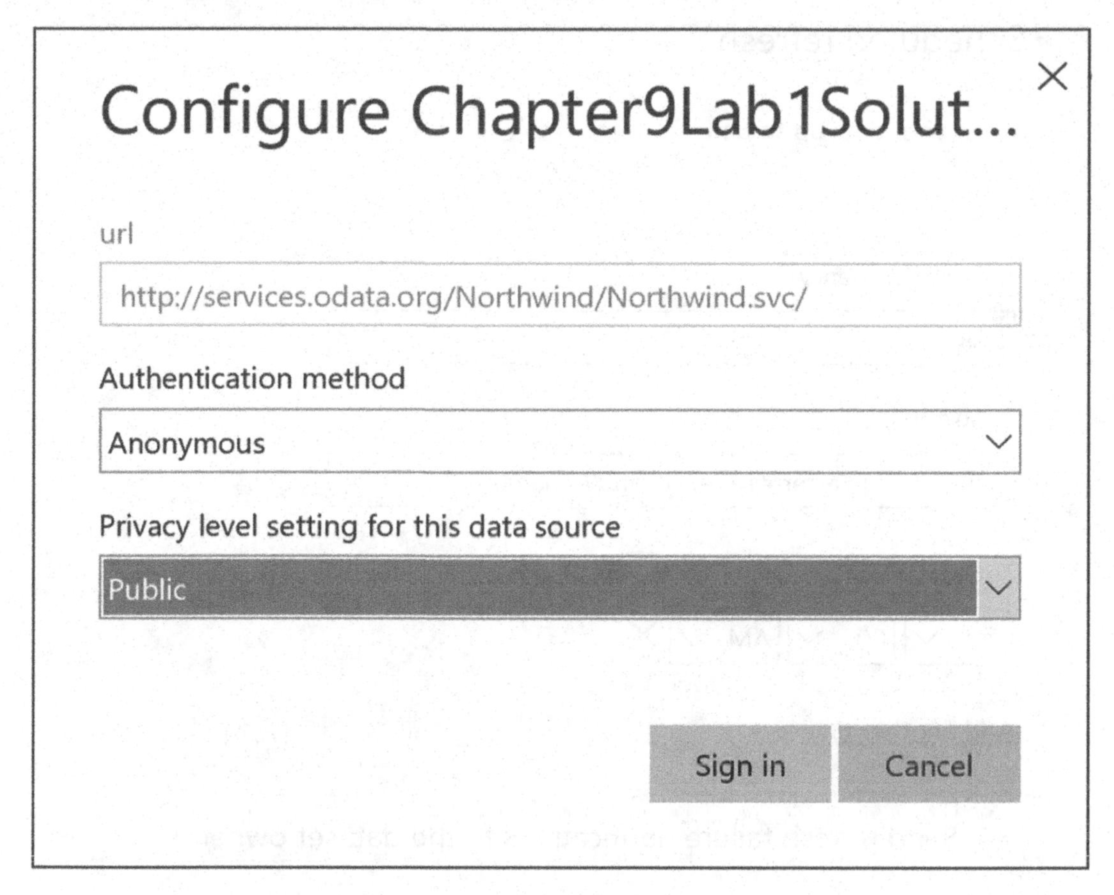

Figure 9-38. *Editing the data source credentials*

15. Create a refresh schedule that runs daily at 7:00 a.m. (see Figure 9-39).

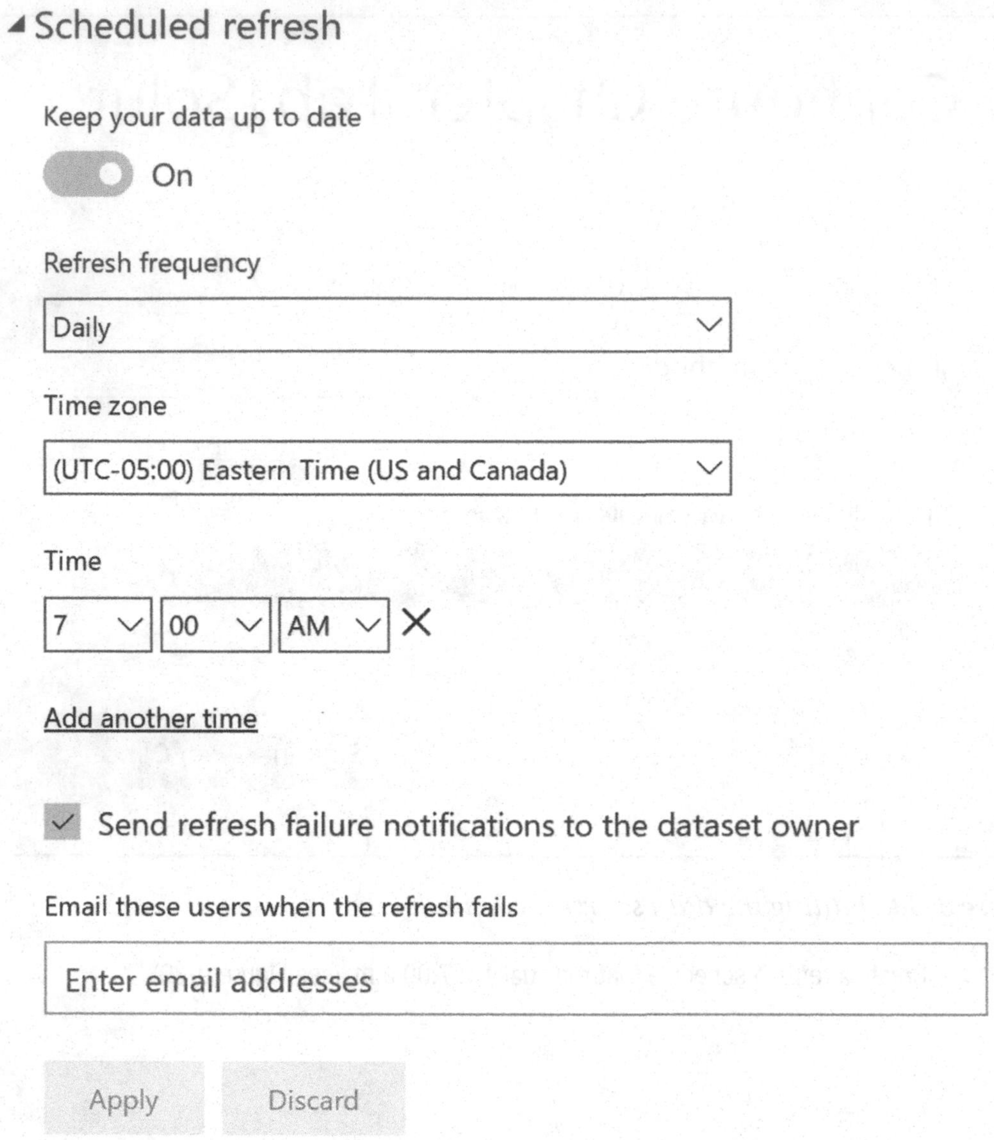

◢ Scheduled refresh

Keep your data up to date

On

Refresh frequency

Daily

Time zone

(UTC-05:00) Eastern Time (US and Canada)

Time

7 00 AM ×

Add another time

Send refresh failure notifications to the dataset owner

Email these users when the refresh fails

Enter email addresses

Apply Discard

Figure 9-39. *Creating a daily refresh*

Summary

In this chapter, you have seen how to publish reports created in Power BI Desktop to the Power BI Service (portal). You created dashboards and saw how to share them with colleagues. You also learned how to set up an automated data refresh schedule.

While Power BI reports and dashboards are excellent tools for sharing data insights with others, they are not the only tools in your arsenal. Excel is one of the most used data analytic tools in the world. If you are a hard-core data analyst, it is probably your preferred tool for discovering insights and trends into your data. The good news is that many of the features of Power BI have been incorporated into Excel. In the next chapter, you will learn how to use these features.

CHAPTER 10

Introducing Power Pivot in Excel

As you have seen in the previous chapters, Power BI is an excellent tool for developing analytic solutions. The Power BI portal is where you can host, share, and secure interactive dashboards and reports with others. Power BI Desktop is where you create the model and visuals on which the dashboards in the portal are based. It is great for sharing the results of your analysis with a broader audience. But where it is lacking (arguably) is when you need to perform pure data discovery. This is where most analysts turn to Excel. The great thing about Excel is that it uses the same tools you've been using in Power BI Desktop. It uses Power Query to get, clean, and shape the data. It then uses a Power Pivot Model Designer to construct a tabular model on top of which you create interactive pivot tables to explore the data.

This chapter introduces you to using Power Query and Power Pivot in Excel. Hopefully, this will be a familiar experience and be intuitive after working with these tools in Power BI Desktop.

After reading this chapter, you will be familiar with the following:

- Setting up the Power Pivot environment
- Getting, cleaning, and shaping data
- Creating table relationships
- Adding calculations and measures
- Incorporating time-based analysis

D. Clark, *Beginning Microsoft Power BI*, https://doi.org/10.1007/978-1-4842-5620-6_10

Setting Up the Power Pivot Environment

Power Pivot is a free add-in to Excel and has been available since Excel 2010. If you are using Excel 2010, you must download and install the add-in from the Microsoft Office web site. If you are using Excel 2013, the add-in is already installed, and you just have to enable it. If you are using Excel 2016 or Excel for Microsoft 365 (the version covered in this book), it is already installed and enabled for you. To check what edition you have installed, click the File menu in Excel and select the Account tab, as shown in Figure 10-1.

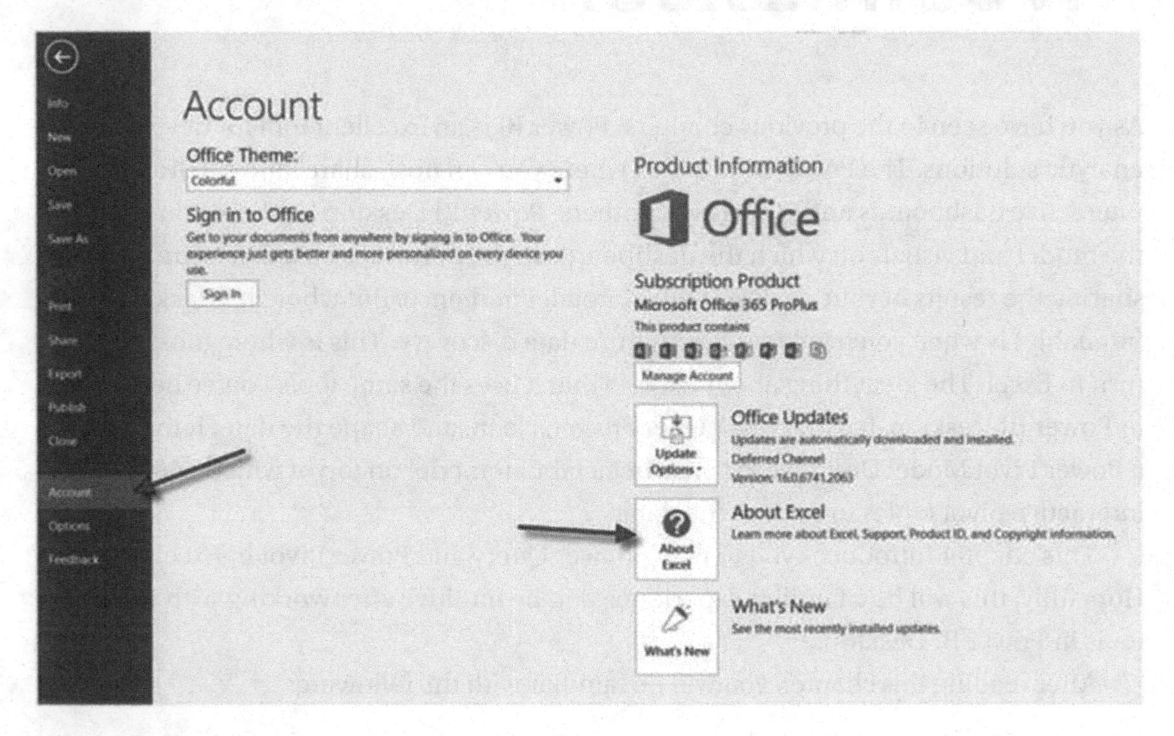

***Figure 10-1.** Checking the Excel version*

On the Account tab, click the About Excel button. You are presented with a screen showing version details, as shown in Figure 10-2. Take note of the edition and the version. Although the 32-bit version will work fine for smaller data sets, to get the optimal performance and experience from Power Pivot, you should use the 64-bit version running on a 64-bit version of Windows with at least 8 GB of RAM.

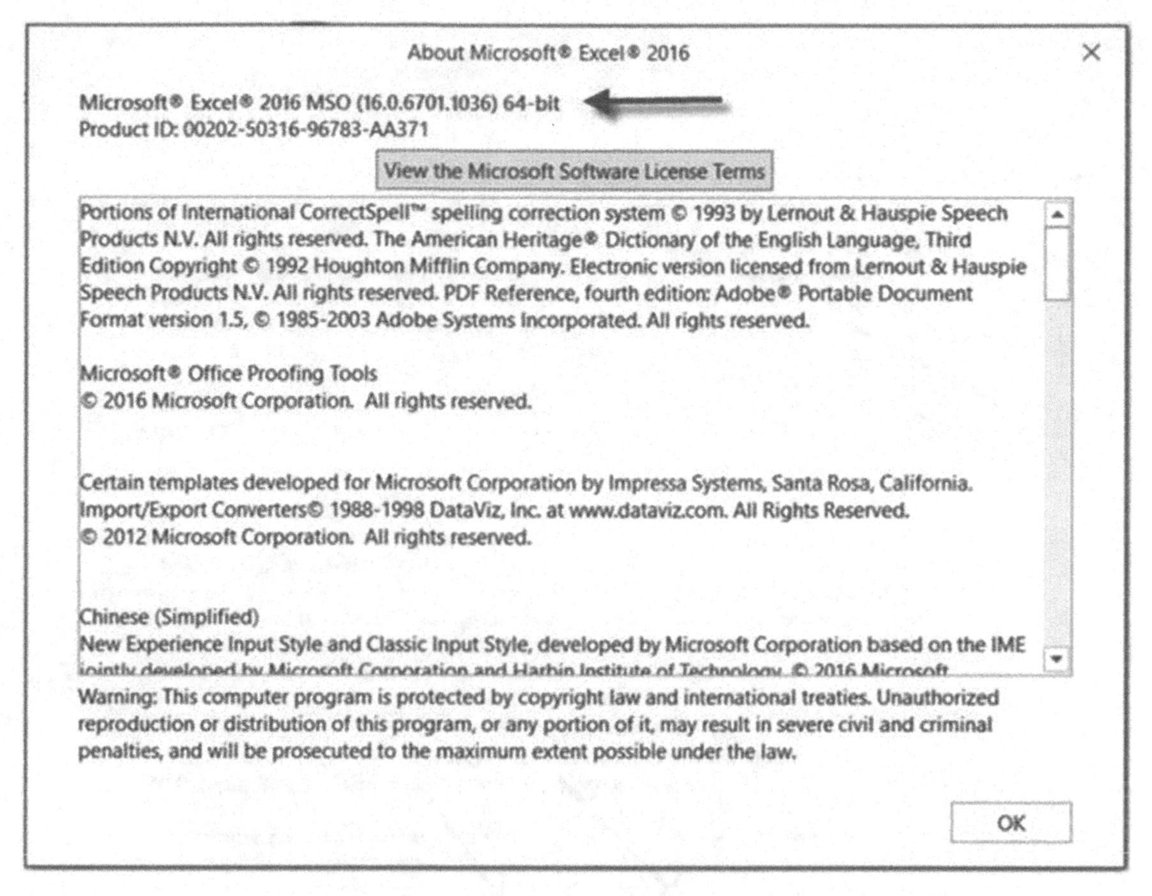

***Figure 10-2.** Checking the Excel edition and version*

Once you have determined that you are running the correct version, you can enable/disable the Power Pivot add-in by clicking the File menu and selecting the Options tab. In the Excel Options window, click the Add-ins tab. In the Manage drop-down, select COM Add-ins and click the Go button (see Figure 10-3).

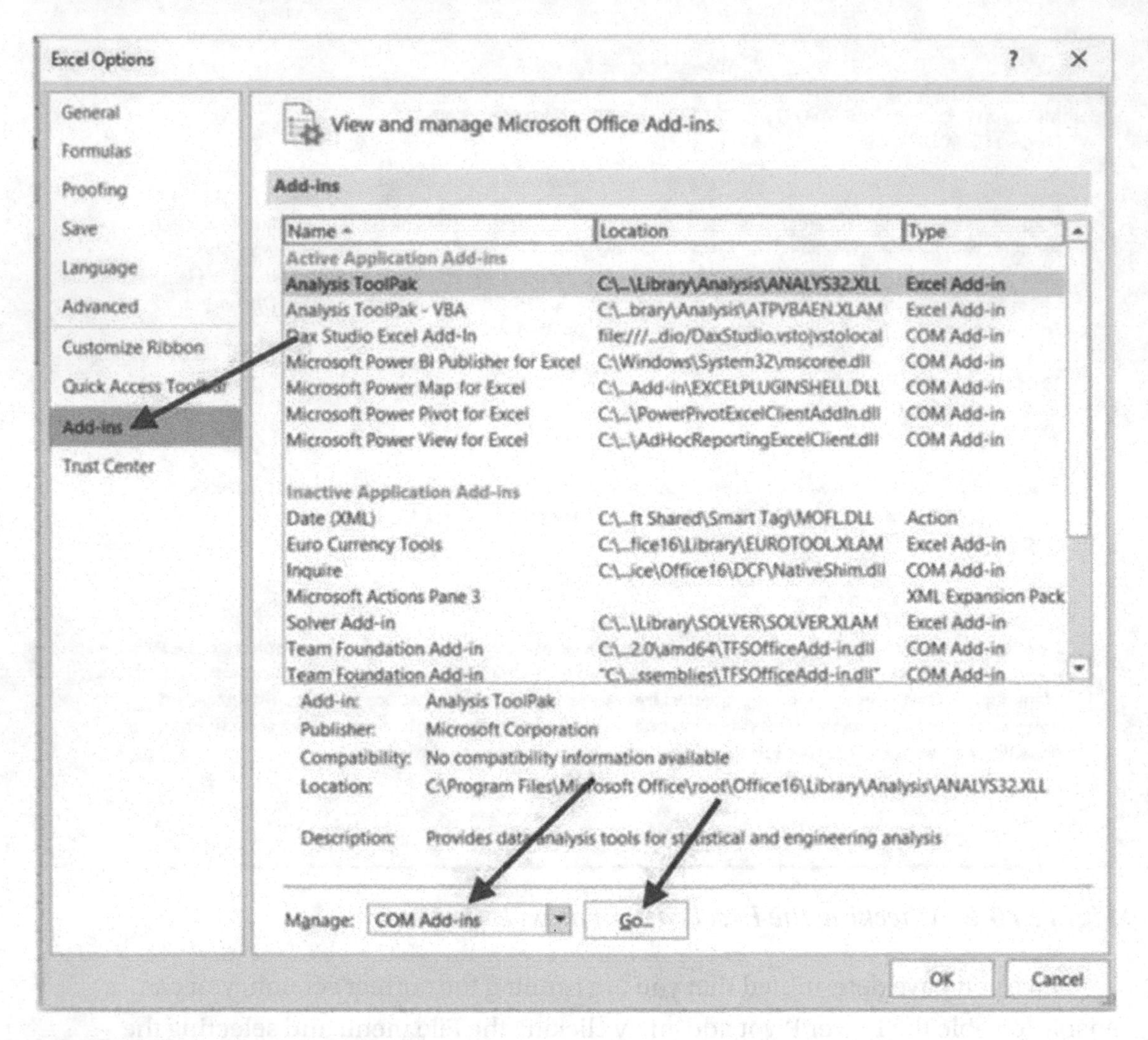

Figure 10-3. Managing COM Add-ins

You are presented with the COM Add-ins window (see Figure 10-4). Make sure Microsoft Power Pivot for Excel is checked and click OK.

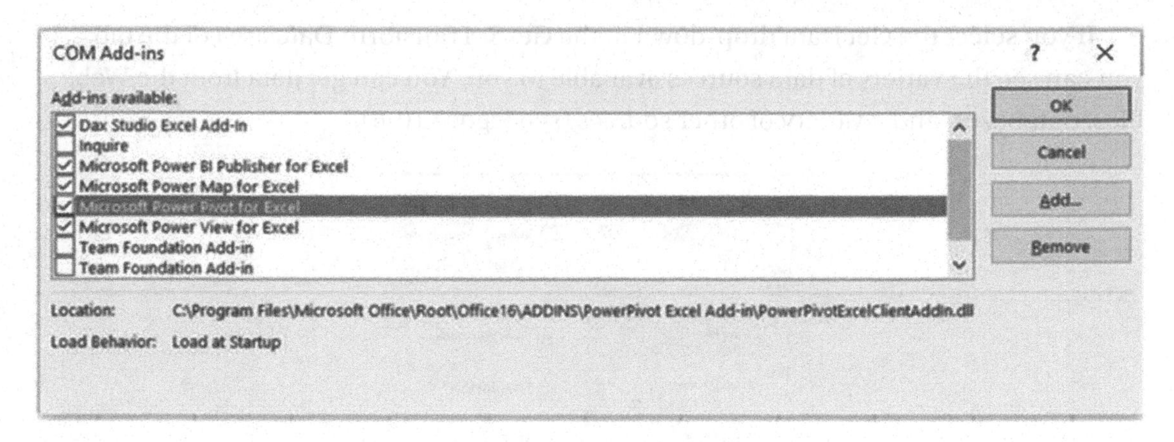

Figure 10-4. *Selecting the Power Pivot add-in*

Now that you have ensured that the Power Pivot add-in for Excel is enabled, you are ready to get some data.

Getting, Cleaning, and Shaping Data

The Power Query interface in Excel is very similar to the Power Query interface used in Power BI Desktop (discussed in Chapter 3). As discussed in that chapter, the first step in creating a data analytics solution is importing data. When you open Excel, you will see a Data tab where you can start importing the data (see Figure 10-5).

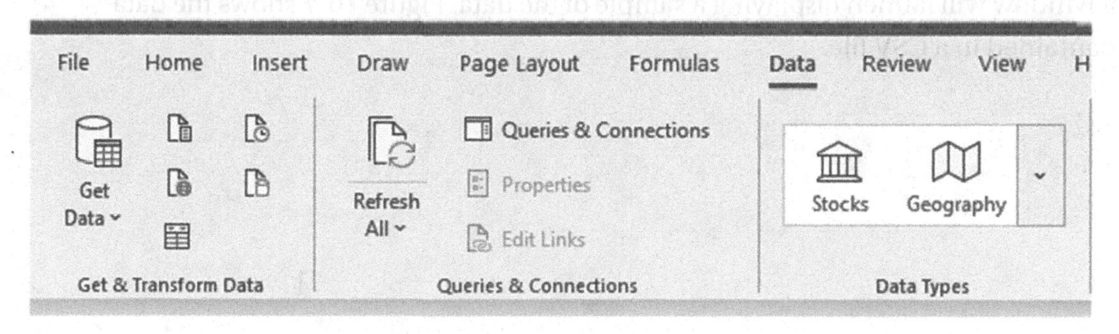

Figure 10-5. *The Data tab in Excel*

If you select the Get Data drop-down in the Get & Transform Data area of the tab, you can see the variety of data sources available to you. You can get data from the Web, files, databases, and a variety of other sources (see Figure 10-6).

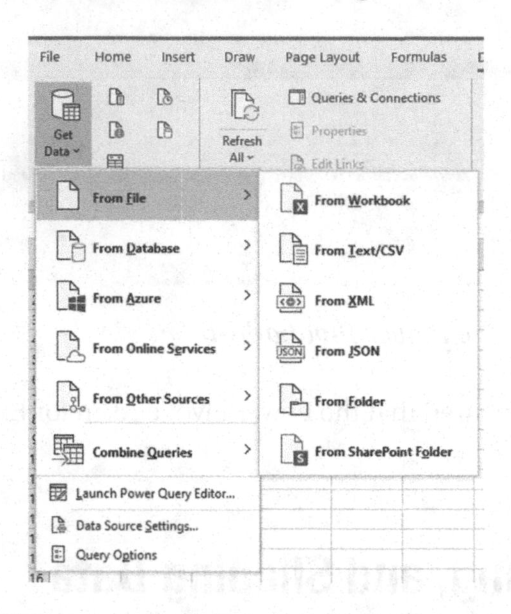

***Figure 10-6.** Some of the many data sources available in Excel*

The type of data source you choose will dictate what information you need to supply to gain access to the data source. For example, an SQL Server database requires log-in credentials, whereas a CSV file requires the file path. Once you connect to a data source, a window will launch displaying a sample of the data. Figure 10-7 shows the data contained in a CSV file.

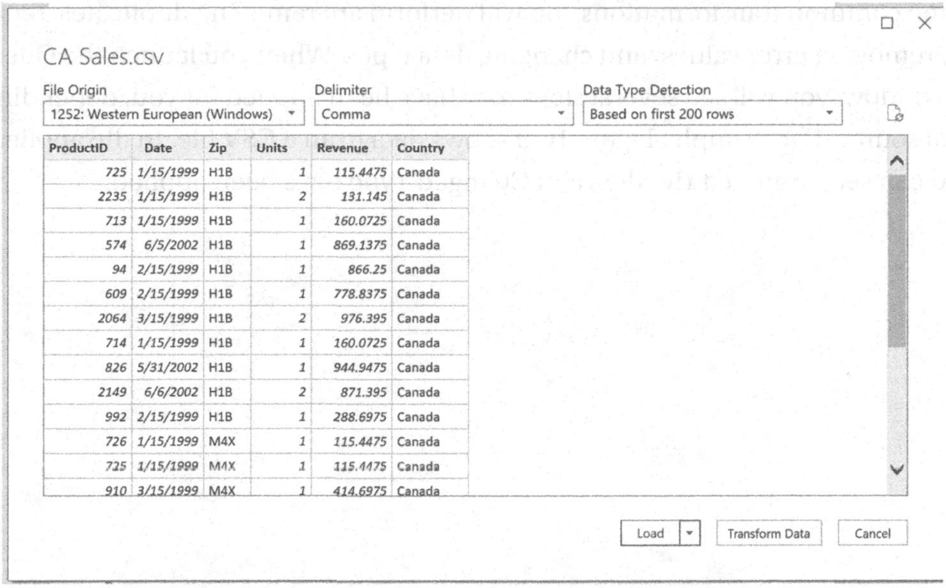

CA Sales.csv

File Origin: 1252: Western European (Windows)
Delimiter: Comma
Data Type Detection: Based on first 200 rows

ProductID	Date	Zip	Units	Revenue	Country
725	1/15/1999	H1B	1	115.4475	Canada
2235	1/15/1999	H1B	2	131.145	Canada
713	1/15/1999	H1B	1	160.0725	Canada
574	6/5/2002	H1B	1	869.1375	Canada
94	2/15/1999	H1B	1	866.25	Canada
609	2/15/1999	H1B	1	778.8375	Canada
2064	3/15/1999	H1B	2	976.395	Canada
714	1/15/1999	H1B	1	160.0725	Canada
826	5/31/2002	H1B	1	944.9475	Canada
2149	6/6/2002	H1B	2	871.395	Canada
992	2/15/1999	H1B	1	288.6975	Canada
726	1/15/1999	M4X	1	115.4475	Canada
725	1/15/1999	M4X	1	115.4475	Canada
910	3/15/1999	M4X	1	414.6975	Canada

Load | Transform Data | Cancel

Figure 10-7. *Data from a CSV file*

After previewing the data, you can either load it directly into the data model or edit the query before loading the data. Clicking Transform Data will launch the familiar Power Query Editor where you can transform, cleanse, and filter the data before importing it into the data model (see Figure 10-8).

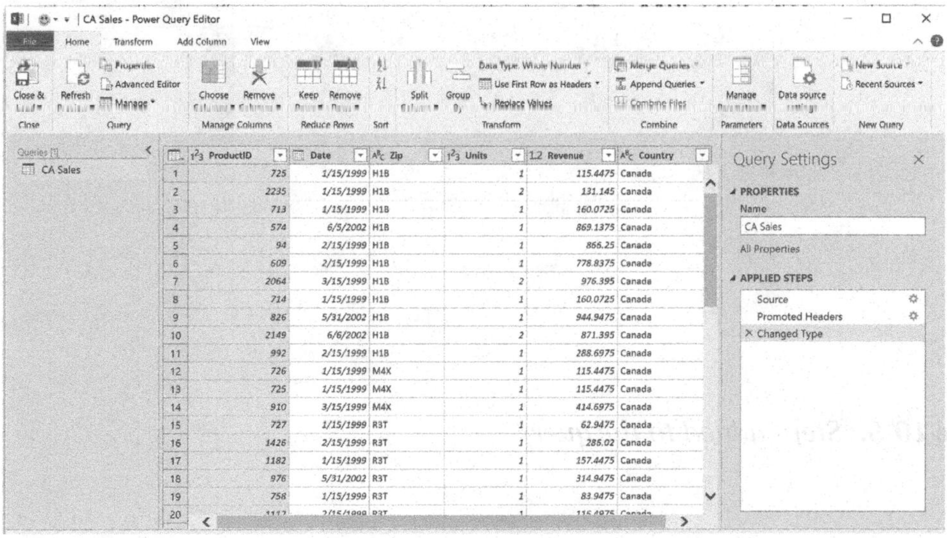

CA Sales - Power Query Editor

File | Home | Transform | Add Column | View

Close & Load | Refresh Preview | Properties | Advanced Editor | Manage | Choose Columns | Remove Columns | Keep Rows | Remove Rows | Split Column | Group By | Data Type: Whole Number | Use First Row as Headers | Replace Values | Merge Queries | Append Queries | Combine Files | Manage Parameters | Data source settings | New Source | Recent Sources

Close | Query | Manage Columns | Reduce Rows | Sort | Transform | Combine | Parameters | Data Sources | New Query

Queries [1]

CA Sales

	ProductID	Date	Zip	Units	Revenue	Country
1	725	1/15/1999	H1B	1	115.4475	Canada
2	2235	1/15/1999	H1B	2	131.145	Canada
3	713	1/15/1999	H1B	1	160.0725	Canada
4	574	6/5/2002	H1B	1	869.1375	Canada
5	94	2/15/1999	H1B	1	866.25	Canada
6	609	2/15/1999	H1B	1	778.8375	Canada
7	2064	3/15/1999	H1B	2	976.395	Canada
8	714	1/15/1999	H1B	1	160.0725	Canada
9	826	5/31/2002	H1B	1	944.9475	Canada
10	2149	6/6/2002	H1B	2	871.395	Canada
11	992	2/15/1999	H1B	1	288.6975	Canada
12	726	1/15/1999	M4X	1	115.4475	Canada
13	725	1/15/1999	M4X	1	115.4475	Canada
14	910	3/15/1999	M4X	1	414.6975	Canada
15	727	1/15/1999	R3T	1	62.9475	Canada
16	1426	2/15/1999	R3T	1	285.02	Canada
17	1182	1/15/1999	R3T	1	157.4475	Canada
18	976	5/31/2002	R3T	1	314.9475	Canada
19	758	1/15/1999	R3T	1	83.9475	Canada
20	[illegible]	[illegible]	[illegible]	[illegible]	[illegible]	[illegible]

Query Settings

PROPERTIES

Name: CA Sales

All Properties

APPLIED STEPS

Source

Promoted Headers

Changed Type

Figure 10-8. *The Power Query Editor*

Some common transformations you will perform are removing duplicates, replacing values, removing error values, and changing data types. When you launch the Query Editor window, you will see several steps may have been applied for you, depending on the data source. For example, Figure 10-9 shows data from a CSV file. In the applied steps list, you can see Promoted Headers and Changed Type have been applied.

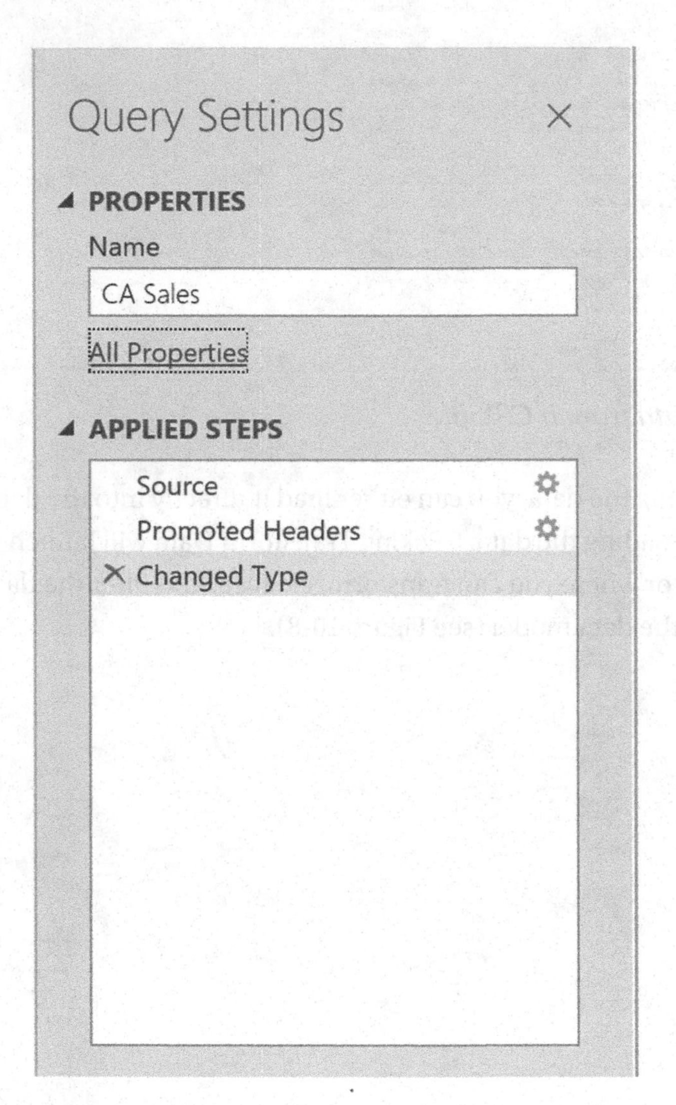

Figure 10-9. *Steps added to the query*

Often you need to replace values from source systems so that they are consistent. For example, some rows have a country abbreviation, and some have the full name. You can easily replace these values as the data is imported by selecting the column and then selecting the Replace Values transformation in the menu. This launches a window to enter the values to find and what to replace them with (see Figure 10-10).

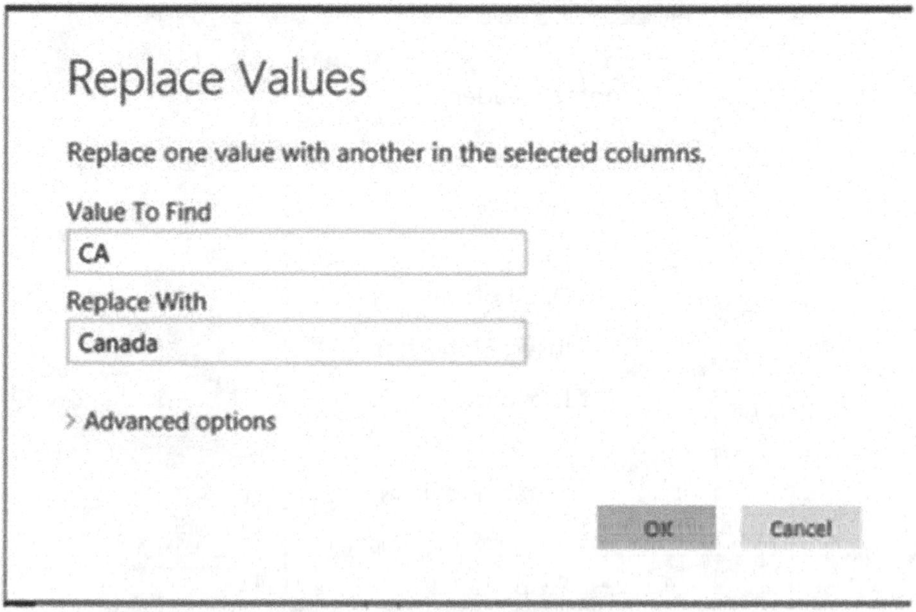

Figure 10-10. *Replacing values in a column*

As you apply the data transformations and filtering, the Query Editor lists the steps you have applied. This allows you to organize and track the changes made to the data. You can rename, rearrange, and remove steps by right-clicking the step in the list (see Figure 10-11).

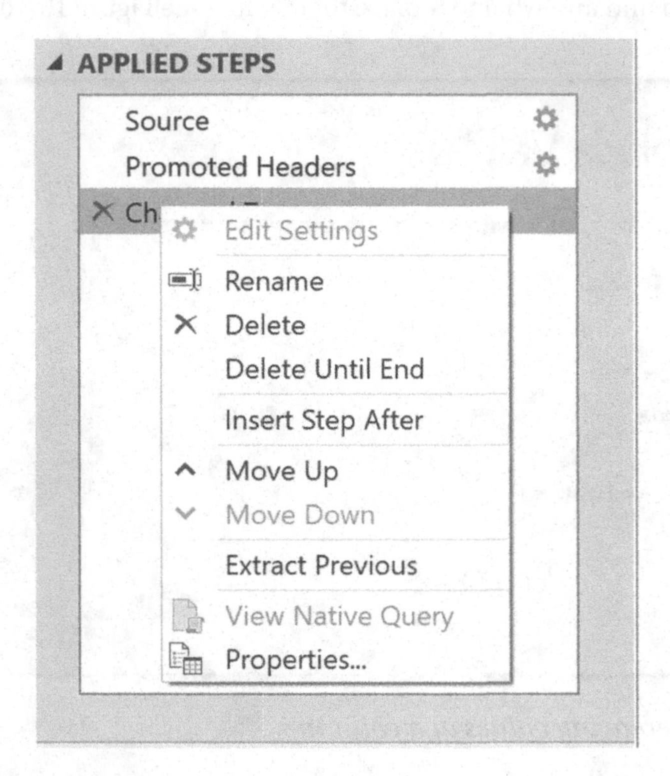

***Figure 10-11.** Managing the query steps*

Sometimes a source may provide you with data in a column that needs to be split up among several columns. For example, you may need to split the city and state, or the first name and last name. To do this, select the column in the Query Editor, and on the Home tab, choose Split Column. You can either split the column by a delimiter or by the number of characters (see Figure 10-12).

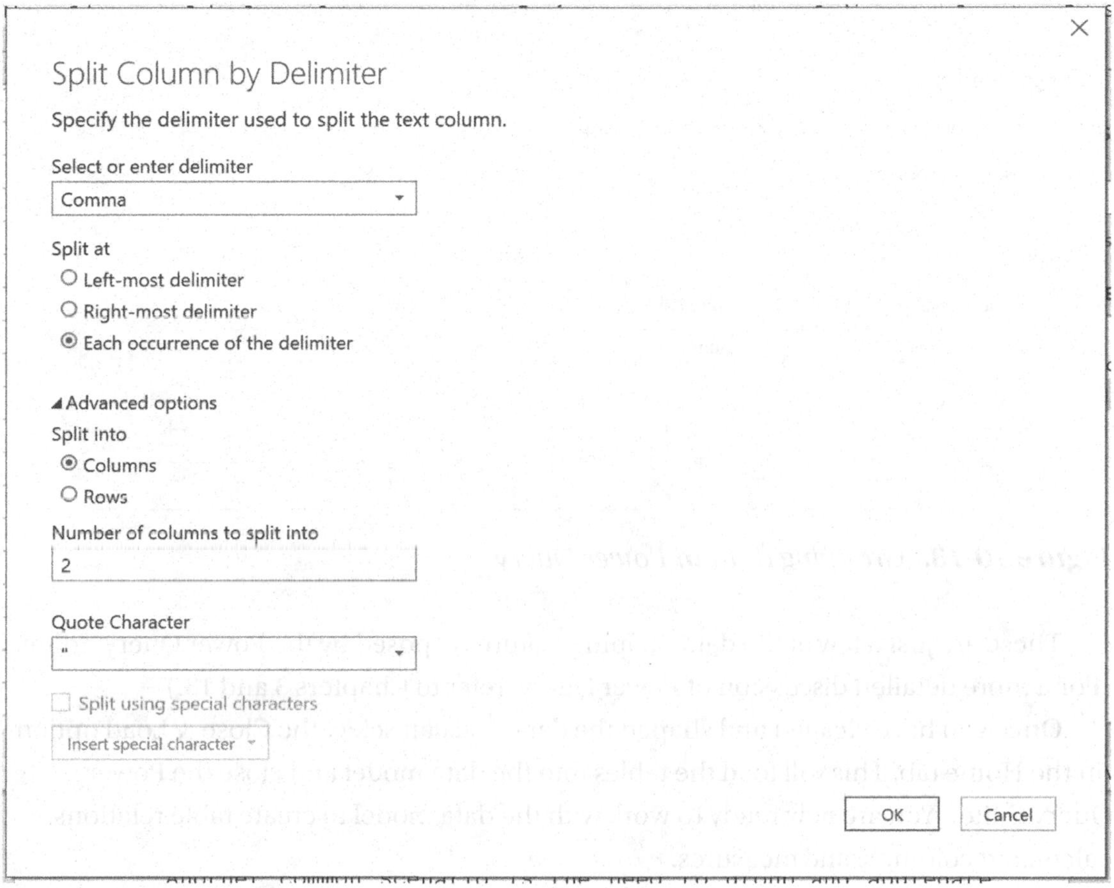

Split Column by Delimiter

Specify the delimiter used to split the text column.

Select or enter delimiter

Comma

Split at

- Left-most delimiter
- Right-most delimiter
- Each occurrence of the delimiter

Advanced options

Split into

- Columns
- Rows

Number of columns to split into

2

Quote Character

"

Split using special characters

Insert special character

OK Cancel

***Figure 10-12.** Splitting a column using a delimiter*

Another common scenario is the need to group and aggregate data. For example, you may need to roll the data up by month or sales territory, depending on the analysis. To aggregate and group the data in the Query Editor, select the column you want to group by and select the Group By transform in the Home tab. You are presented with a Group By Editor window (see Figure 10-13).

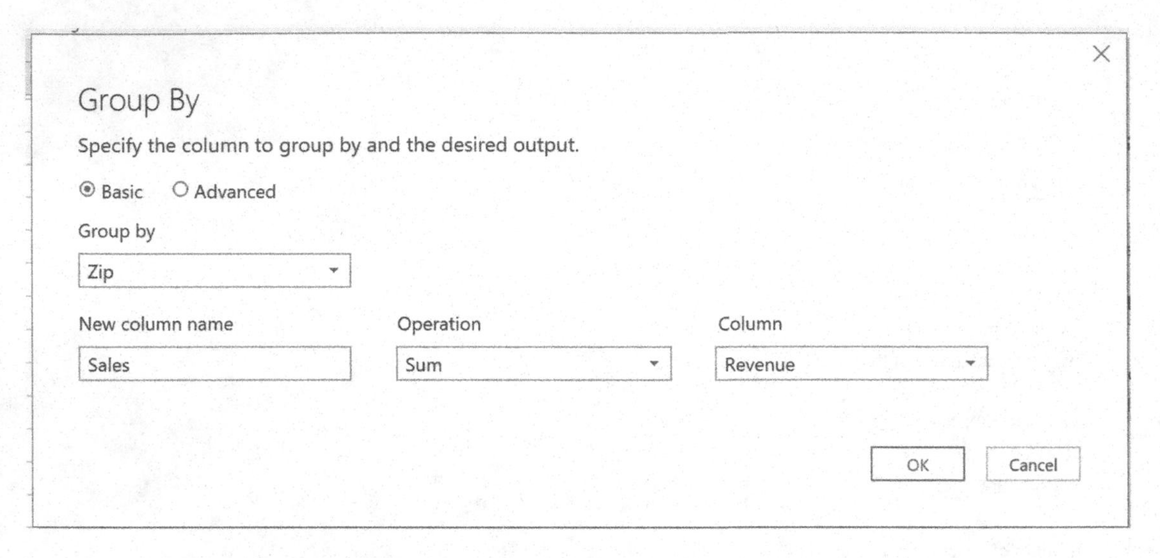

Figure 10-13. *Grouping data in Power Query*

These are just a few of the data-shaping features exposed by the Power Query Editor. (For a more detailed discussion of Power Query, refer to Chapters 3 and 13.)

Once you have cleaned and shaped the data, you can select the Close & Load option on the Home tab. This will load the tables into the data model and close the Power Query Editor. You are now ready to work with the data model to create table relations, calculated columns, and measures.

Creating Table Relationships

After the data tables are imported into the data model, you are ready to create the relationships between the tables. As discussed in Chapter 4, typically you want to set up your model in a star schema where the fact table is in the center of the star, and the dimension tables are related to the fact table by keys (see Figure 10-14).

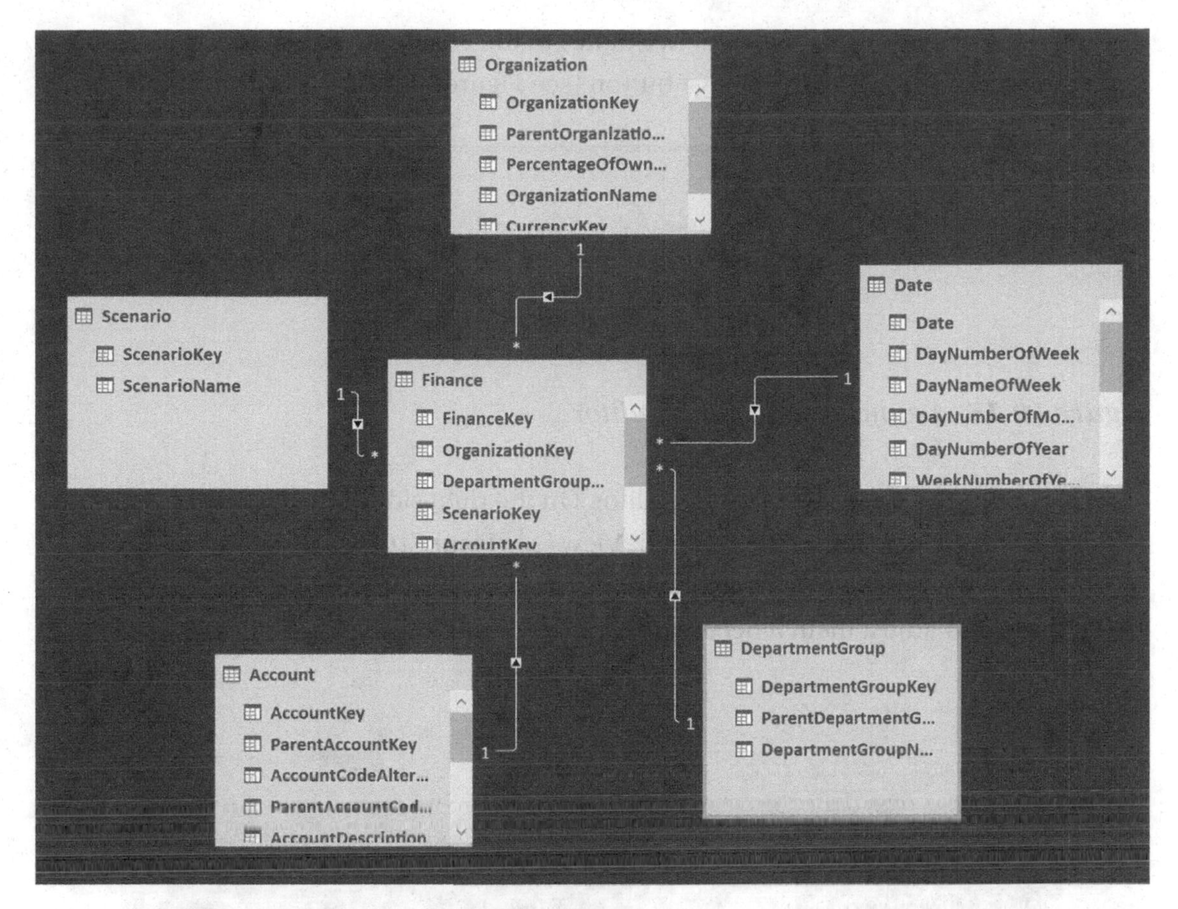

Figure 10-14. *A typical star schema*

The fact table contains numbers that you need to aggregate. For example, in the finance table, you have the amount, which is a monetary value that needs to be aggregated. In a sales fact table, you may have sales amount and item counts. In a human resources system, you might have hours worked. The dimension tables contain the attributes that you are using to categorize and roll up the measures. For example, the financial measures are classified as profit, loss, and forecasted. You want to roll the values up to the department and organization level and you want to compare values between months and years.

To create a relationship between two tables in the Power Pivot model, select the Power Pivot tab and click the Manage button (see Figure 10-15).

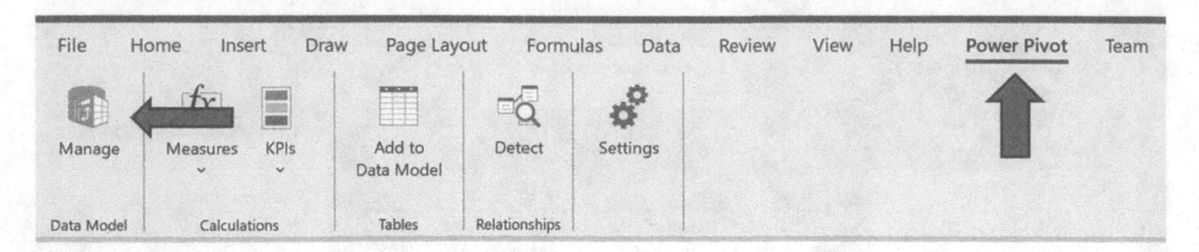

Figure 10-15. Launching the model editor

This opens the Power Pivot model editor. On the right side of the Home tab, you can switch from the Data view to the Diagram View (see Figure 10-16). By default, Power Pivot will try to autodetect new relationships when data is loaded. Make sure to double check these and adjust them if necessary.

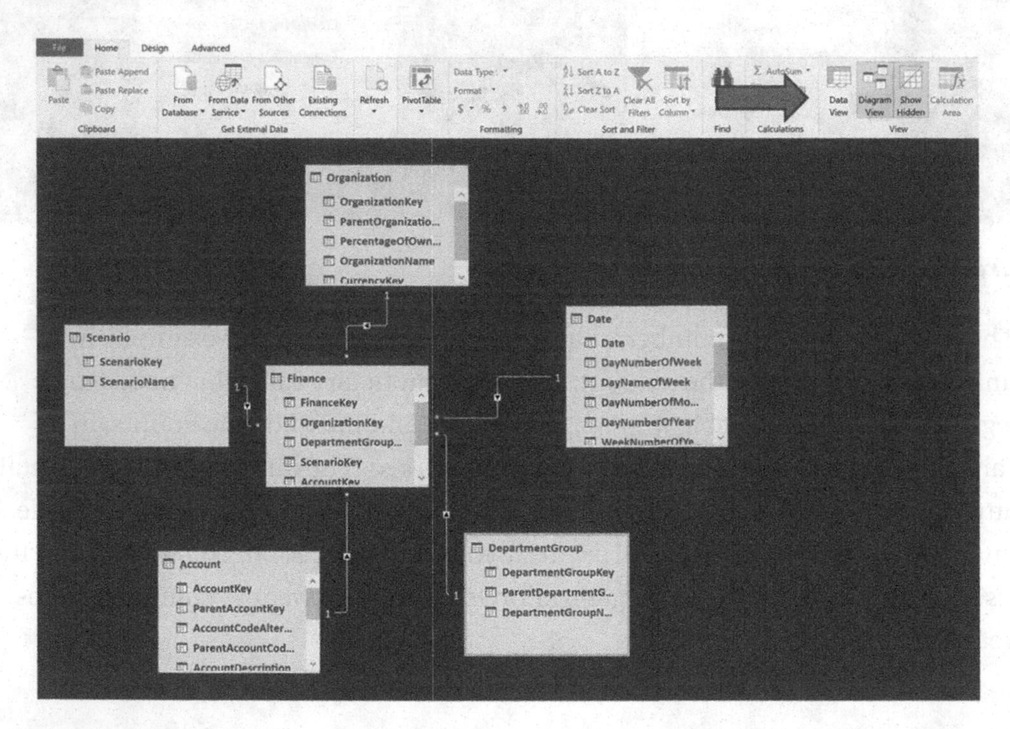

Figure 10-16. Switching to the Diagram view

To manage the relationships between the tables, select the Manage Relationships button in the Design tab. This launches the Manage Relationships window where you can create new relationships, edit existing relationships, and delete existing relationships (see Figure 10-17).

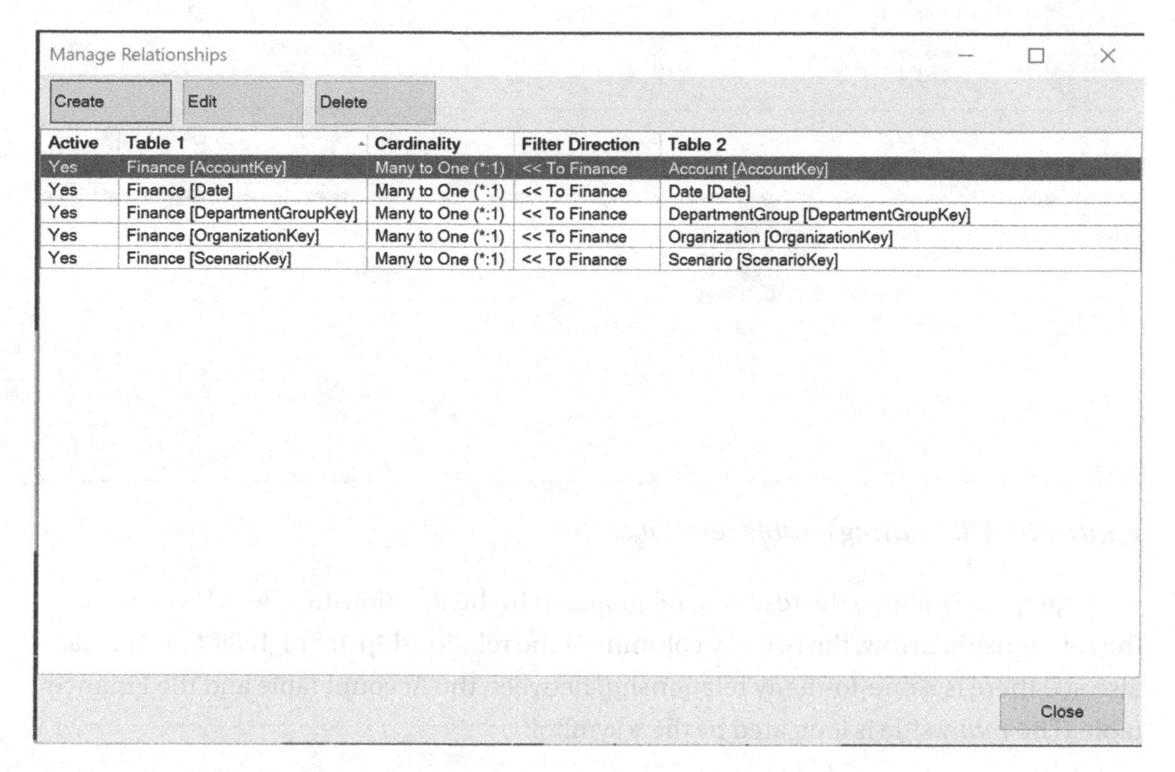

Manage Relationships
Create
Edit
Delete
Active
Table 1
Cardinality
Filter Direction
Table 2
Yes
Finance [AccountKey]
Many to One (*:1)
<< To Finance
Account [AccountKey]
Yes
Finance [Date]
Many to One (*:1)
<< To Finance
Date [Date]
Yes
Finance [DepartmentGroupKey]
Many to One (*:1)
<< To Finance
DepartmentGroup [DepartmentGroupKey]
Yes
Finance [OrganizationKey]
Many to One (*:1)
<< To Finance
Organization [OrganizationKey]
Yes
Finance [ScenarioKey]
Many to One (*:1)
<< To Finance
Scenario [ScenarioKey]
Close

***Figure 10-17.** Managing table relationships*

Clicking the Edit button presents you with the Edit Relationship window (see Figure 10-18). This is where you set the tables and key columns used that define the relationship.

Edit Relationship ? ×

Select tables and columns that relate to one another.

Finance

AccountKey	Amount	Date	DepartmentGroupKey	FinanceKey	OrganizationKey	ScenarioKey
52	782	12/29/2010 12:00:00 AM	6	850	7	1
53	782	12/29/2010 12:00:00 AM	6	853	7	1
56	28863	12/29/2010 12:00:00 AM	6	854	7	1
57	2607	12/29/2010 12:00:00 AM	6	857	7	1
60	28840	12/29/2010 12:00:00 AM	6	858	7	1

Account

AccountCodeAlternateKey	AccountDescription	AccountKey	AccountType	CustomMemberOptions	CustomMembers	Operator	ParentAccountCodeAlternateKey	Pare
10	Assets	2	Assets			+	1	1
110	Current Assets	3	Assets			+	10	2
1110	Cash	4	Assets			+	110	3
1120	Receivables	5	Assets			+	110	3
1130	Trade Receivables	6	Assets			+	1120	5

☑ Active

OK Cancel

Figure 10-18. *Editing a table relationship*

Figure 10-19 shows the resulting relationship in the Relationship view. If you click the relationship arrow, the two key columns of the relationship are highlighted. You can also see there is a one-to-many relationship between the Account table and the Finance table. The *many* side is indicated by the * symbol.

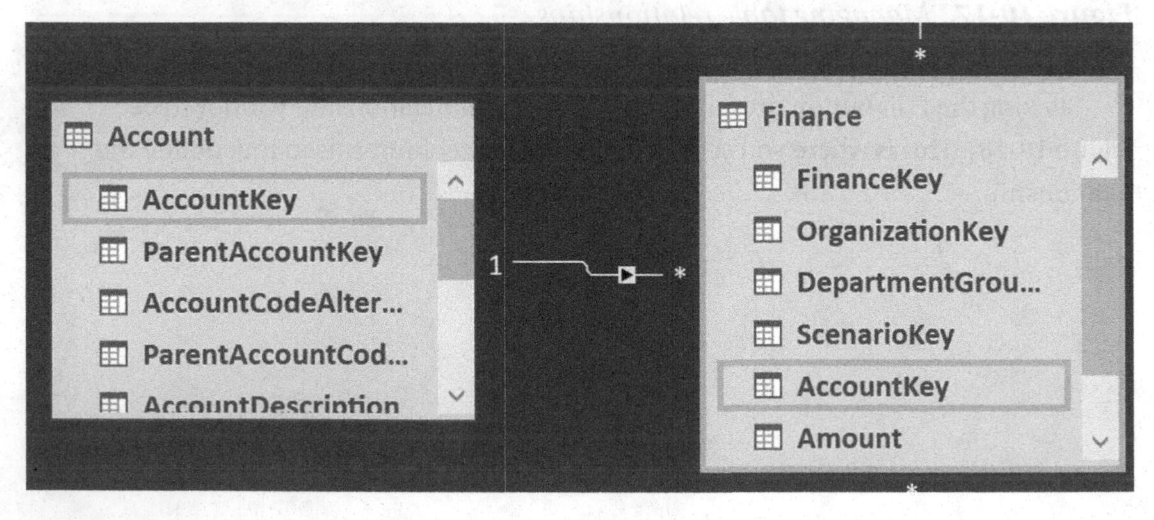

Figure 10-19. *Viewing a relationship in the Relationship view*

The arrow on the relationship line indicates the direction filtering works. When you create a filter on the Account table, it will filter the corresponding rows of the Sales table.

After setting up the relationships between the tables in the model, you are ready to add any calculations and measures that will aid in the data analysis.

Adding Calculations and Measures

As you saw in Chapters 5 and 6, one of the most important aspects of creating a solid data model involves adding calculated columns and measures that aid in the analysis of the data. For example, you may need to calculate years of service for employees or concatenate first and last names. Just like Power BI Desktop, Power Pivot in Excel uses DAX to create calculated columns and measures.

To create a calculated column, select the Data view button in the Home tab. Next, select the table to add the calculated column to. The table tabs are along the bottom on the left (see Figure 10-20).

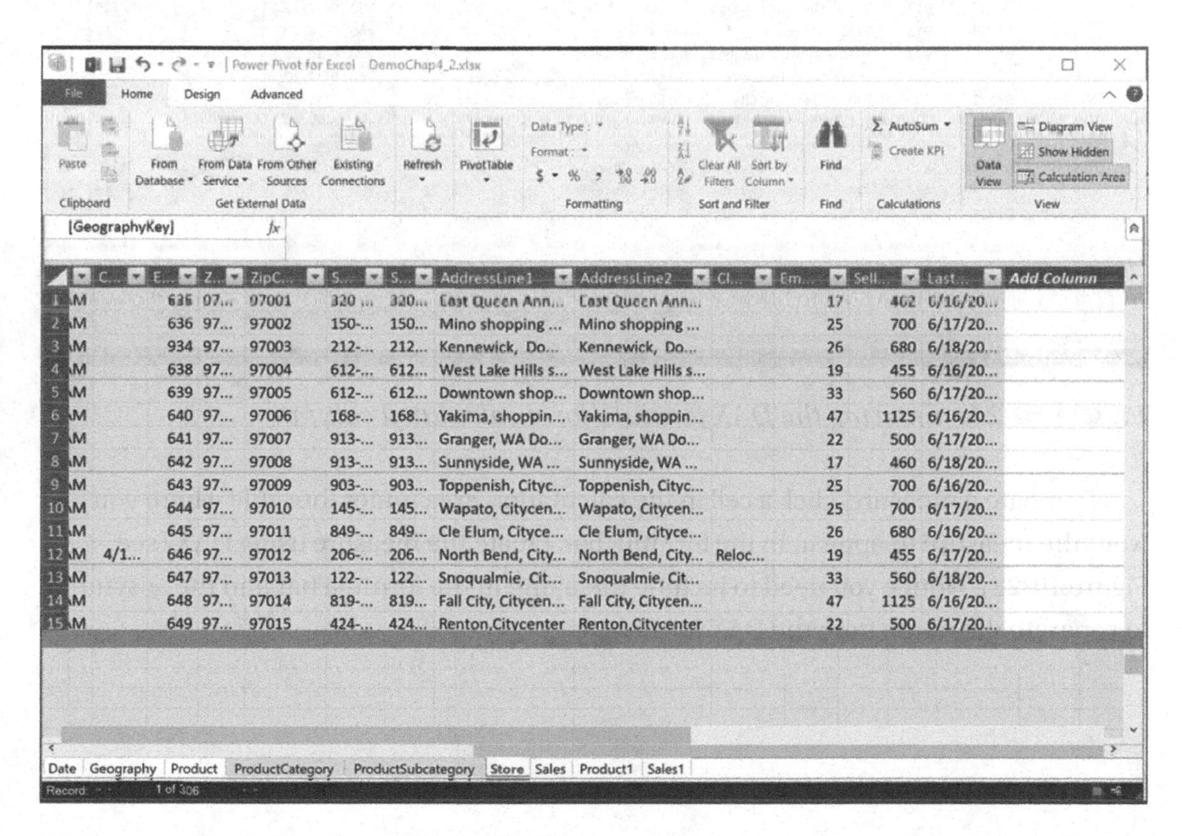

Figure 10-20. Selecting tables in the Data view

Scroll to the last column in the table and you should see the Add Column column. Select the column and in the formula bar enter the DAX formula for the column and select the check mark to complete entering the formula (see Figure 10-21). One slight difference between Power Pivot in Excel and the Power BI Desktop is that you don't include the name of the column in the formula bar; instead you have to rename the column in the table. Remember to start your DAX formula with an = sign.

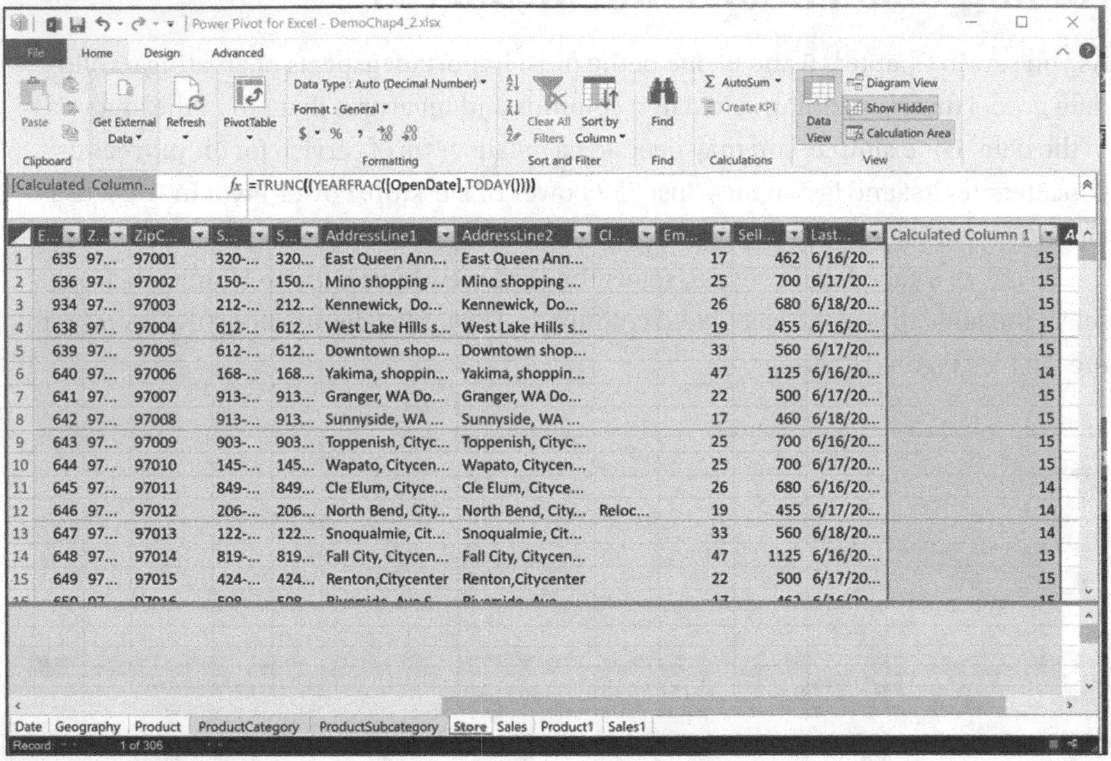

	E...	Z...	ZipC...	S...	S...	AddressLine1	AddressLine2	Cl...	Em...	Sell...	Last...	Calculated Column 1
1	635	97...	97001	320-...	320...	East Queen Ann...	East Queen Ann...		17	462	6/16/20...	15
2	636	97...	97002	150-...	150...	Mino shopping ...	Mino shopping ...		25	700	6/17/20...	15
3	934	97...	97003	212-...	212...	Kennewick, Do...	Kennewick, Do...		26	680	6/18/20...	15
4	638	97...	97004	612-...	612...	West Lake Hills s...	West Lake Hills s...		19	455	6/16/20...	15
5	639	97...	97005	612-...	612...	Downtown shop...	Downtown shop...		33	560	6/17/20...	15
6	640	97...	97006	168-...	168...	Yakima, shoppin...	Yakima, shoppin...		47	1125	6/16/20...	14
7	641	97...	97007	913-...	913...	Granger, WA Do...	Granger, WA Do...		22	500	6/17/20...	15
8	642	97...	97008	913-...	913...	Sunnyside, WA ...	Sunnyside, WA ...		17	460	6/18/20...	15
9	643	97...	97009	903-...	903...	Toppenish, Cityc...	Toppenish, Cityc...		25	700	6/16/20...	15
10	644	97...	97010	145-...	145...	Wapato, Citycen...	Wapato, Citycen...		25	700	6/17/20...	15
11	645	97...	97011	849-...	849...	Cle Elum, Cityce...	Cle Elum, Cityce...		26	680	6/16/20...	14
12	646	97...	97012	206-...	206...	North Bend, City...	North Bend, City...	Reloc...	19	455	6/17/20...	14
13	647	97...	97013	122-...	122...	Snoqualmie, Cit...	Snoqualmie, Cit...		33	560	6/18/20...	14
14	648	97...	97014	819-...	819...	Fall City, Citycen...	Fall City, Citycen...		47	1125	6/16/20...	13
15	649	97...	97015	424-...	424...	Renton,Citycenter	Renton,Citycenter		22	500	6/17/20...	15

Figure 10-21. Entering the DAX formula for a calculated column

To create a measure, click a cell in the calculation area under the table where you want the measure to appear. In the formula bar, create the measure using DAX (see Figure 10-22). Notice you need to include the name in the formula bar and the := symbol to separate the name from the DAX formula.

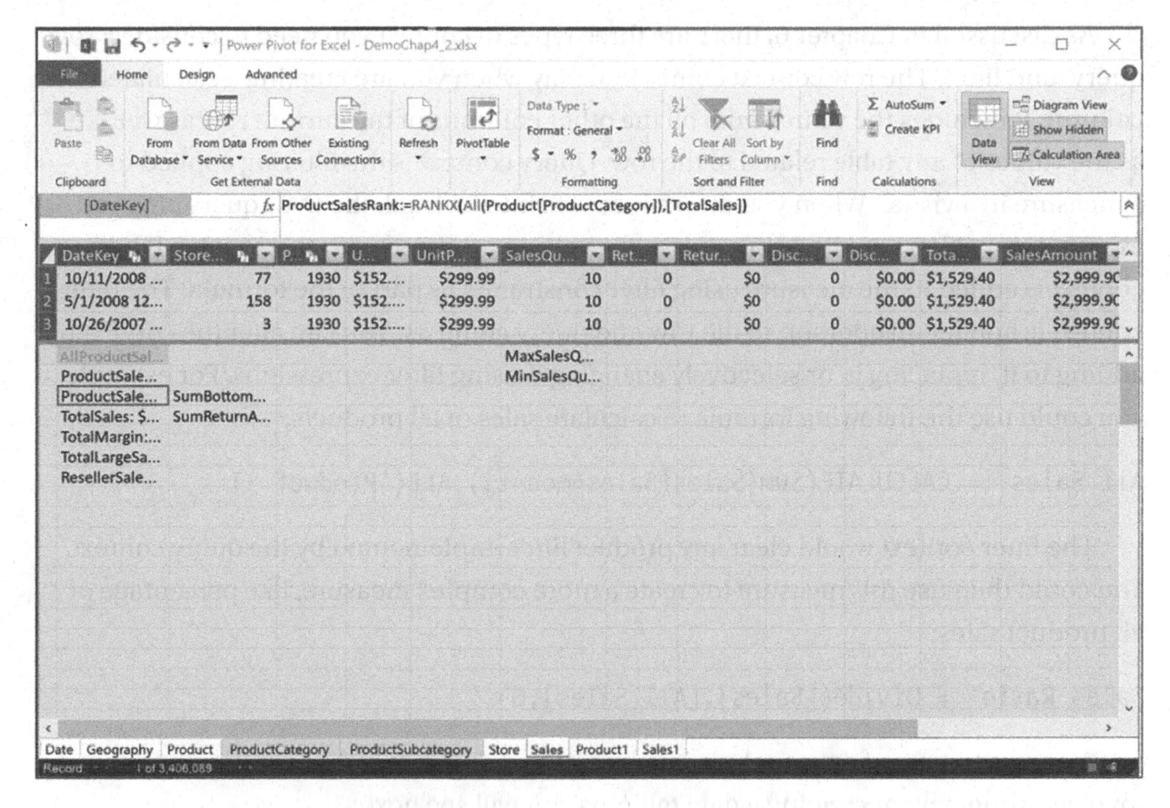

***Figure 10-22.** Creating a measure*

```
ProductSalesRank:=RANKX(All(Product[ProductCategory]),[TotalSales])
```

Remember, the difference between a calculated column and a measure is that calculated columns are precalculated and stored in the model. Measures are calculated when filters are applied to them and are recalculated every time different filters are applied. So, the more calculated columns you have, the greater the size of your Power BI Excel file, and the more measures you have, the greater the memory needed to update the data visuals when different filters are applied. Recall, however, that the Power Pivot model is based on an in-memory analytics engine and columnar storage (discussed in Chapter 1). It is designed to load all the data into memory and make calculations on the fly. As such, you are better off limiting the number of calculated columns in favor of measures.

Context plays an important role when creating measures in the Power Pivot model. Unlike static reports, Excel visuals are designed for dynamic analysis. When the user interacts with the visual, the context changes, and the values are recalculated. Knowing how the context changes and how it affects the results is essential to being able to build and troubleshoot DAX formulas.

As discussed in Chapter 6, there are three types of context you need to consider: row, query, and filter. The row context comes into play when you are creating a calculated column. It includes the values from all the other columns of the current row as well as the values of any table related to the row. Query context is the filtering applied to a measure in a visual. When you drop a measure into a visual, the DAX query engine examines any filters applied and returns the value associated with the context. Filter context is added to the measure using filter constraints as part of the formula. The filter context is applied in addition to the row and query contexts. You can alter the context by adding to it, replacing it, or selectively changing it using filter expressions. For example, you could use the following formula to calculate sales of all products:

```
All Sales := CALCULATE(Sum(Sales[SalesAmount]),ALL('Product'))
```

The filter context would clear any product filter implemented by the query context. You could then use this measure to create a more complex measure, like percentage of all product sales:

```
Sales Ratio := Divide([Sales],[All Sales],0)
```

A common type of data analysis involves comparing measures over time, and that involves importing or creating a date table, as you will see next.

Incorporating Time-Based Analysis

As discussed in Chapter 7, one of the most common types of data analysis is comparing values over time. For example, you may want to look at sales this quarter compared to sales last quarter or month-to-date (MTD) help desk tickets compared to last month. DAX contains many functions that help you create the various datetime-based analysis you may need. For example, you can use the following expressions to calculate the sum of the sales and the sales year-to-date (YTD) values:

```
Total Sales := SUM([SalesAmount])
YTD Sales := TOTALYTD([Total Sales],'Date'[Datekey])
```

To use the built-in time intelligence functions in DAX, you need to have a date table in your model for the functions to reference. The only requirement for the table is that it needs a distinct row for each day in the date range you are interested in looking at. Each of these rows needs to contain the full date of the day. The date table can and often does have more columns, but it doesn't have to. You can either import a date table from the source if one is available, create one in Power Query, or use a template provided by Power Pivot. To use the template, on the Design tab, select the Date Table drop-down and click the New button. This creates a Calendar table with a date range spanning the minimum date in your model to the maximum date (see Figure 10-23).

	Date	Year	Month Number	Month	MMM-YYYY	Day Of Week Number	Day Of Week	Add Column
1	1/1/2003 12:00:00 AM	2003	1	January	Jan-2003	4	Wednesday	
2	1/2/2003 12:00:00 AM	2003	1	January	Jan-2003	5	Thursday	
3	1/3/2003 12:00:00 AM	2003	1	January	Jan-2003	6	Friday	
4	1/4/2003 12:00:00 AM	2003	1	January	Jan-2003	7	Saturday	
5	1/5/2003 12:00:00 AM	2003	1	January	Jan-2003	1	Sunday	
6	1/6/2003 12:00:00 AM	2003	1	January	Jan-2003	2	Monday	
7	1/7/2003 12:00:00 AM	2003	1	January	Jan-2003	3	Tuesday	
8	1/8/2003 12:00:00 AM	2003	1	January	Jan-2003	4	Wednesday	
9	1/9/2003 12:00:00 AM	2003	1	January	Jan-2003	5	Thursday	
10	1/10/2003 12:00:00 AM	2003	1	January	Jan-2003	6	Friday	
11	1/11/2003 12:00:00 AM	2003	1	January	Jan-2003	7	Saturday	
12	1/12/2003 12:00:00 AM	2003	1	January	Jan-2003	1	Sunday	
13	1/13/2003 12:00:00 AM	2003	1	January	Jan-2003	2	Monday	
14	1/14/2003 12:00:00 AM	2003	1	January	Jan-2003	3	Tuesday	
15	1/15/2003 12:00:00 AM	2003	1	January	Jan-2003	4	Wednesday	
16	1/16/2003 12:00:00 AM	2003	1	January	Jan-2003	5	Thursday	
17	1/17/2003 12:00:00 AM	2003	1	January	Jan-2003	6	Friday	
18	1/18/2003 12:00:00 AM	2003	1	January	Jan-2003	7	Saturday	
19	1/19/2003 12:00:00 AM	2003	1	January	Jan-2003	1	Sunday	
20	1/20/2003 12:00:00 AM	2003	1	January	Jan-2003	2	Monday	

Figure 10-23. *Creating a calendar table*

After creating the calendar table, you can use DAX to create additional calculated columns in the table if you need them. You can also update the date range by selecting Update Range under the Date Table drop-down (see Figure 10-24).

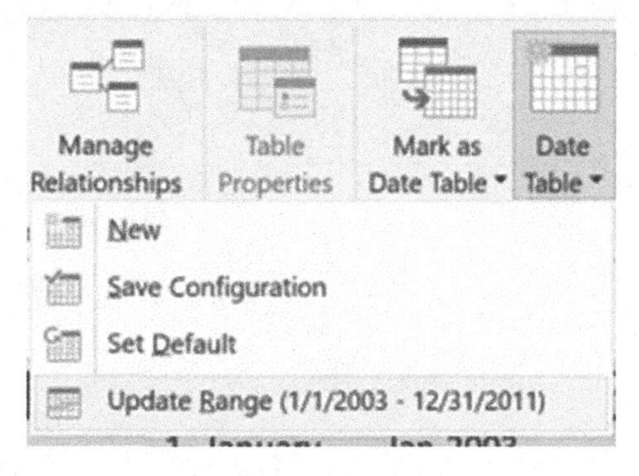

Figure 10-24. *Updating the date range*

The final step is to create a relationship between the date table and the table that contains the values you want to analyze.

Now that you are familiar with using Power Query and the Power Pivot Model Designer in Excel, it's time to get your hands dirty and complete the following hands-on lab. This lab will help you become familiar with working with the Power Pivot in Excel.

HANDS-ON LAB: CREATING THE DATA MODEL IN POWER PIVOT

In the following lab, you will

- Import data
- Clean and shape the data
- Create table relationships
- Add calculated columns and measures

1. If not already installed, download and install Excel 2016 or Excel 365.
2. Open Excel and create a blank spreadsheet.
3. On the Data tab, click the Get Data drop-down and choose the From Text/CSV option. Navigate to the SF311Calls.csv in the LabStarterFiles\Chapter10Lab1 folder. Open the file and click Transform Data to open it up in the Query Editor. You should see the Query Editor window with San Francisco call center data, as shown in Figure 10-25.

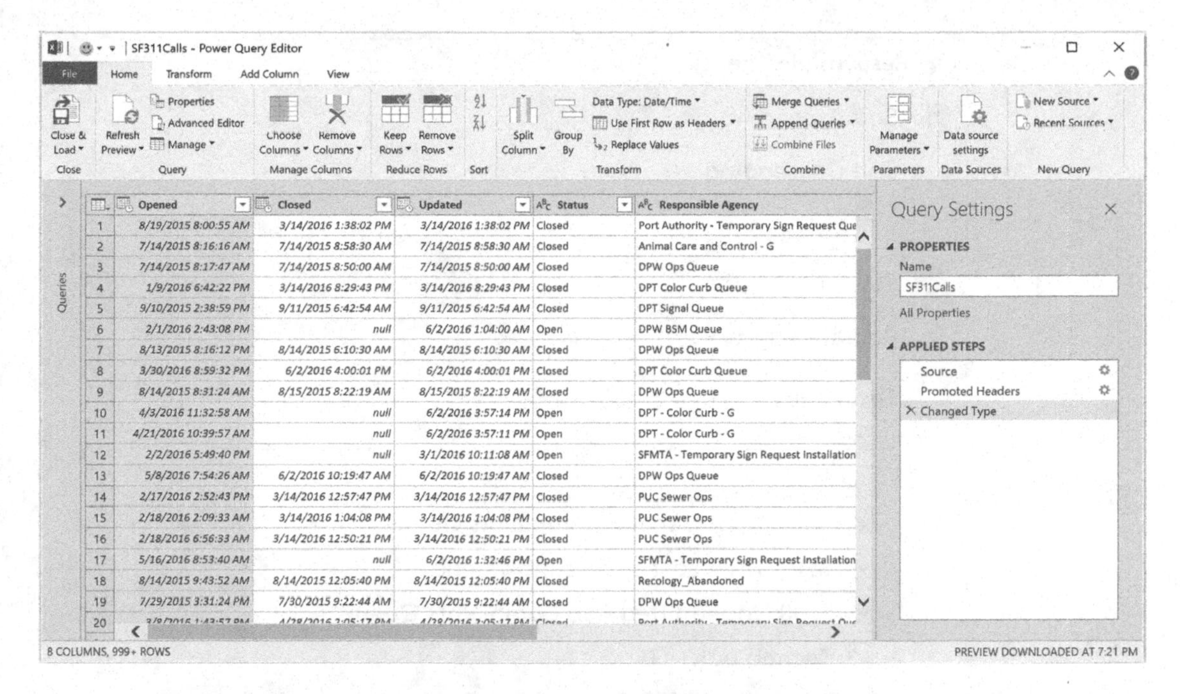

***Figure 10-25.** The Query Editor window with call center data*

4. In the Applied Steps list, if the Query Editor didn't automatically add the transform to set the first row as headers, add it now.

5. Check the types of each column to see whether the Query Editor updated the Opened, Closed, and Updated columns to a datetime data type. The rest of the columns should be the text data type.

6. Filter the data so that it doesn't include data from the Test Queue, zzRPD, and zzTaxi Commission agencies (see Figure 10-26). If you don't see those agencies, you may need to click "Load more" at the bottom of the filter list.

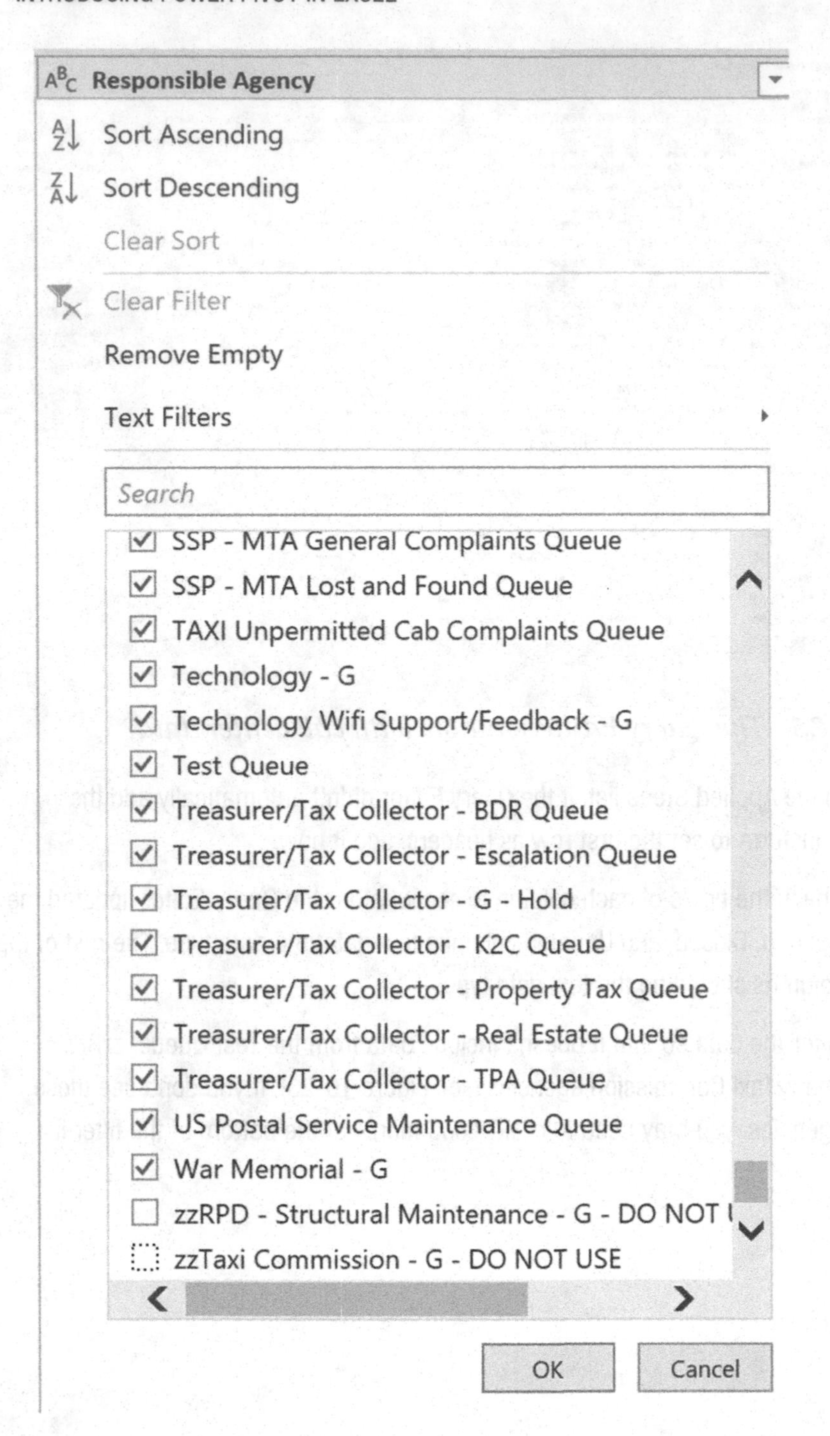

Figure 10-26. *Filtering out agencies*

7. Select the Opened column and filter the rows so that Opened is after 12/31/2014.

8. Select the Opened column again, and on the Add Column tab, select the Time drop-down. Select Hour and then Start Of Hour. Rename the new column to Hour Opened.

9. Change the data types of the Opened and Closed columns to Date (no time) and the Hour Opened to the Time data type.

10. In the Home tab, select the New Source drop-down and choose the CSV option. Navigate to the file Categories.csv in the LabStarterFiles\Chapter10Lab1 folder. Open the file and click Import and then OK to open it in the Query Editor.

11. Under the Home tab, select Use First Row As Headers.

12. Select the `SF311Calls` query and merge it with the `Categories` query using the Category column (see Figure 10-27).

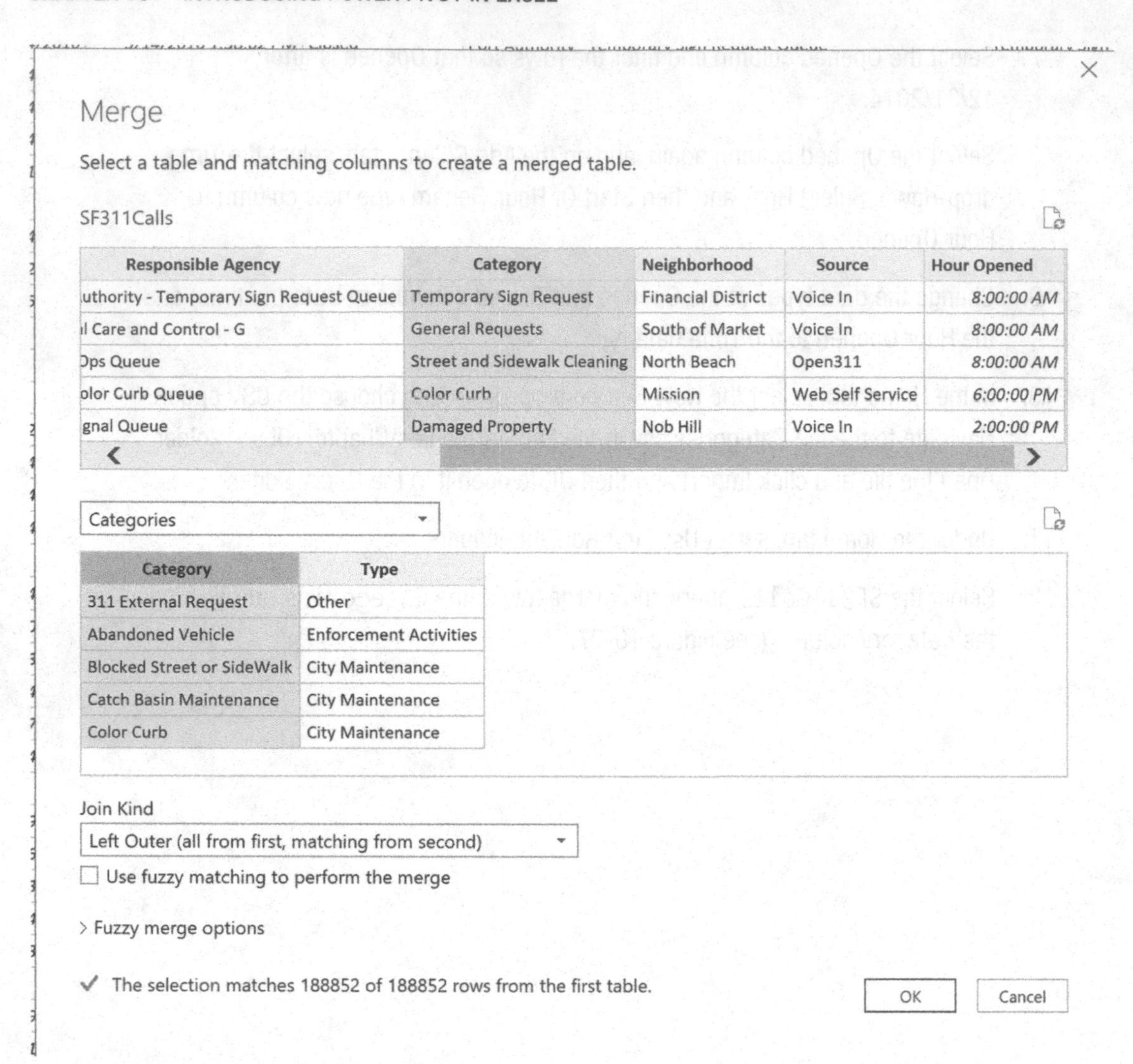

***Figure 10-27.** Merging queries*

13. Expand the resulting column and uncheck the Category field, leaving only the Type field selected (see Figure 10-28).

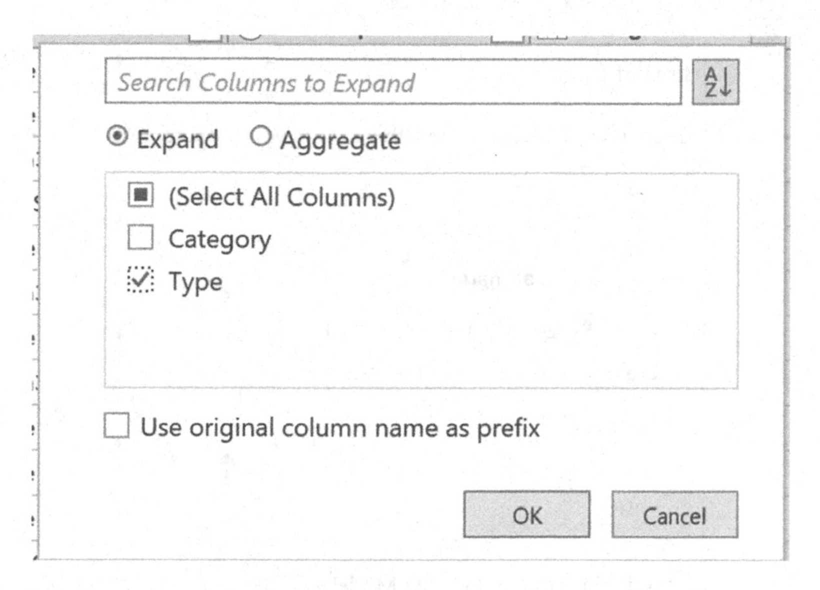

Figure 10-28. *Expanding the new column*

14. On the Home tab, select the Close & Load drop-down and the Close & Load To… option (see Figure 10-29).

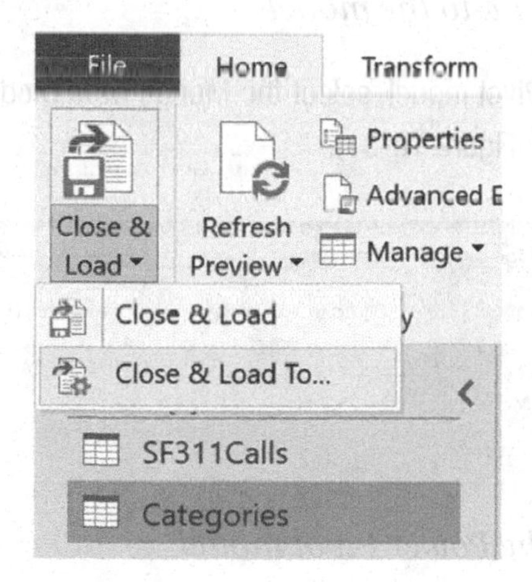

Figure 10-29. *Selecting the Load To option*

15. In the Import Data pop-up widow, select Only Create Connection option and check the Add this data to the Data Model (see Figure 10-30). Click OK.

***Figure 10-30.** Loading data to the model*

16. To open the Power Pivot model, select the Mange Data Model button on the Power Pivot tab (see Figure 10-31).

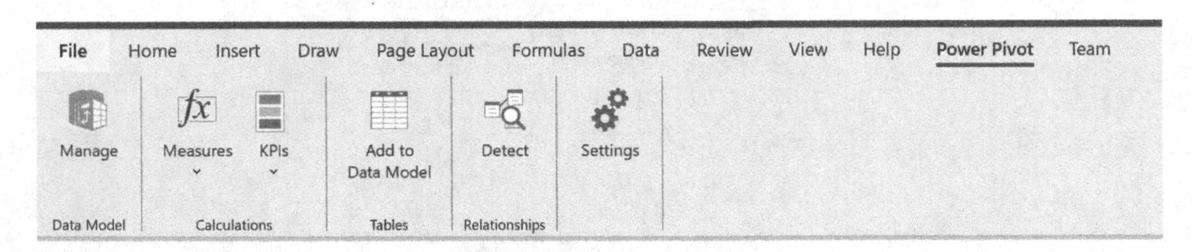

***Figure 10-31.** Opening the Power Pivot model*

17. Select the SF311Calls table, select the Hour Open column, and on the Home tab change the format to show the hour (see Figure 10-32).

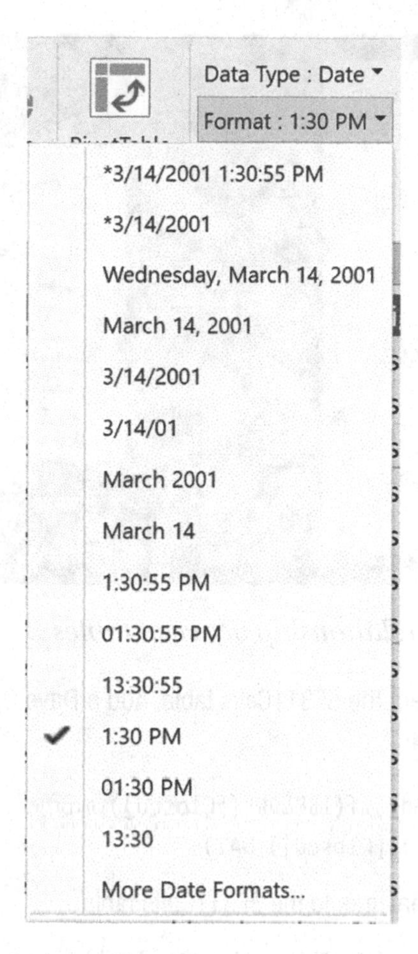

Figure 10-32. *Changing the time format*

18. To create a calendar table, on the Design tab, under the Date Table drop-down, click the New Table button.
19. Once the calendar table is created, select the Date Table drop-down again and update the date range from 1/1/2014 to 12/31/2016.
20. In the Data view, select the Calendar table. Change the Date column's data type to Date and format it to show MM/dd/yyyy.
21. Sort the Month column by the Month No column.
22. Switch to the Diagram View and create a relationship between the SF311Calls table and the Calendar table using the Date Opened and Date columns (see Figure 10-33).

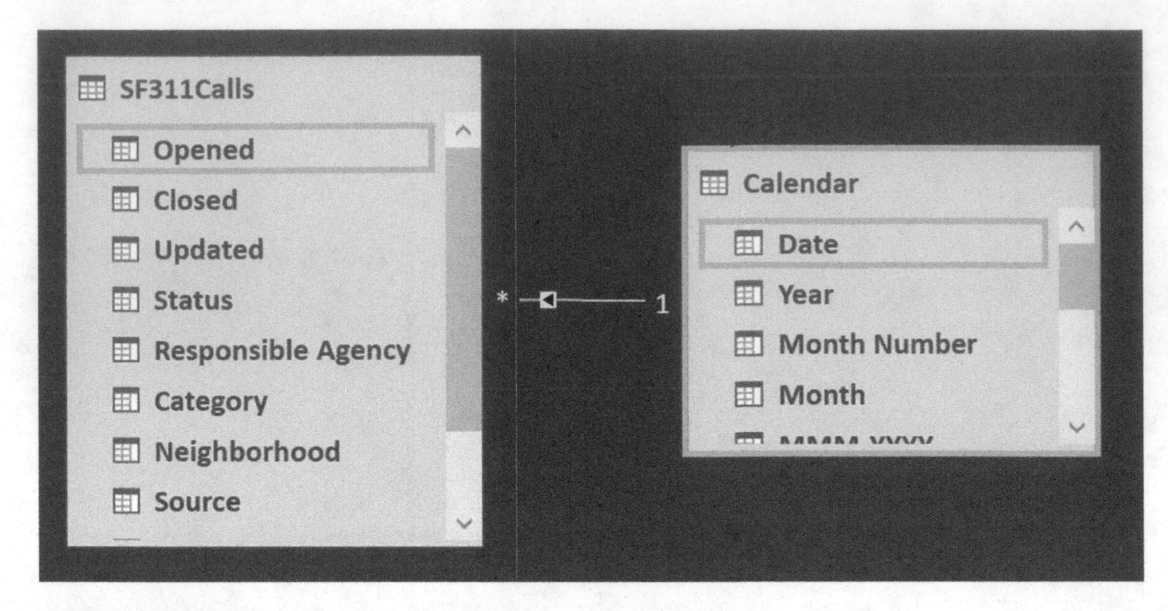

Figure 10-33. *Creating a relationship between tables*

23. In the Data view, select the SF311Calls table. Add a Days Open column using the following formula:

```
= DATEDIFF([Opened],if(ISBLANK([Closed]),max
('Calendar'[Date]),[Closed]),DAY)
```

24. Add the following measures to the SF311Calls table:

```
Number of Cases := COUNTROWS('SF311Calls')
Previous Month Number of Cases := Calculate([Number of Cases],
PARALLELPERIOD('Calendar'[Date],-1,MONTH))
```

25. Close the Power Pivot editor and save the Excel file as Chapter10Lab1.xlsx.

Summary

This chapter introduced you to Power Pivot in Excel. Most of the concepts were a review of topics covered in previous chapters. The purpose of this chapter was for you to gain familiarity with using Power Pivot in Excel and discovering how similar it is to working in Power BI Desktop. Once you have the data model created in Power Pivot, you need to create an interface for users to interact with the data model and perform data analysis using the model. This is where pivot tables and pivot charts in Excel come into play and are the topics covered in Chapter 11.

CHAPTER 11

Data Analysis with Pivot Tables and Charts

Once you have the data model created in Power Pivot, you need to create an interface for users to interact with the data model and perform data analysis using the model. Excel is a feature-rich environment for creating dashboards using pivot tables and pivot charts. Furthermore, it is very easy to share Excel files with colleagues or host the Excel workbook on SharePoint for increased performance and security. This chapter covers the basics of building an interface for analyzing the data contained in a Power Pivot model using pivot tables and pivot charts in Excel.

After completing this chapter, you will be able to

- Use pivot tables to explore the data
- Filter data using slicers
- Add visualizations to a pivot table
- Use pivot charts to explore trends
- Use multiple charts and tables linked together
- Use cube functions to query the data model

Note This chapter contains references to color figures. If you are reading this book in print, or in a black-and-white e-book edition, you can find copies of color figures in the Source Code/Downloads package for the book at `https://github.com/Apress/beginningpower-bi-3ed`.

D. Clark, *Beginning Microsoft Power BI*, https://doi.org/10.1007/978-1-4842-5620-6_11

Pivot Table Fundamentals

One of the most widely used tools for analyzing data is the pivot table. The pivot table allows you to easily detect patterns and relationships from the data. For example, you can determine what products sell better during certain times during the year. Or you can see how marketing campaigns affect the sales of various products. Figure 11-1 shows the various areas of a pivot table.

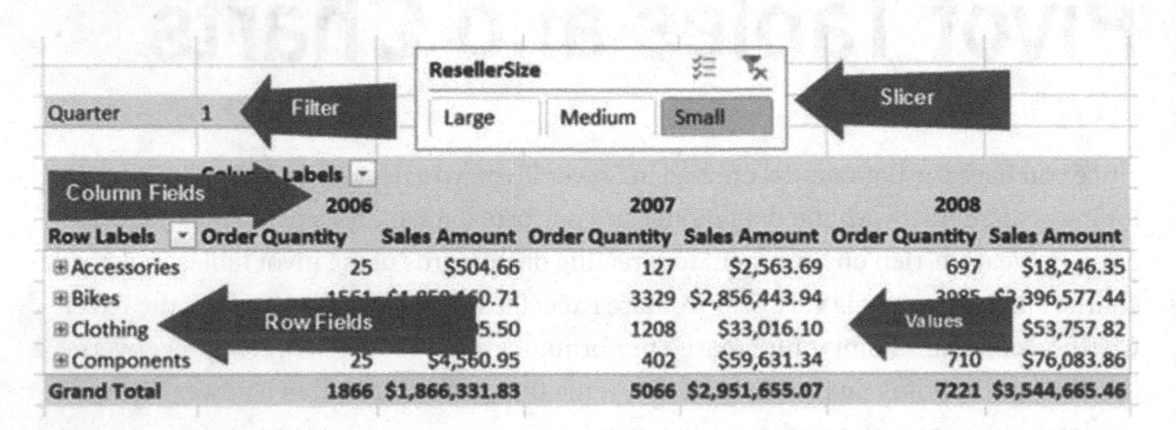

Figure 11-1. The parts of a pivot table

The row and column fields contain the attributes that you are interested in using to summarize the data. For example, the pivot table in Figure 11-1 is aggregating the values by product categories and years. The filter is used to filter the values in the pivot table by some attribute. The filter in Figure 11-1 is limiting the results to the first quarter of each year. The slicer works the same as the filter in this case, limiting the values to small resellers. Slicers have the added advantage of filtering multiple pivot tables and pivot charts.

To construct the pivot table, on the Insert tab, select the Pivot Table button. In the Create PivotTable window, select the option to use the workbook's data model (see Figure 11-2).

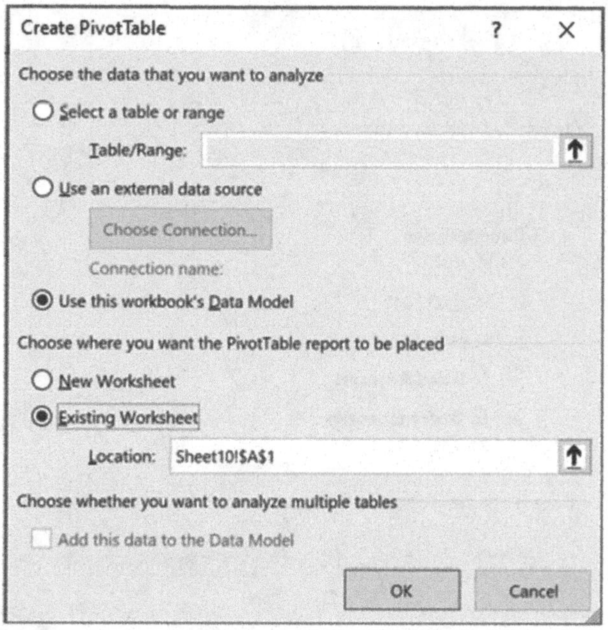

***Figure 11-2.** Inserting a pivot table*

You should see a field list and a pivot table on a new Excel sheet. Drag and drop the fields from the field list to the drop areas below the field list, as shown in Figure 11-3. If you just click the check box to select the fields, it will place text fields in the rows and any numeric values in the Values area. This can become quite annoying because it will place a field like Years in the Values area and treat it as a set of values to be summed.

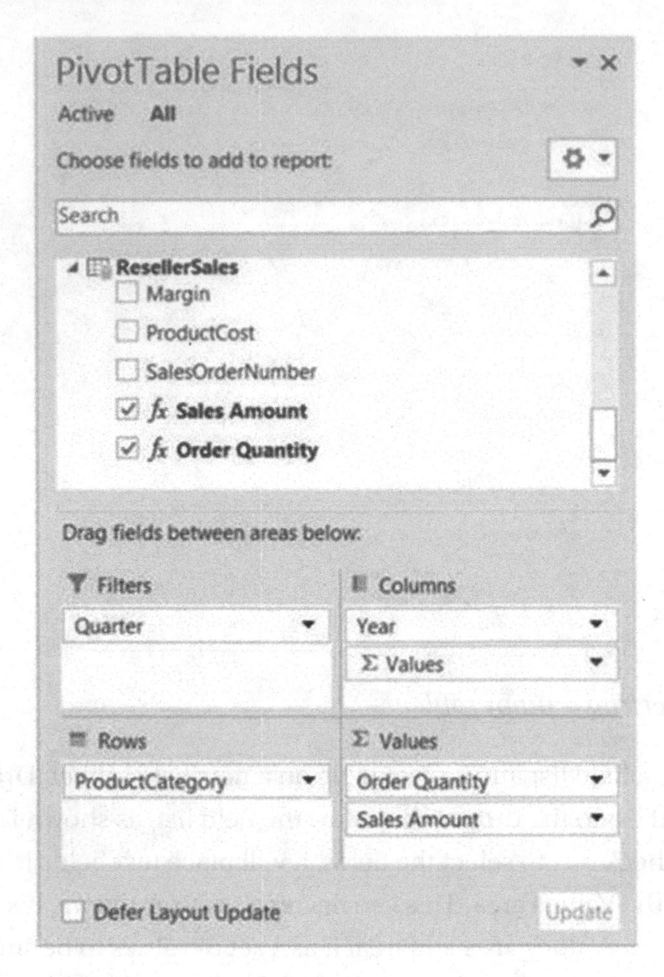

Figure 11-3. *Adding fields to the pivot table*

To add slicers to filter the pivot table, you need to go to the Insert tab in Excel and select the Slicer button. The next section looks at adding slicers and controlling multiple pivot tables with the same slicer.

Slicing the Data

To add a slicer to filter a pivot table, click the pivot table, and on the Insert tab, select the Slicer button. You should see a pick list containing the fields in the Power Pivot data model (see Figure 11-4).

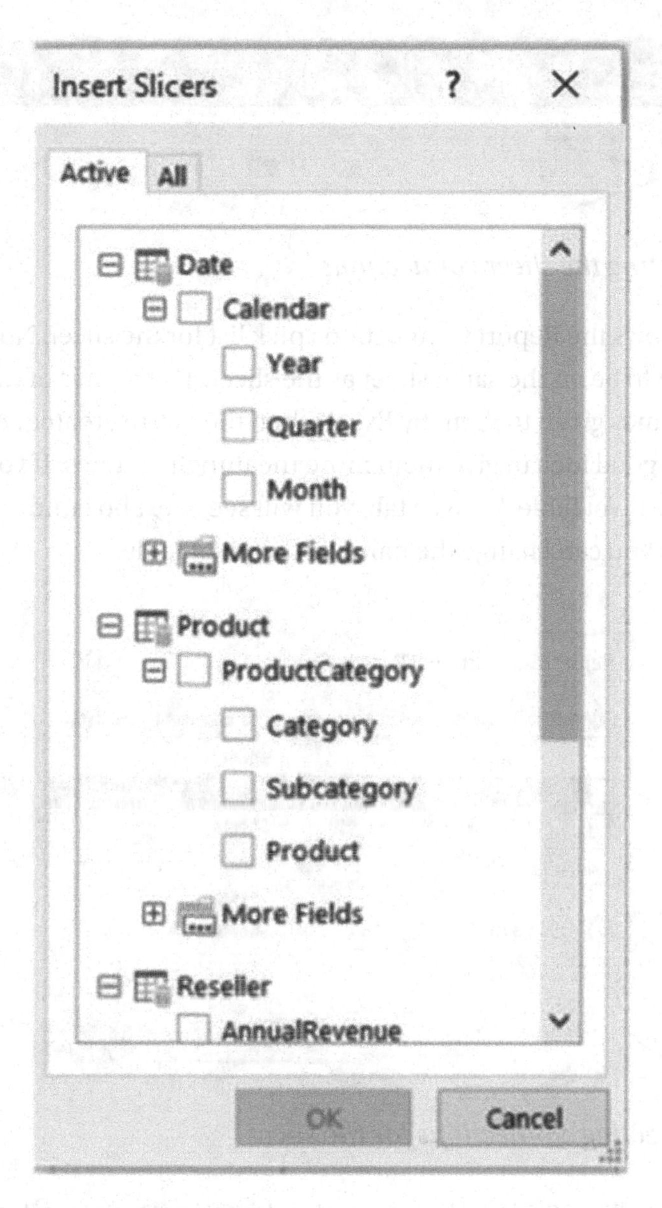

Figure 11-4. *Selecting the slicer field*

If you had the pivot table selected when you inserted the slicer, it is automatically wired up to the pivot table. You can verify this by clicking the slicer and selecting the Slicer Tools Options tab (see Figure 11-5). You can use this tab to format the slicer and choose the Report Connections for the slicer.

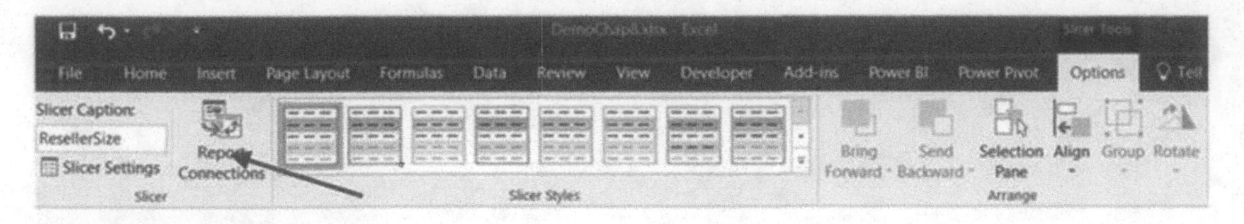

Figure 11-5. Setting the slicer connections

Figure 11-6 shows the Report Connections pick list for the slicer. Notice that the pivot table doesn't have to be on the same sheet as the slicer. The names of the pivot tables are the generic names given to them by Excel when they were created. As you add more pivot tables, it is a good idea to give them more meaningful names. If you click the pivot table and go to the PivotTable Analyze tab, you will see a text box under the PivotTable drop-down where you can change the name of the pivot table.

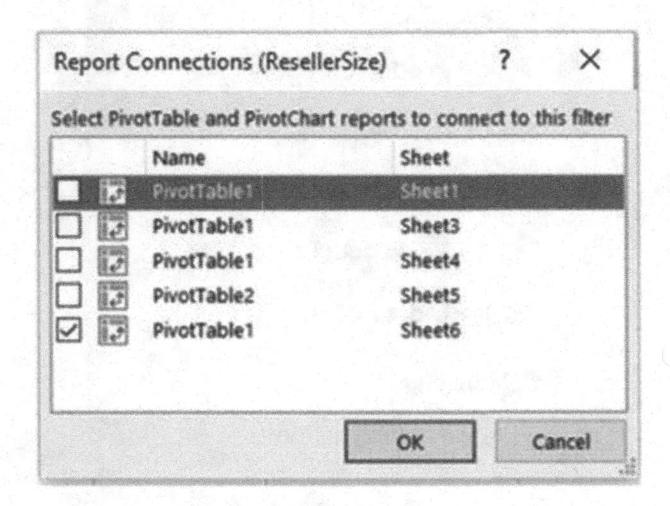

Figure 11-6. Selecting connections for the slicer

If you select the Slicer Settings button under the Slicer Options tab, you launch the Slicer Settings window (see Figure 11-7) where you can change the name, caption, and sort order of the slicer. You can also choose how to show items with no data.

Slicer Settings ? ×

Source Name: ResellerSize
Name to use in formulas: Slicer_ResellerSize2
Name: ResellerSize 2
Header
☑ Display header
Caption: ResellerSize
Item Sorting and Filtering
◉ Data source order
○ Ascending (A to Z)
○ Descending (Z to A)
☐ Hide items with no data
☑ Visually indicate items with no data
☑ Show items with no data last
OK Cancel

***Figure 11-7.** Changing the slicer settings*

There are times when you need *cascading filters*, where one filter limits what can be chosen in the next filter. This is easy with slicers based on fields that are related in the model. When you select the related fields, the slicers are linked automatically for you. For example, Figure 11-8 shows a product category and product subcategory filter. If you select a product category, the corresponding product subcategories are highlighted.

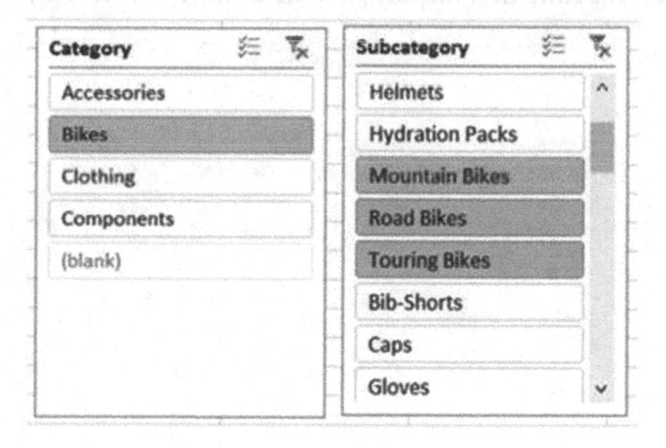

***Figure 11-8.** Creating cascading filters using slicers*

Adding Visualizations to a Pivot Table

To help identify trends and outliers and provide insight into the data, you can add many types of visualizations to a pivot table. These include conditional formatting, data bars, and trend lines. Figure 11-9 shows a pivot table with data bars and conditional formatting.

Row Labels	Order Quantity	Sales Amount	Margin
Accessories	2740	$60,086.05	$17,312.40
Bikes	4669	$3,699,096.26	($502,489.41)
Clothing	4967	$129,921.56	($3,853.94)
Components	5680	$1,453,875.07	$66,463.69
Grand Total	18056	$5,342,978.94	($422,567.26)

Figure 11-9. *Using conditional formatting and data bars*

To create the visual formatting, select the data you want to format in the pivot table, and in the Home tab, click the Conditional Formatting drop-down (see Figure 11-10). As you can see, you have a lot of options for creating conditional formatting. (In Figure 11-9, there is a Highlight Cells Rule that displays negative numbers in red.)

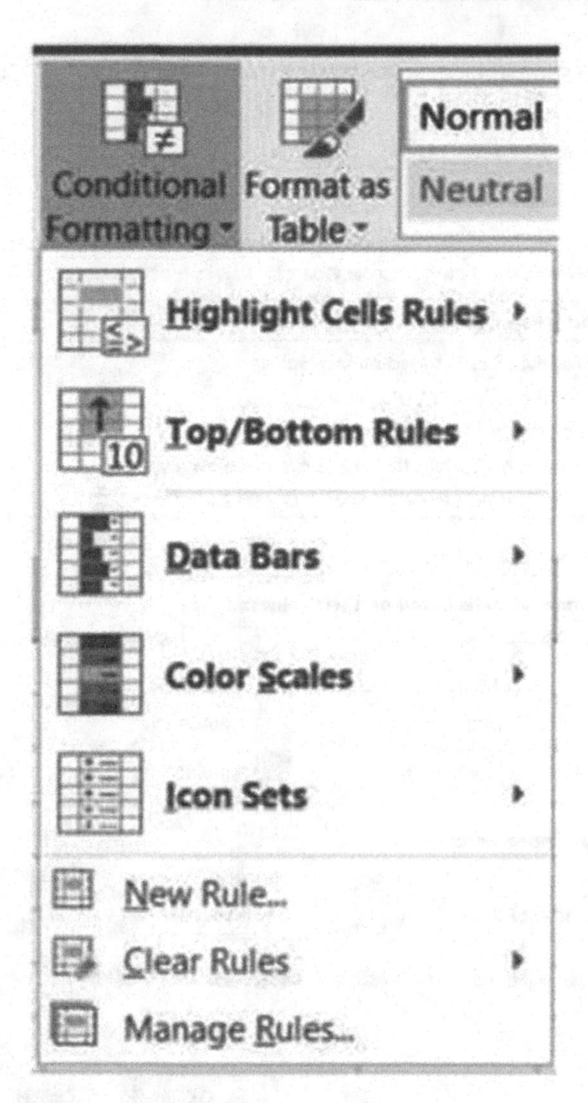

***Figure 11-10.** Setting up conditional formatting*

Once you establish a rule, you can edit it by clicking the Manage Rules option and selecting the rule you want to edit. Figure 11-11 shows the various options you can set for a data bar formatting rule.

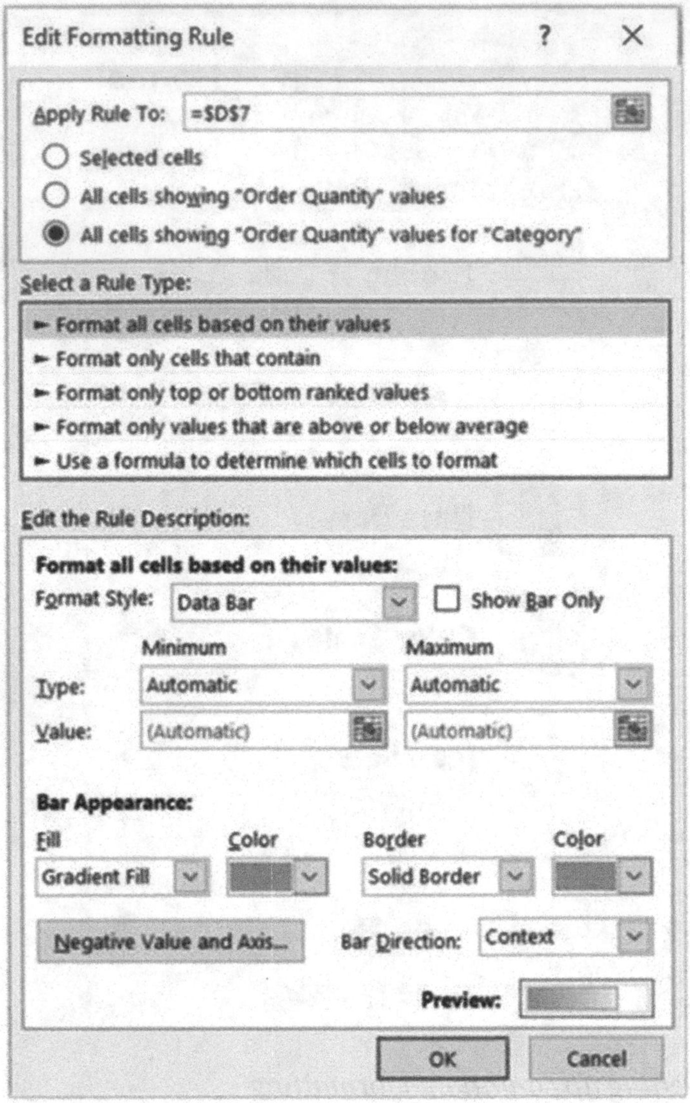

Figure 11-11. *Editing a data bar rule*

You can produce some interesting effects with the formatting rules. For example, the pivot table in Figure 11-12 shows a heat map used to quickly determine which months have good sales and which are bad.

Order Quantity	Column Labels				
Row Labels	Australia	Canada	France	Germany	United Kingdom
April	115	2861	628	541	750
August	415	5269	3252	1184	1350
December	901	3981	652	482	1505
February	110	2313	1444	504	562
January	103	1726	371	359	507
July	124	3794	890	756	1073
June	893	3827	625	427	1513
March	509	2428	376	286	1013
May	209	3824	2177	783	960
November	224	3885	2229	836	968
October	110	2848	659	615	829
September	1235	5005	1045	607	2163

***Figure 11-12.** Creating a heat map with conditional formatting*

Another popular feature associated with pivot tables is the spark lines. *Spark lines* are mini graphs that show the trend of a series of data. Figure 11-13 shows spark lines that display the sales trend across the four quarters of the year.

Year	2006				
Sum of SalesAmount	Column Labels				
Row Labels	1	2	3	4	Sales Trend
Large	$1,512,840.61	$1,673,185.81	$4,031,239.79	$2,895,748.58	
Medium	$690,013.60	$685,873.20	$1,253,354.23	$822,995.20	
Small	$1,866,331.83	$1,794,761.41	$3,595,645.41	$3,322,439.98	

***Figure 11-13.** Using spark lines to show trends*

To create a spark line, highlight the values in the pivot table that contain the data and select the Sparklines button on the Insert tab. You then select the location of the spark line (see Figure 11-14). You can choose between a line, a column, and a win/loss spark line.

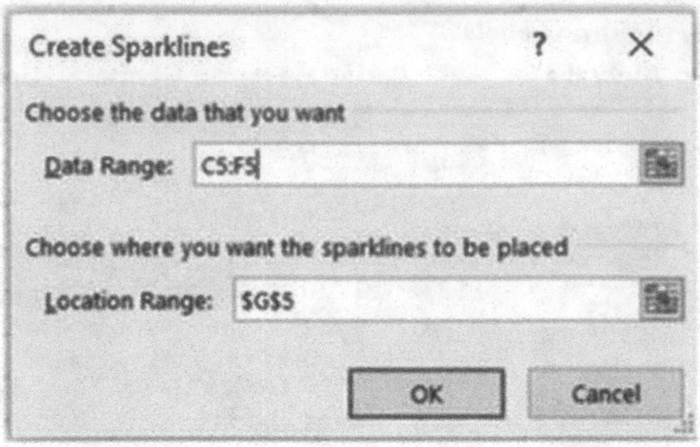

Figure 11-14. *Setting up a spark line*

Although adding visualizations to pivot tables can enhance your ability to analyze the data, many times the best way to spot trends and compare and contrast the data is through the use of charts and graphs. In the next section, you will see how charts and graphs are useful data analysis tools and how to add them to your analysis dashboards.

Working with Pivot Charts

Along with pivot tables, Excel has a robust set of charts and graphs available for you to use to analyze your data. Adding a pivot chart is very similar to adding a pivot table. On Excel's Insert tab, click the PivotChart drop-down. You have the option of selecting a single PivotChart or a PivotChart and PivotTable that are tied together (see Figure 11-15).

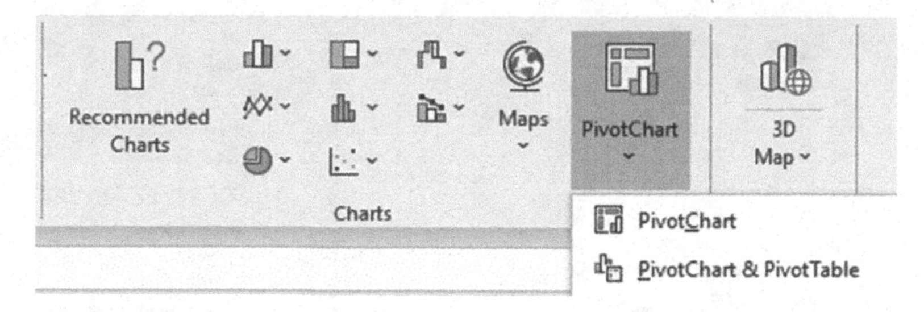

***Figure 11-15.** Adding a pivot chart*

After selecting the pivot chart, you are presented with a window to select a data source and choose where you want to put the pivot chart (see Figure 11-16).

Create PivotChart
Choose the data that you want to analyze
Select a table or range
Table/Range:
Use an external data source
Choose Connection...
Connection name:
Use this workbook's Data Model
Choose where you want the PivotChart to be placed
New Worksheet
Existing Worksheet
Location: Sheet7!A1
Choose whether you want to analyze multiple tables
Add this data to the Data Model
OK
Cancel

***Figure 11-16.** Selecting a data source*

Once you select the data source, you are presented with a blank chart and the field selection box where you can drag and drop fields for the chart values and axis (see Figure 11-17).

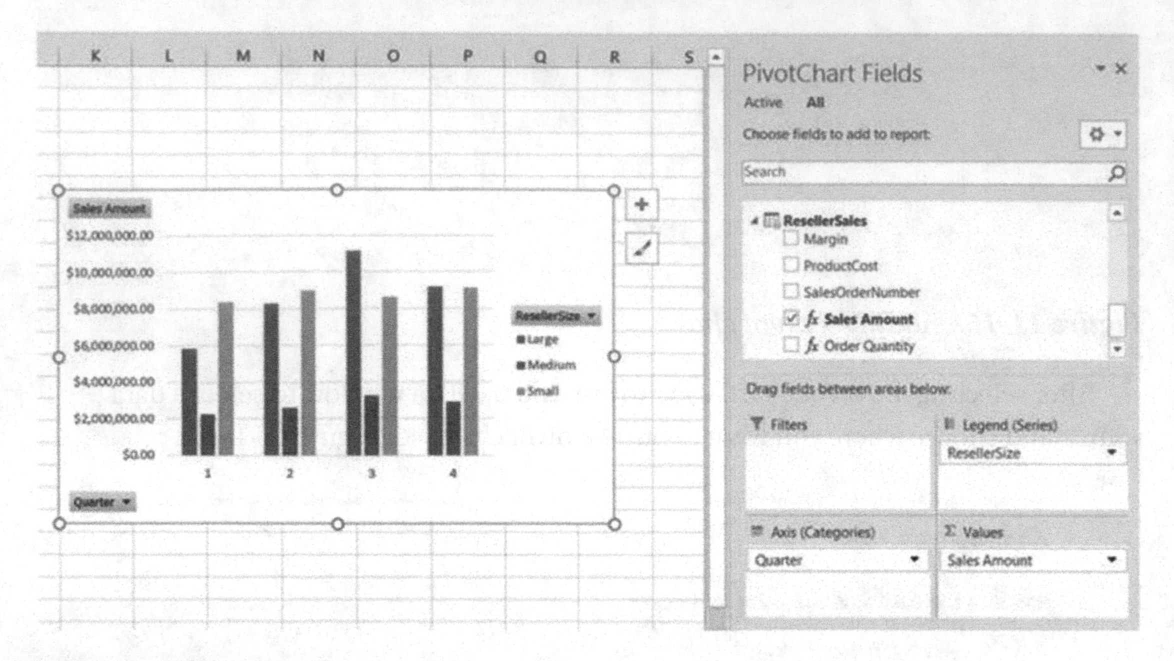

***Figure 11-17.** Creating a chart*

By default, the chart is a column chart, but you can choose from many different types of charts. If you click the chart and select the Design tab, you can select the Change Chart Type button and choose from a variety of types (see Figure 11-18).

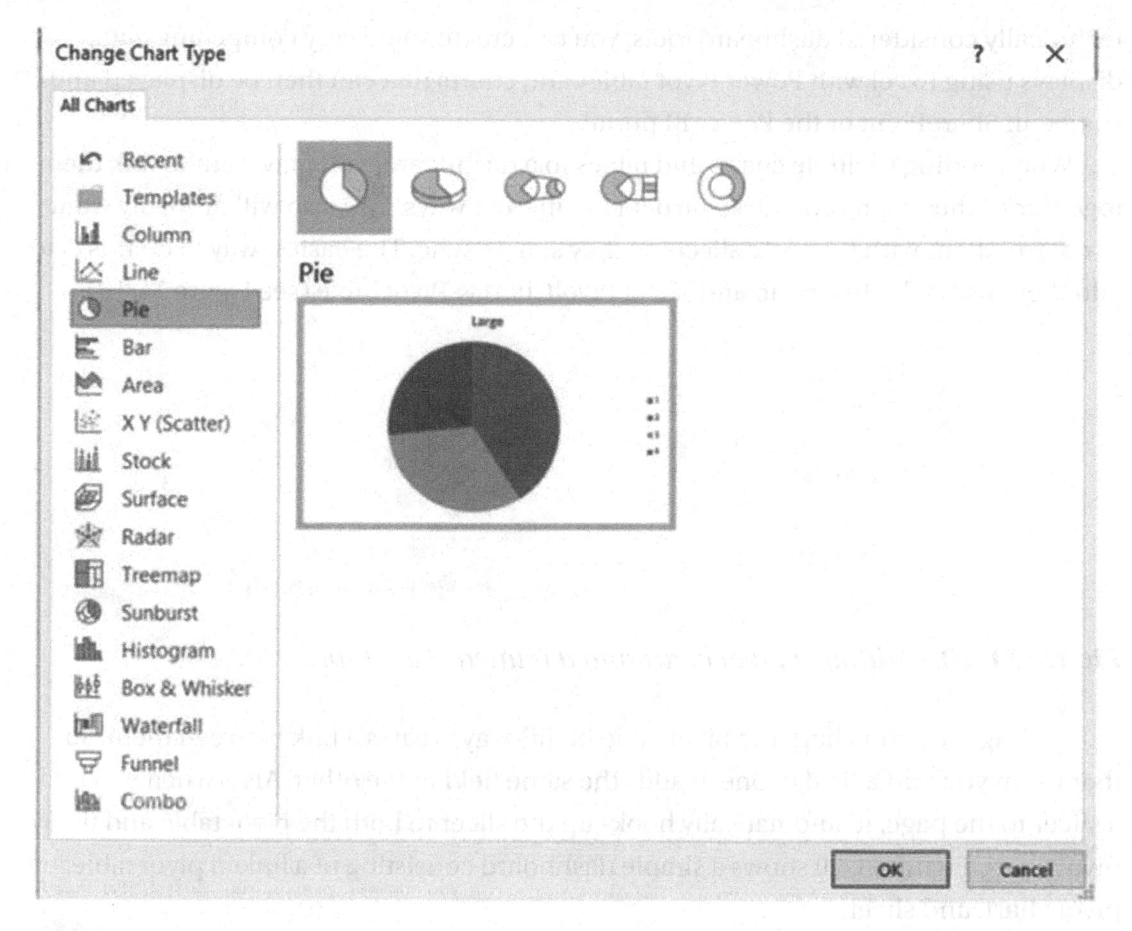

Figure 11-18. *Selecting a chart type*

Along with changing the chart type, the PivotChart Tools tabs offer you a vast assortment of design and formatting options you can use to customize the look and feel of your charts and graphs. It is worth your time to play around with the tools and create various graphs to gain a better idea of the various options available.

Using Multiple Charts and Tables

When creating pivot tables and charts to display and make sense of the data, you often want to create a dashboard that easily allows you to determine performance. You may be interested in sales performance, network performance, or assembly-line performance. Dashboards combine visual representations, such as key performance indicators (KPIs), graphs, and charts, into one holistic view of the process. Although they are not technically considered dashboard tools, you can create some very compelling data displays using Excel with Power Pivot tables and charts that can then be displayed and shared in SharePoint or the Power BI portal.

When adding multiple charts and tables to a dashboard, you may want to link them together so they represent the same data in different ways. You also will probably want to control them with the same slicers so they stay in sync. The easiest way to do this is to add them using the Insert tab and select PivotChart & PivotTable (see Figure 11-19).

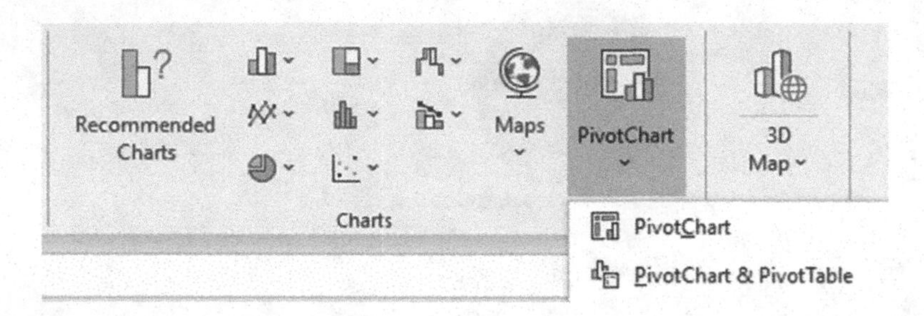

***Figure 11-19.** Adding a pivot chart and a related pivot table*

Adding the pivot chart and pivot table in this way creates a link between them, so that when you add a field to one, it adds the same field to the other. Also, when you add a slicer to the page, it automatically hooks up the slicer to both the pivot table and the pivot chart. Figure 11-20 shows a simple dashboard consisting of a linked pivot table, pivot chart, and slicer.

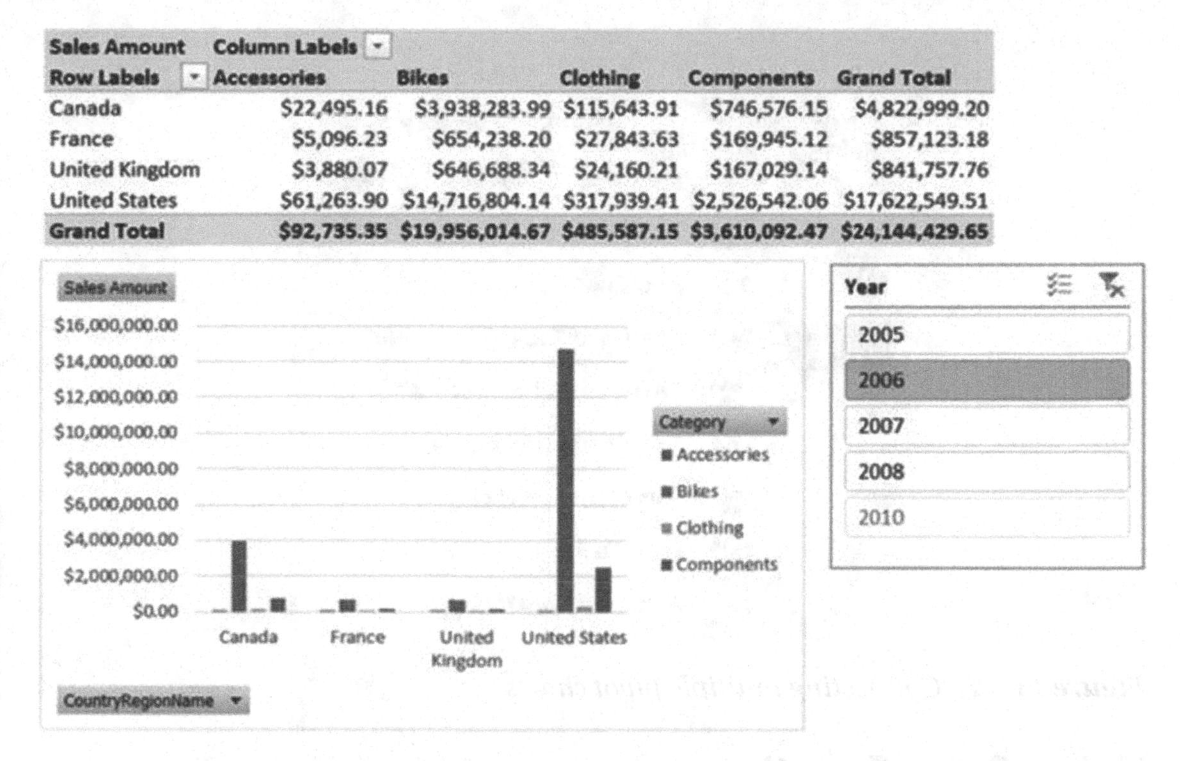

Sales Amount	Column Labels				
Row Labels	Accessories	Bikes	Clothing	Components	Grand Total
Canada	$22,495.16	$3,938,283.99	$115,643.91	$746,576.15	$4,822,999.20
France	$5,096.23	$654,238.20	$27,843.63	$169,945.12	$857,123.18
United Kingdom	$3,880.07	$646,688.34	$24,160.21	$167,029.14	$841,757.76
United States	$61,263.90	$14,716,804.14	$317,939.41	$2,526,542.06	$17,622,549.51
Grand Total	$92,735.35	$19,956,014.67	$485,587.15	$3,610,092.47	$24,144,429.65

Figure 11-20. Creating a simple dashboard in Excel

If you need to connect multiple charts together, open the Model Designer by selecting the Manage button on the PowerPivot tab. Select the PivotTable drop-down on the Home tab (see Figure 11-21). This also allows you to insert a flattened pivot table, which is useful for printing.

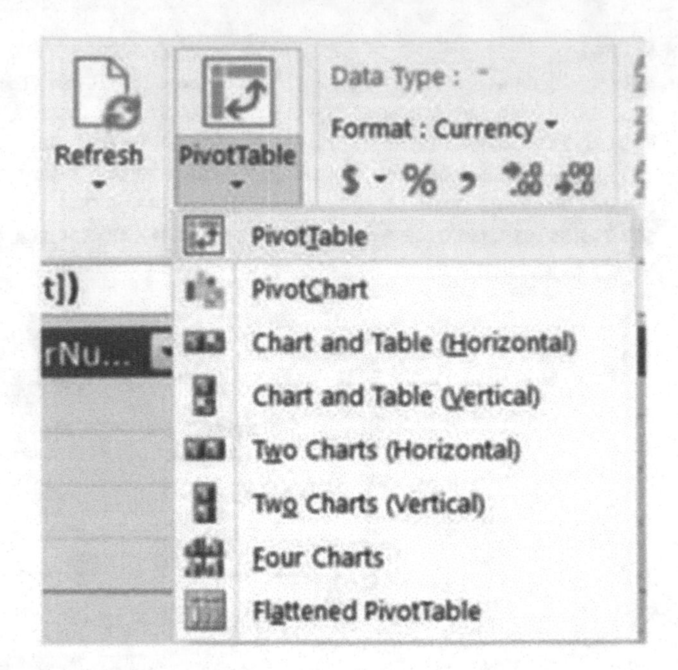

Figure 11-21. *Connecting multiple pivot charts*

Using Cube Functions

Using the built-in PivotChart and PivotTable layouts in Excel allows you to create compelling dashboards and provide great interfaces for browsing the data. There are times, though, when you may find yourself frustrated with some of the limitations inherent with these structures. For example, you can't insert your own columns inside the pivot table to create a custom calculation. You may also want to display the data in a non-tabular format for a customized report. This is where the Excel cube functions are useful.

The Excel cube functions allow you to connect directly to the Power Pivot data model without needing to use a pivot table. The cube functions are Excel functions (as opposed to DAX functions) and can be found on the Excel Formulas tab under the More Functions drop-down (see Figure 11-22).

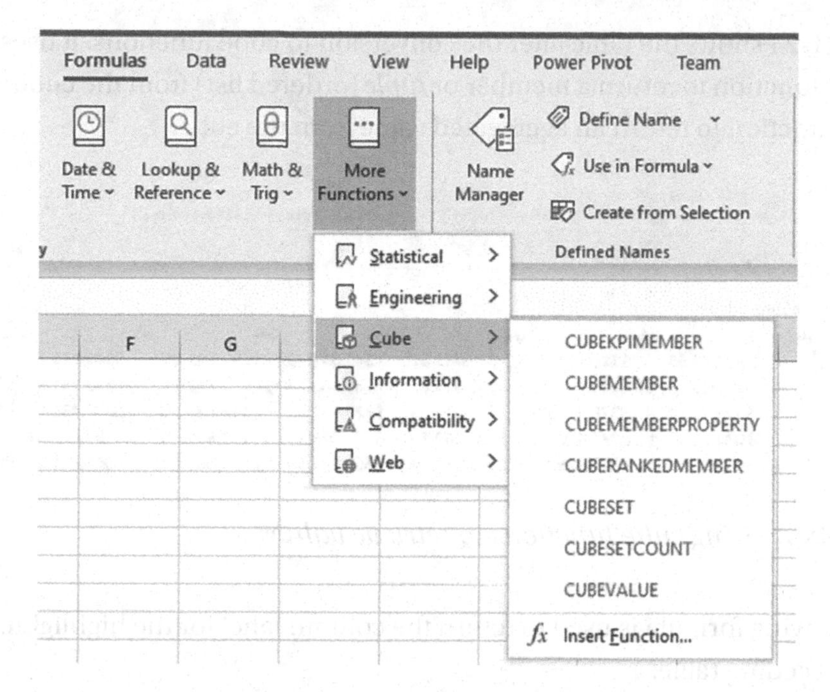

Figure 11-22. *Using cube functions to connect to the model*

The easiest way to see how these functions are used is to create a pivot table. On the PivotTable Analyze tab, click the OLAP Tools drop-down and select the Convert to Formulas option (see Figure 11-23).

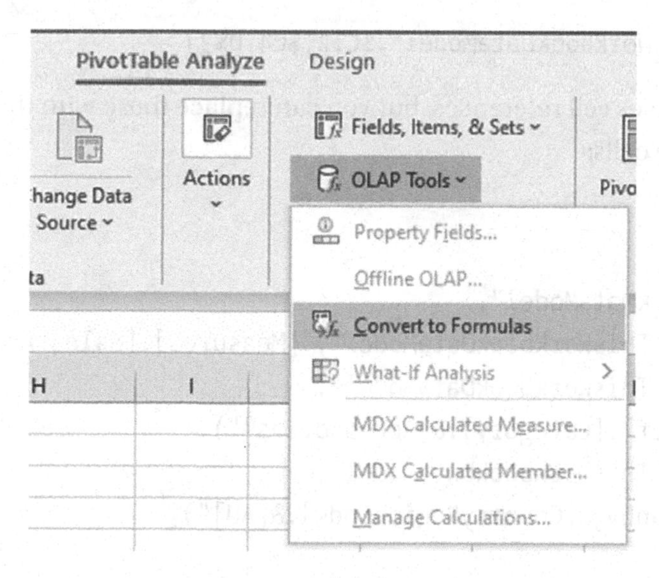

Figure 11-23. *Converting from a pivot table to cube functions*

Figure 11-24 shows the table after the conversion to cube functions. It uses the CUBEMEMBER function to return a member or *tuple* (ordered list) from the cube and the CUBEVALUE function to return an aggregated value from the cube.

=CUBEMEMBER("ThisWorkbookDataModel","[Product].[Category].&[Accessories]")

C	D	E	F	G	H	I	J
Sales Amount	Column Labels						
Row Labels	AU	CA	DE	FR	GB	US	Grand Total
Accessories	$23,947.53	$118,127.35	$35,083.07	$48,031.73	$42,593.03	$303,515.23	$571,297.93
Bikes	$1,323,820.73	$11,636,380.59	$1,543,015.65	$3,560,665.65	$3,405,747.21	$44,832,751.73	$66,302,381.56
Clothing	$42,915.80	$378,947.63	$71,619.43	$128,092.22	$118,828.80	$1,037,436.95	$1,777,840.84
Components	$203,651.31	$2,244,470.02	$334,269.89	$870,748.34	$711,839.79	$7,434,097.31	$11,799,076.66
Grand Total	$1,594,335.38	$14,377,925.60	$1,983,988.04	$4,607,537.94	$4,279,008.83	$53,607,801.21	$80,450,596.98

Figure 11-24. *Using cube functions to retrieve values*

The following formula is used to return the column label for the highlighted cell C4 in the preceding table:

```
=CUBEMEMBER("ThisWorkbookDataModel","[Product].[Category].&[Accessories]")
```

The first parameter is the name of the connection to the data model, and the second parameter is the member expression.

The value in cell D4 is returned using the following formula:

```
=CUBEVALUE("ThisWorkbookDataModel",$C$2,$C4,D$3)
```

This formula uses cell references, but you can replace these with the cube functions contained in these cells:

```
=CUBEVALUE
(
    "ThisWorkbookDataModel",
    CUBEMEMBER("ThisWorkbookDataModel","[Measures].[Sales Amount]"),
    CUBEMEMBER("ThisWorkbookDataModel",
        "[Product].[Category].&[Accessories]"),
    CUBEMEMBER("ThisWorkbookDataModel",
        "[Geography].[CountryRegionCode].&[AU]")
)
```

Note that one of the parameters is the CUBEMEMBER function returning the aggregation you want the value of; the other parameters use the CUBEMEMBER to define the portion of the cube used in the aggregation.

You can now rearrange the values and labels to achieve the layout and formatting that you need (see Figure 11-25).

Sales	US	CA	GB
Total	$53,607,801	$14,377,926	$4,279,009
Bikes	$44,832,752	$11,636,381	$3,405,747
Components	$7,434,097	$2,244,470	$711,840
Total	$52,266,849	$13,880,851	$4,117,587
Others			
Accessories	$303,515	$118,127	$42,593
Clothing	$1,037,437	$378,948	$118,829

***Figure 11-25.** Rearranging the data*

HANDS-ON LAB: CREATING THE BI INTERFACE IN EXCEL

In the following lab, you will

- Add conditional formatting to a pivot table
- Create a chart to help analyze data
- Link together pivot tables and pivot charts
- Use cube functions to display model data

1. In the LabStarterFiles\Chapter11Lab1 folder, open the Chapter11Lab1.xlsx file. This file contains inventory and sales data from the test AdventureWorksDW database.

2. In Excel Sheet1, insert a Power Pivot table using the PivotTable button on the Insert tab. Make sure the Use this workbook's Data Model is selected. If you don't see the PivotTable button, you may need to click the Table button first.

3. Add TotalQuantity, TotalSales, and TotalMargin from the ResellerSales table to the Values drop area in the Field List window. Add the Calendar hierarchy to the Rows drop area.
4. Note that some of the margins are negative. To bring attention to the negative values, you are going to format them in red.
5. Select a TotalMargin cell in the pivot table. On the Home tab, click the Conditional Formatting drop-down. Select the Highlight Cells, Less Than option. Format the cells that are less than 0 with red text (see Figure 11-26).

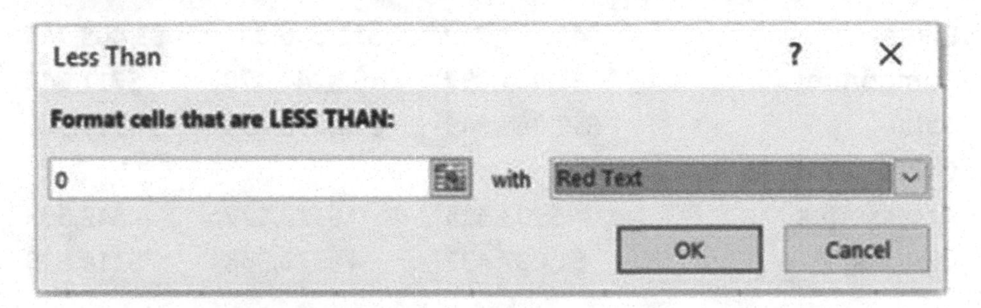

Figure 11-26. Setting conditional formatting

6. When finished, you should see an icon next to the selected cell. Click it and select the All Cells Showing TotalMargin Values option.
7. Select one of the TotalSales cells. This time select the Data Bars under the Conditional Formatting drop-down. Select one of the gradient styles. After selecting the data bar, you should see a small icon next to the values. Click it and select the All Cells Showing TotalSales Values.
8. Your pivot table should look like Figure 11-27. Expand the years and note that the formatting shows up for the quarters, months, and years.

Row Labels	TotalQuantity	TotalSales	TotalMargin
⊞2005	10835	$8,065,435.31	$328,927.08
⊞2006	58241	$24,144,429.65	$323,401.79
⊟2007	100172	$32,202,669.43	($168,557.73)
⊞1	12307	$5,266,343.51	$193,324.41
⊞2	19466	$6,733,903.82	$298,556.44
⊟3	39784	$10,926,196.09	($710,603.34)
July	9871	$2,665,650.54	($181,857.11)
August	15139	$4,212,971.51	($250,821.21)
September	14774	$4,047,574.04	($277,925.03)
⊞4	28615	$9,276,226.01	$50,164.75
⊞2008	45130	$16,038,062.60	($13,288.53)
Grand Total	214378	$80,450,596.98	$470,482.60

***Figure 11-27.** Adding conditional formatting to the pivot table*

9. Open Sheet2 and on the Insert tab, and select the PivotChart drop-down. From the drop-down, choose the pivot chart. Insert the chart on the current sheet using the Use this workbook's Data Model option.

10. In the Field List window, add the TotalSales from the ResellerSales table to the Values drop area and add the CountryRegionName from the Geography table to the Axis Fields drop area. You should see a column chart showing sales by country (see Figure 11-28).

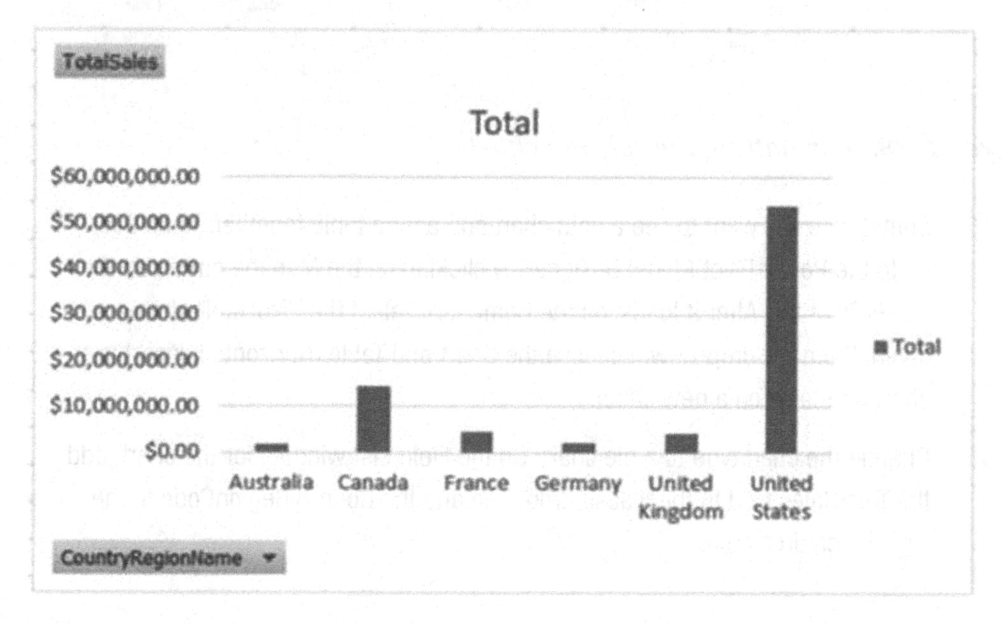

***Figure 11-28.** Adding a column chart*

11. Select the column chart. In the Design tab, you can change the chart colors, layout, and chart type.

12. The Format tab lets you format the shapes and text in the chart. The Analyze tab lets you show/hide the field buttons.

13. Rename the title of the chart to *Sales By Country*. Hide the field buttons and delete the legend.

14. Right-click the vertical axis and select Format Axis. Under Axis Options, change the Display Units to Millions. Under the Number node, change the Category to Currency with zero decimal places. Your chart should look like Figure 11-29.

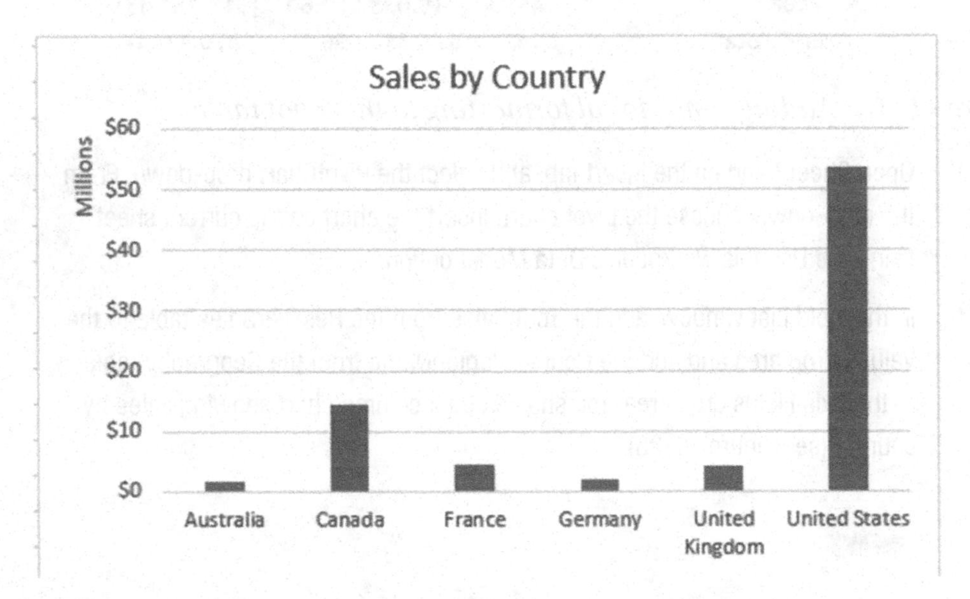

Figure 11-29. *Formatting the column chart*

15. Sometimes you want to see a data chart and a data table together. To do this, go to the Power Pivot Model Designer by clicking on the Manage button on the PowerPivot tab. After it loads, on the Home tab, select the PivotTable drop-down. From the drop-down, choose the Chart and Table (Horizontal). Insert the chart and table on a new sheet.

16. Change the chart type to a pie chart. On the Field List window for the chart, add the TotalSales field to the Values drop area and the CountryRegionCode to the Axis Fields drop area.

17. Change the title of the chart to Sales By Country and remove the field button.
18. Change the Chart Layout on the Design tab to show values as % of total (see Figure 11-30).

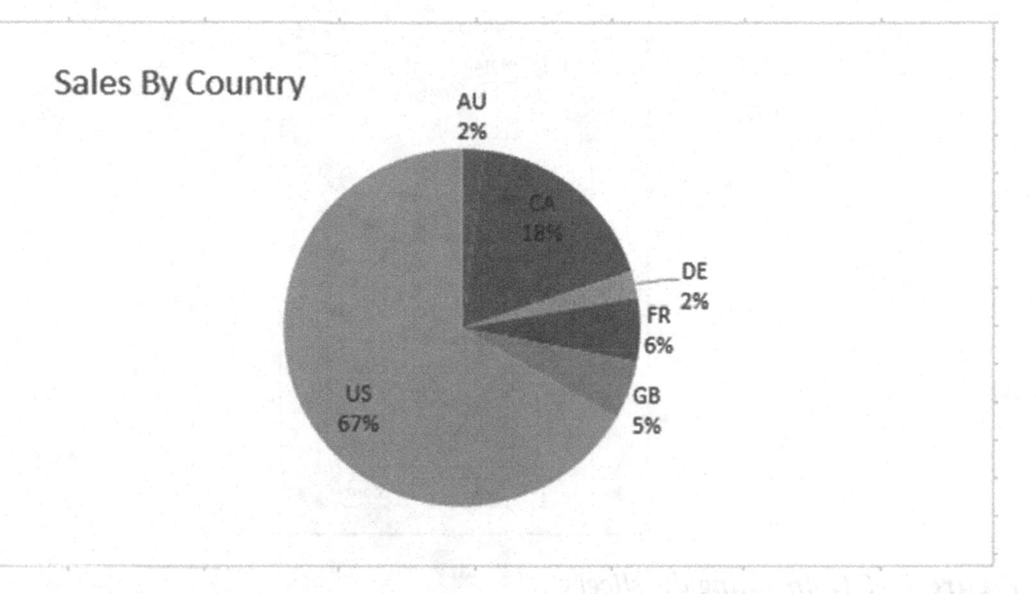

Figure 11-30. Creating the sales pie chart

19. Click the pivot table to bring up its Field List window. Add the TotalSales field from the ResellerSales table to the Values drop area. Add the Subcategory field from the Product table to the Rows drop area.
20. With the pie chart selected, insert a slicer for the product category and one for the year (see Figure 11-31). Verify that the slicers filter the pie chart but not the pivot table.

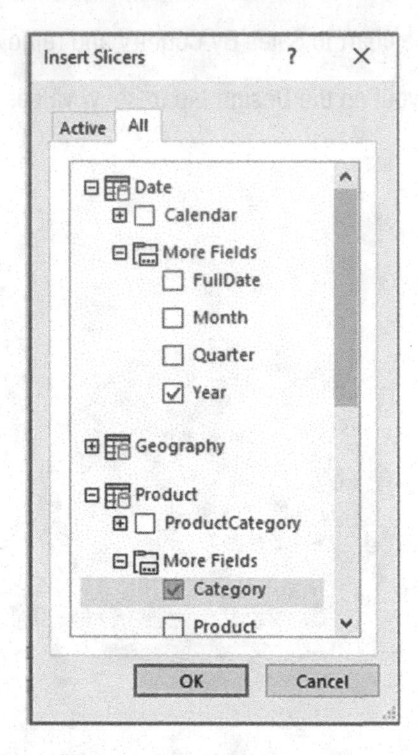

Figure 11-31. *Inserting the slicers*

21. Select the Year slicer and on the Slicer tab, click the Report Connections button. Add the pivot table located on the same sheet as the pie chart. Repeat this for the category slicer.

22. Verify that the slicers filter both the chart and the pivot table. Your dashboard should look like Figure 11-32.

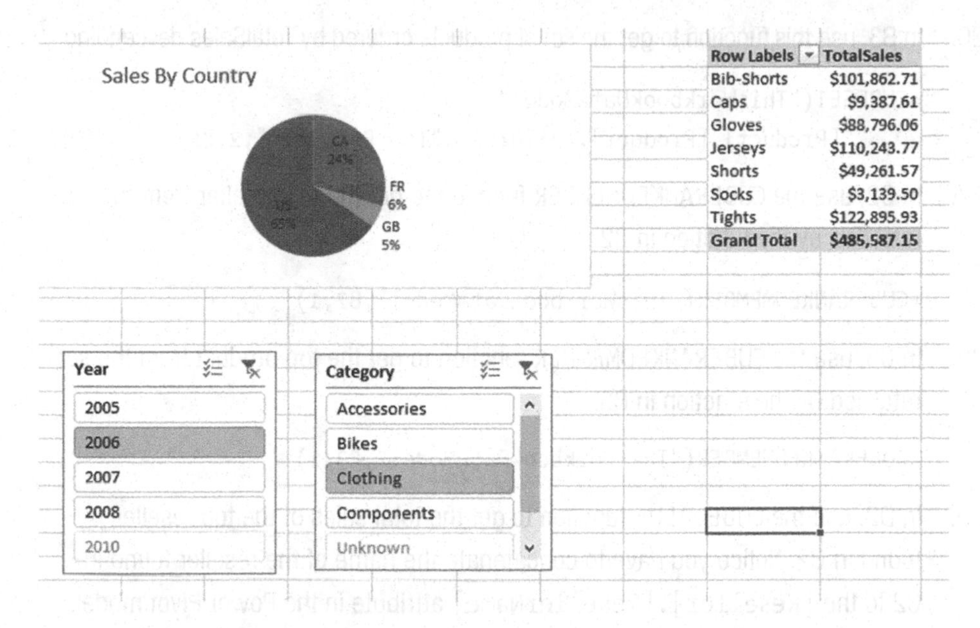

***Figure 11-32.** Creating a dashboard*

23. Add a new sheet to the workbook. You are going to use cube functions to calculate the top product and top reseller in total sales. The result should look like Figure 11-33.

	A	B	C	D
1		TotalSales		
2		Top Reseller	Brakes and Gears	$877,107.19
3		Top Products	Mountain-200 Black, 38	$3,105,726.66

***Figure 11-33.** Using cube functions*

24. Add the following code in B1 to select the TotalSales measure:

```
=CUBEMEMBER("ThisWorkbookDataModel","[Measures].[TotalSales]")
```

25. In B2, use the following function to get the set of resellers ordered by TotalSales descending—you will use this set to select the top reseller:

```
=CUBESET("ThisWorkbookDataModel",
    "[Reseller].[ResellerName].children","Top Reseller",2,B1)
```

26. In B3, use this function to get the set of products ordered by TotalSales descending:

```
=CUBESET("ThisWorkbookDataModel",
     "[Product].[Product].children","Top Products",2,B1)
```

27. In C2, use the CUBERANKEDMEMBER function to get the top reseller from the set returned by the function in B2:

```
=CUBERANKEDMEMBER("ThisWorkbookDataModel",B2,1)
```

28. In C3, use the CUBERANKEDMEMBER function to get the top product from the set returned by the function in B3:

```
=CUBERANKEDMEMBER("ThisWorkbookDataModel",B3,1)
```

29. In D2, use the CUBEVALUE function to get the total sales of the top reseller found in C2. Notice you have to concatenate the name of the reseller found in C2 to the [Reseller].[ResellerName] attribute in the Power Pivot model:

```
=CUBEVALUE("ThisWorkbookDataModel",
    "[Reseller].[ResellerName].&["& C2 &"]","[Measures].[TotalSales]")
```

30. Using a similar CUBEVALUE function, you can get the total sales value for the top product in cell D3:

```
=CUBEVALUE("ThisWorkbookDataModel",
      "[Product].[Product].&[" & C3 &"]","[Measures].[TotalSales]")
```

31. You can verify the results by building a pivot table.

Summary

As you saw in this chapter, Excel is a feature-rich environment for creating dashboards using pivot tables and pivot charts. At this point, you should feel comfortable creating pivot tables and pivot charts using your workbook model as a data source. You also used Excel cube functions to query the model directly without needing to use a pivot table. Although this chapter covered the basics to get you started, you have a lot more to learn about Excel and how it can help you analyze your data. I encourage you to dig deeper into these features.

In the next chapter, you will look at several realistic case studies to solidify the concepts of the previous chapters. By working through these case studies, you will be able to gauge which areas you have mastered and which you need to study further.

CHAPTER 12

Creating a Complete Solution

So far in this book, you've gained experience working with each of the pieces of Microsoft's self-service BI toolset. You've used Power Query, Power BI Desktop, Power Pivot, and Excel. This chapter provides you with several use cases to solidify the concepts of the previous chapters. By working through these use cases, you will gauge which areas you have mastered and which you need to spend more time studying. Because this is sort of like your final exam, I have deliberately not included step-by-step instructions as I did for earlier exercises. Instead, I have given you general directions that should be sufficient to get you started. If you get stuck, refer back to the previous chapters to remind yourself of how to accomplish the task.

This chapter contains the following use cases:

- Sales quota analysis
- Reseller sales analysis
- Sensor analysis

Use Case 1: Sales Quota Analysis

For this scenario, you work for a bike equipment company and have been asked to analyze the sales data. Using Power BI Desktop, you need to create a dashboard that allows the sales manager to track the performance of the sales team. You will compare actual sales to the sales quotas for the sales team.

D. Clark, *Beginning Microsoft Power BI*, https://doi.org/10.1007/978-1-4842-5620-6_12

Load the Data

Create a new Power BI Desktop file named SalesRepAnalysis.pbix. In the Chapter12Labs folder is a folder named UseCase1; this folder contains a SalesRepAnalysis.accdb Access database with sales data extracted from the sales data warehouse. Using this as a data source, select the Employee, SalesTerritory, and ResellerSales table. Then select the Transform Data button to launch Power Query (see Figure 12-1).

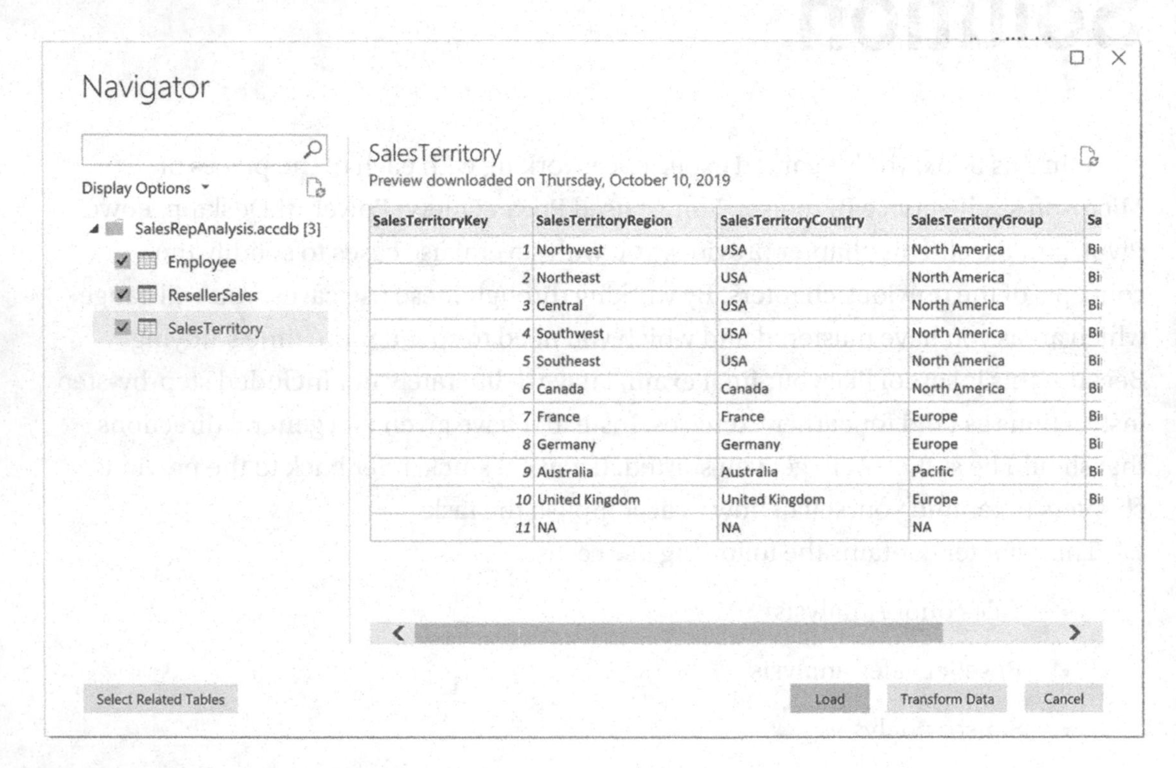

***Figure 12-1.** Selecting the tables*

Using Power Query, select the columns listed in Table 12-1.

Table 12-1. Columns to Import

Source Table	Query Name	Fields
Employee	Rep	EmployeeKey, EmployeeNationalID, FirstName, LastName, MiddleName, Title
SalesTerritory	Territory	SalesTerritoryKey, SalesTerritoryRegion, SalesTerritoryCountry, SalesTerritoryGroup
ResellerSales	Sales	OrderDateKey, EmployeeKey, SalesTerritoryKey, SalesAmount

Filter the Employee table to only include salespeople. Add a Name Column that combines the FirstName and Last Name columns. In the Sales table, use the OrderDateKey to create an OrderDate column. Change OrderDateKey's data type from whole number to text. Because the OrderDateKey is in the format YYYYMMDD, you can now use the `Date.FromText()` function to create an OrderDate column. Make sure the new column is a date format type. Once you have the OrderDate column, you can use this to create a Year and a Quarter column. Then group the SalesAmount by EmployeeKey, SalesTerritoryKey, OrderYear, and Quarter. Add a leading Q to the quarter numbers. Create a TimePeriod column that concatenates the Year and Quarter columns. After creating the TimePeriod column, you can delete the Year and Quarter columns. The result should look like Figure 12-2.

	EmployeeKey	SalesTerritoryKey	SalesAmount	TimePeriod
1	285	5	2024.994	2013Q3
2	285	5	6074.982	2013Q3
3	285	5	2024.994	2013Q3
4	285	5	2039.994	2013Q3
5	285	5	2039.994	2013Q3
6	285	5	4079.988	2013Q3
7	285	5	2039.994	2013Q3
8	285	5	86.5212	2013Q3
9	285	5	28.8404	2013Q3
10	285	5	34.2	2013Q3
11	285	5	10.373	2013Q3
12	285	5	80.746	2013Q3
13	285	5	419.4589	2013Q3
14	285	5	874.794	2013Q3
15	288	6	809.76	2013Q3
16	288	6	714.7043	2013Q3
17	288	6	1429.4086	2013Q3
18	288	6	20.746	2013Q3

***Figure 12-2.** The final Sales query data*

The sales quotas are kept in an Excel workbook called SalesQuotas.xlsx in the UseCase1 folder. Import the data using the Excel source and select the sheets for each year. Once the data is imported into Power Query using the Transform Data button, you should see a query for each year. You need to add the year column to each query. Now you can combine the queries by appending them into a new query called `Quotas`. Rearrange the column order to SalesPerson, Year, Q1, Q2, Q3, and Q4. Select the SalesPerson and Year columns. Unpivot the other columns. Rename the Value column to Quota. Add a TimePeriod column that concatenates the year and quarter. After adding the TimePeriod, remove the Year and Quarter columns. The result should look like Figure 12-3.

	SalesPerson	TimePeriod	Quota
1	David Campbell	2013Q3	180800
2	David Campbell	2013Q4	282400
3	Garrett Vargas	2013Q3	195200
4	Garrett Vargas	2013Q4	284800
5	Jillian Carson	2013Q3	452000
6	Jillian Carson	2013Q4	697600
7	José Saraiva	2013Q3	420000
8	José Saraiva	2013Q4	613600
9	Linda Mitchell	2013Q3	509600
10	Linda Mitchell	2013Q4	624800
11	Michael Blythe	2013Q3	293600
12	Michael Blythe	2013Q4	444800
13	Pamela Ansman-Wolfe	2013Q3	132000
14	Pamela Ansman-Wolfe	2013Q4	375200

***Figure 12-3.** The final Quotas query data*

Right-click each of the year queries and uncheck the Enable load option (see Figure 12-4).

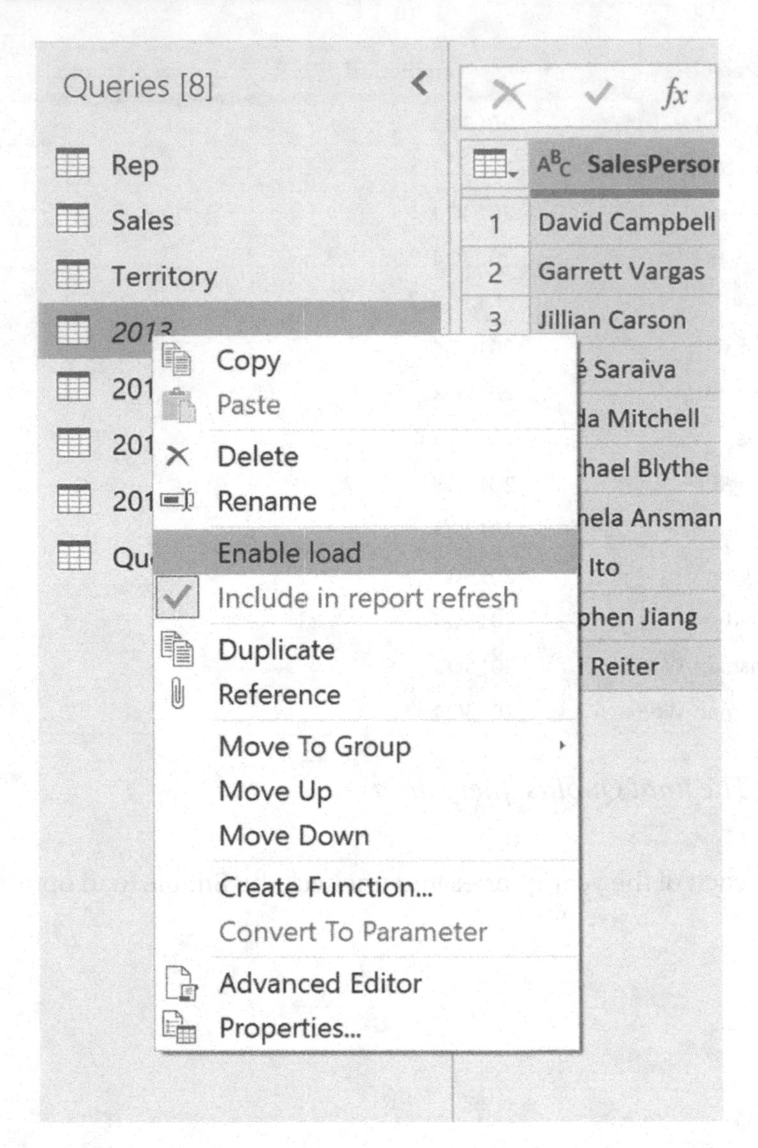

Figure 12-4. *Disabling the load option*

To create the appropriate relationships between the tables, you need to create a Period table that contains the calendar year and quarter. To get this table, right-click the `Quotas` query and select Duplicate. Rename the query to `Period`. Delete the step that removed the Year and Quarter columns. Remove the SalesPerson and Quota columns. With the TimePeriod column selected, click the Remove Rows drop-down button on the Home tab and choose Remove Duplicates. Finally, reorder the columns and order the rows by TimePeriod. The result should look like Figure 12-5.

	TimePeriod	Year	Quarter
1	2013Q3	2013	Q3
2	2013Q4	2013	Q4
3	2014Q1	2014	Q1
4	2014Q2	2014	Q2
5	2014Q3	2014	Q3
6	2014Q4	2014	Q4
7	2015Q1	2015	Q1
8	2015Q2	2015	Q2
9	2015Q3	2015	Q3
10	2015Q4	2015	Q4
11	2016Q1	2016	Q1
12	2016Q2	2016	Q2

Figure 12-5. *The results of the* `Period` *query*

Click Close & Apply to load the data into the model. You are now ready to create the data model.

Create the Model

After importing the tables, rename the table columns and show or hide the columns from client tools according to the information listed in Table 12-2.

Table 12-2. *Table and Column Property Settings*

Table	Column	New Name	Hide in Report view
Rep	EmployeeKey	—	X
	EmployeeNationalID	Employee ID	
	FirstName	First Name	
	LastName	Last Name	
	MiddleName	Middle Name	
	Name	Sales Person	

(continued)

Table 12-2. *(continued)*

Table	Column	New Name	Hide in Report view
Territory	SalesTerritoryKey	—	X
	SalesTerritoryRegion	Region	
	SalesTerritoryCountry	Country	
	SalesTerritoryGroup	Group	
Quota	TimePeriod	—	
Sales	SalesPerson	Salesperson	
	Quota	—	
	EmployeeKey	—	X
	SalesTerritoryKey	—	X
	TimePeriod	—	X
	SalesAmount	Sales Amount	

After renaming and hiding the key fields, format the Sales Amount and quota columns as currency.

The next step is to verify the table relationships in the model. Figure 12-6 shows what the model should look like.

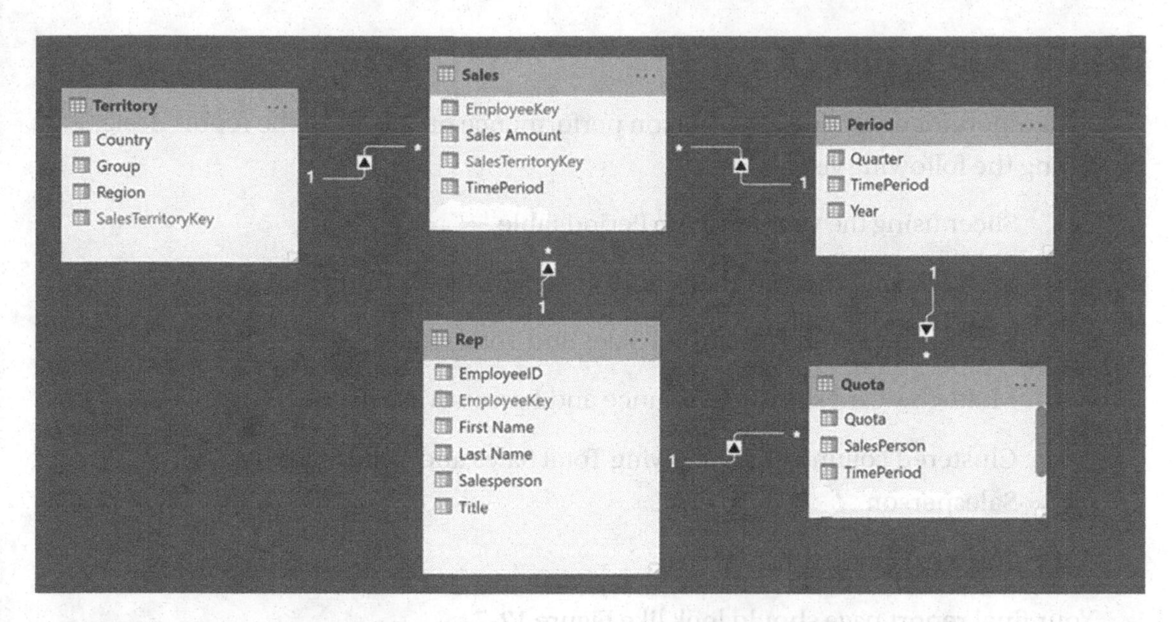

Figure 12-6. *The Sales Rep Analysis model with relationships defined*

You are now ready to add some measures to the model.

Create Measures

Add the following measure to the Sales table:

```
Total Sales = Sum(Sales[Sales Amount])
```

Add the following measures to the Quota table:

```
Total Quota = Sum(Quota[Quota])

Variance = Sales[Total Sales] -Quota[Total Quota]

Percent Variance = DIVIDE(Quota[Variance],Quota[Total Quota],BLANK())
```

Make sure you format the measures appropriately.

Create the Report

Create a report for analyzing sales person performance on Page 1 of the report designer by adding the following visuals:

- Slicer using the year from the Period table
- Stacked column chart that shows Total Sales by Country and Quarter
- Multi-row card showing Total Sales and Total Quota
- Multi-row card showing Variance and Percent Variance
- Clustered column chart showing Total Sales and Total Sales by Salesperson
- Text box for the title

Your final report page should look like Figure 12-7.

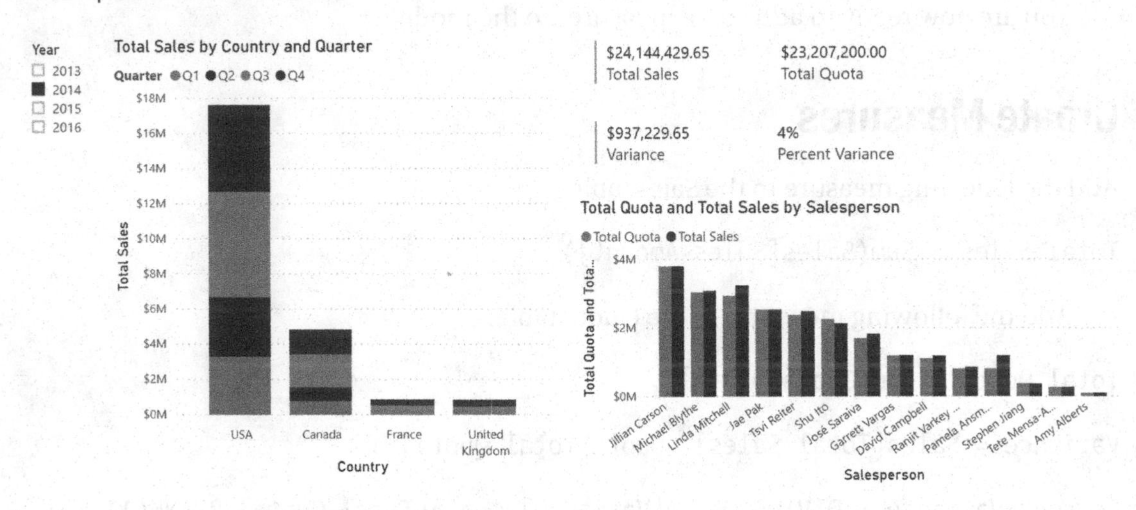

Figure 12-7. *The final Salesperson Performance report*

Add another page to the report and add the following visuals to the page:

- Slicer with the Year field from the Period table
- Slicer with the Salesperson field from the Rep table
- Two tables with the Salesperson, Total Sales, Total Quota, and Percent Variance fields
- A line chart showing Total Sales and Total Quota by TimePeriod

Filter the tables so that one shows the top five Salesperson by Percent Variance and the other shows the bottom five Salespersons by Percent Variance. Edit the visual interactions so that the Year slicer affects the tables but not the line chart. The Salesperson slicer should affect the line chart but not the tables. The final report page should look like Figure 12-8.

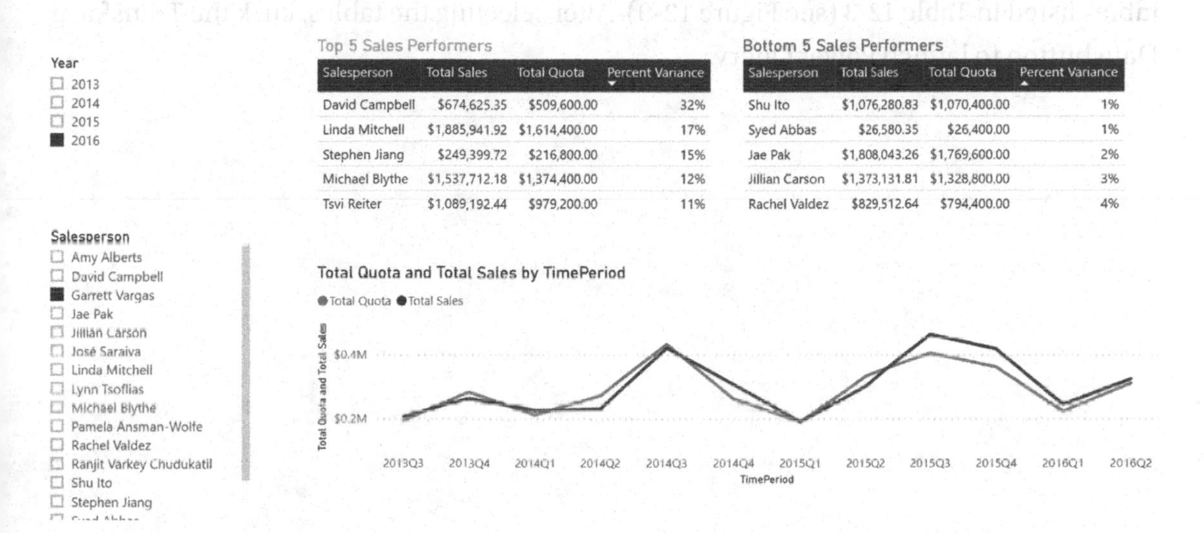

Figure 12-8. *The final report, Page 2*

After completing and experimenting with the report, close and save the desktop file.

Use Case 2: Reseller Sales Analysis

In this scenario, you work for a bike equipment company and have been asked to analyze the sales data. You need to compare monthly store sales. An added constraint is that you only want to compare resellers who have been open for at least a year when you make the comparison.

Load the Data

Create a new Excel workbook named StoreSalesAnalysis.xlsx. In the Chapter12Labs folder, find the UseCase2 folder, which contains a StoreSales.accdb Access database. This database houses the sales data you need to analyze. On the Data tab, select the Get Data drop-down to connect to the Access database. In the navigator pane, select the source tables listed in Table 12-3 (see Figure 12-9). After selecting the tables, click the Transform Data button to launch Power Query.

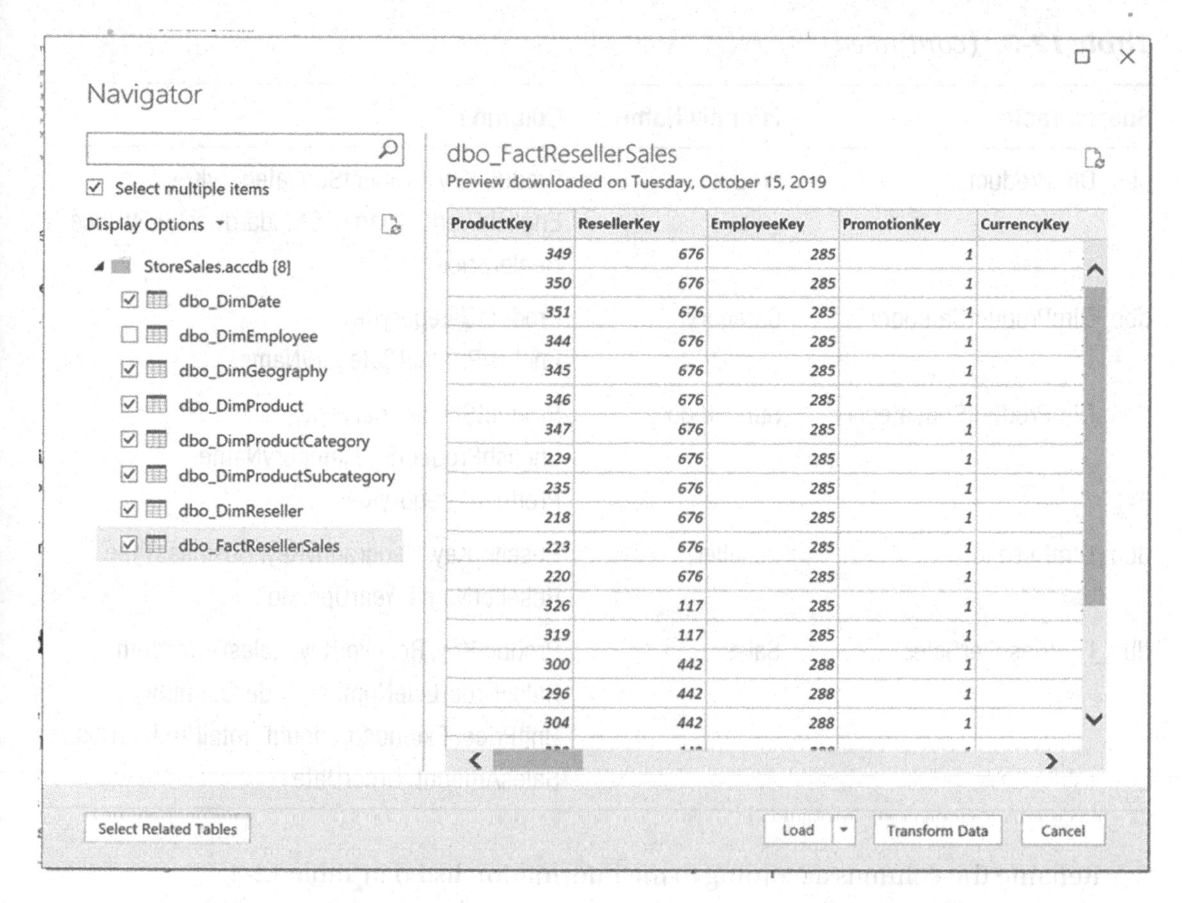

Figure 12-9. *Selecting the source tables*

Using Power Query, rename the tables and select columns listed in Table 12-3.

Table 12-3. *Store Sales Tables and Columns to Import*

Source Table	Friendly Name	Columns
dbo_DimDate	Date	FullDate, MonthName, MonthNumberOfYear, CalendarQuarter, CalendarYear
dbo_DimGeography	Location	GeographyKey, City, StateProvinceCode, StateProvinceName, CountryRegionCode, EnglishCountryRegionName, PostalCode

(*continued*)

Table 12-3. (*continued*)

Source Table	Friendly Name	Columns
dbo_DimProduct	Product	ProductKey, ProductSubcategoryKey, EnglishProductName, StandardCost, ListPrice, DealerPrice
dbo_DimProductCategory	Category	ProductCategoryKey, EnglishProductCategoryName
dbo_DimProductSubcategory	Subcategory	ProductSubcategoryKey, EnglishProductSubcategoryName, ProductCategoryKey
dbo_DimReseller	Reseller	ResellerKey, GeographyKey, BusinessType, ResellerName, YearOpened
dbo_FactResellerSales	Sales	ProductKey, ResellerKey, SalesOrderNumber, SalesOrderLineNumber, OrderQuantity, UnitPrice, ExtendedAmount, TotalProductCost, SalesAmount, OrderDate

Rename the columns according to the information listed in Table 12-4.

Table 12-4. *Friendly Column Names*

Table	Column	New Name
Date	FullDate	Full Date
	MonthName	Month
	MonthNumberOfYear	Month Number
	CalendarQuarter	Calendar Quarter
	CalendarYear	Calendar Year

(*continued*)

Table 12-4. *(continued)*

Table	Column	New Name
Product	ProductKey	—
	ProductSubcategoryKey	—
	EnglishProductName	Product
	StandardCost	Standard Cost
	ListPrice	List Price
	DealerPrice	Dealer Price
Reseller	ResellerKey	—
	GeographyKey	—
	BusinessType	Business Type
	ResellerName	Reseller Name
	YearOpened	Year Opened
Location	GeographyKey	—
	StateProvinceCode	State Provlnce Code
	StateProvinceName	State Province
	CountryRegionCode	Country Code
	EnglishCountryRegionName	Country
	PostalCode	Postal Code
Sales	ProductKey	—
	ResellerKey	—
	SalesOrderNumber	Order Number
	SalesOrderLineNumber	Order Line Number
	OrderQuantity	Quantity
	UnitPrice	Unit Price
	ExtendedAmount	Extended Amount
	TotalProductCost	Product Cost
	SalesAmount	Sales Amount
	OrderDate	—

After renaming the tables, select Close & Load To... on the Home tab in Power Query. In the Import Data window, select Only Create Connection and check the Add this data to the Data Model checkbox (see Figure 12-10).

***Figure 12-10.** Importing the data*

Create the Model

After importing the tables, go to the PowerPivot ribbon and select the Manage button to open the data model. In the model designer, create the table relationships using the appropriate keys. Your model should now look like Figure 12-11.

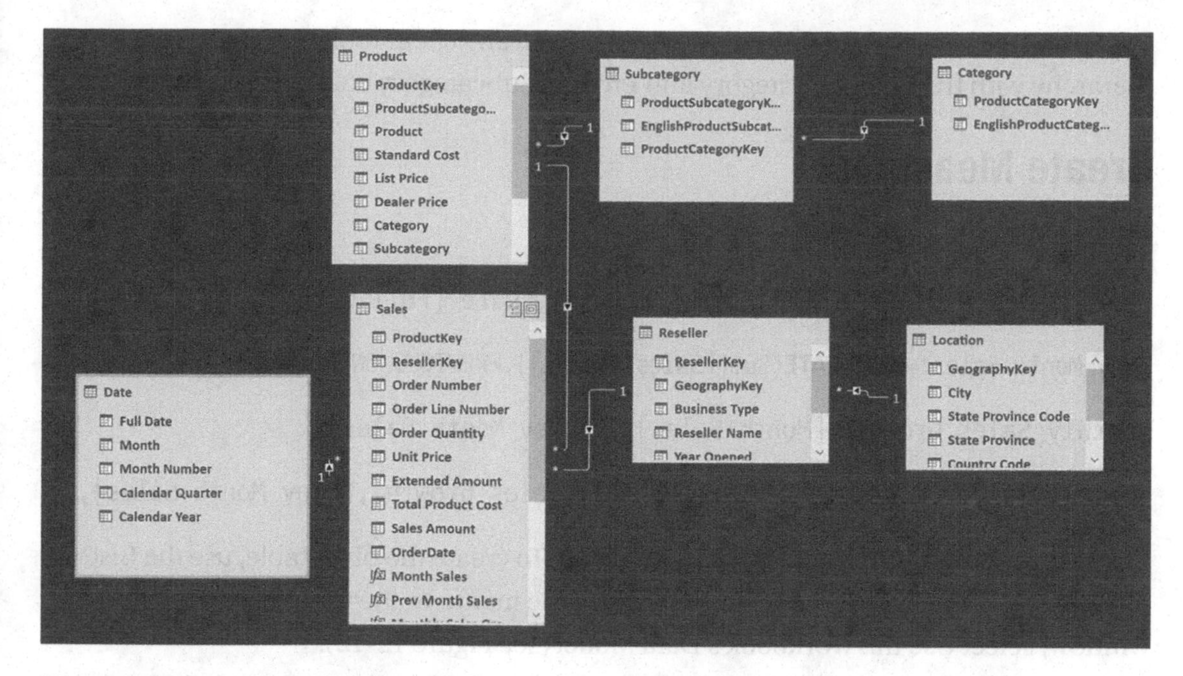

Figure 12-11. *Store sales model*

Right-click the Date table in the diagram and select create hierarchy. Create a calendar hierarchy of Year-Quarter-Month. Mark the table as the Date table with the Full Date column as the key.

Return to the Data view and sort the Month column by the Month Number column. Format the Full Date column to only show the date and not the time. Hide the key columns in the tables from the client tools.

Create Calculated Columns

Using the DAX `RELATED` function, create a calculated column in the Product table for Product Subcategory and Product Category. If the ProductSubcategoryKey is blank, fill in the Category and Subcategory columns with `"Misc"`.

```
[Category]
=IF(ISBLANK([ProductSubcategoryKey]),"Misc",
    RELATED(Category[EnglishProductCategoryName]))

[Subcategory]
=IF(ISBLANK([ProductSubcategoryKey]),"Misc",
    RELATED(Subcategory[EnglishProductSubcategoryName]))
```

Hide the Category and Subcategory tables from any client tools. Finally, create a hierarchy with the Product Category and Product Subcategory columns named Prd Cat.

Create Measures

Add the following measures to the Sales table:

```
Month Sales:=TOTALMTD(SUM([Sales Amount]),'Date'[Full Date])

Prev Month Sales:=CALCULATE(Sum([Sales Amount]),PREVIOUSMONTH('Date'[Full Date]))

Monthly Sales Growth:=[Month Sales] - [Prev Month Sales]

Monthly Sales Growth %:=DIVIDE([Monthly Sales Growth],[Prev Month Sales],0)
```

Test your measures by creating a pivot table. To create the pivot table, use the Insert tab and select the PivotTable on the left side of the menu. In the Create PivotTable window, select Use this workbook's Data Model (see Figure 12-12).

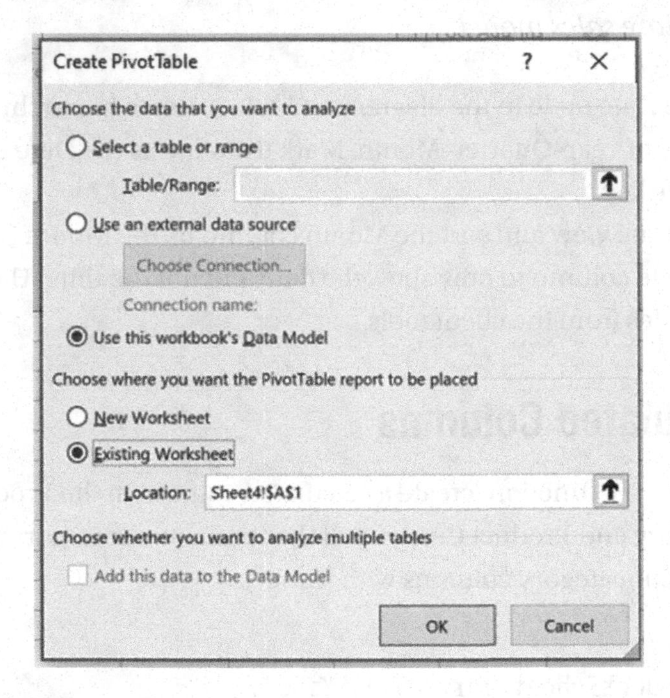

Figure 12-12. *Creating a PivotTable*

Select the PivotTable to see the field list (see Figure 12-13). If you don't see the field list, select the PivotTable Analyze tab and click the show field list option on the right side of the menu.

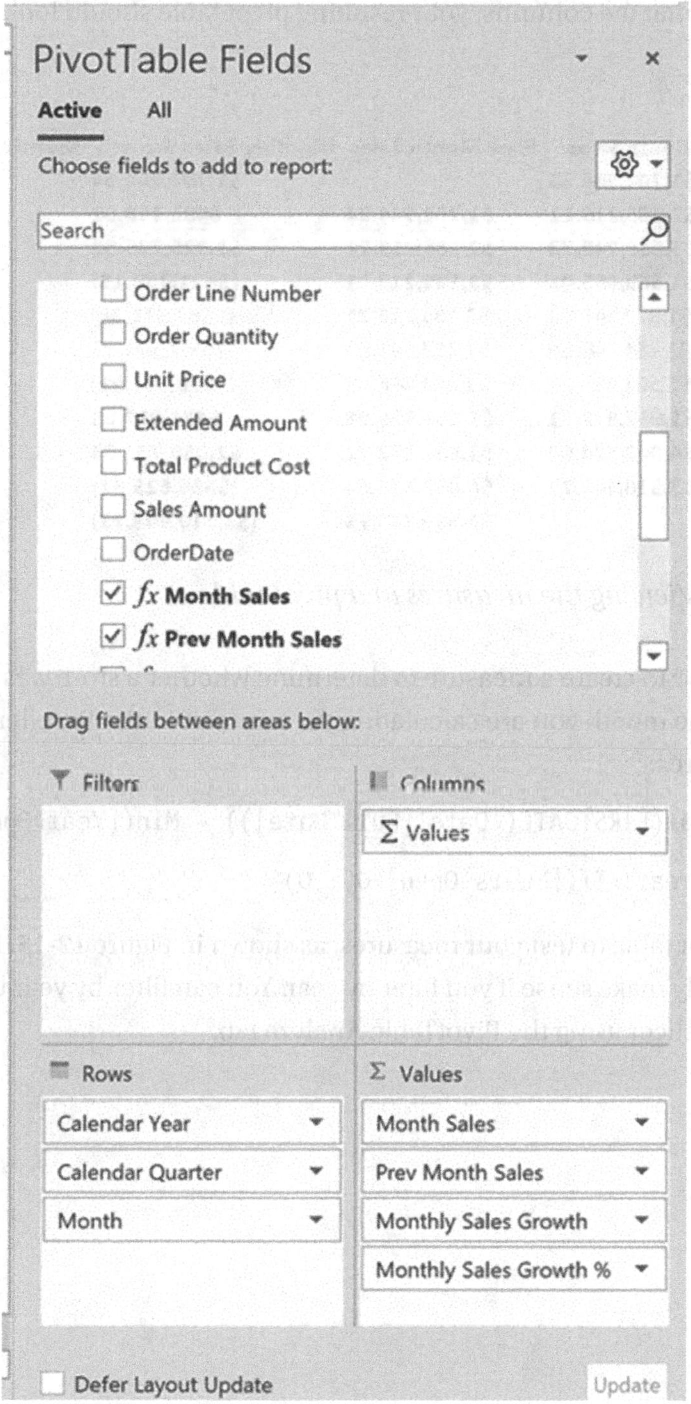

PivotTable Fields

Active All

Choose fields to add to report:

Search

- [ ] Order Line Number
- [ ] Order Quantity
- [ ] Unit Price
- [ ] Extended Amount
- [ ] Total Product Cost
- [ ] Sales Amount
- [ ] OrderDate
- [x] *fx* **Month Sales**
- [x] *fx* **Prev Month Sales**

Drag fields between areas below:

Filters	Columns
	Σ Values

Rows	Σ Values
Calendar Year	Month Sales
Calendar Quarter	Prev Month Sales
Month	Monthly Sales Growth
	Monthly Sales Growth %

Defer Layout Update Update

***Figure 12-13.** Adding fields to the PivotTable*

Once you format the columns, your resulting pivot table should look similar to Figure 12-14.

Row Labels	Month Sales	Prev Month Sales	Monthly Sales Growth	Monthly Sales Growth %
⊞2013	$1,702,944.54		$1,702,944.54	0.00 %
⊞2014	$2,185,213.21	$1,702,944.54	$482,268.67	28.32 %
⊟2015	$3,510,948.73	$2,185,213.21	$1,325,735.52	60.67 %
⊟1	$1,563,955.08	$2,185,213.21	($621,258.13)	-28.43 %
January	$1,317,541.83	$2,185,213.21	($867,671.38)	-39.71 %
February	$2,384,846.59	$1,317,541.83	$1,067,304.76	81.01 %
March	$1,563,955.08	$2,384,846.59	($820,891.51)	-34.42 %
⊞2	$1,987,872.71	$1,563,955.08	$423,917.63	27.11 %
⊞3	$4,047,574.04	$1,987,872.71	$2,059,701.33	103.61 %
⊞4	$3,510,948.73	$4,047,574.04	($536,625.31)	-13.26 %
⊞2016		$3,510,948.73	($3,510,948.73)	-100.00 %

Figure 12-14. *Viewing the measures in a pivot table*

Now you need to create a measure to determine whether a store was open for at least a year for the month you are calculating the sales for. In the Reseller table, add the following measures:

```
Years Open:=Year(FIRSTDATE('Date'[Full Date])) - Min([Year Opened])
```

```
Was Open Prev Year:=If([Years Open]>0,1,0)
```

Create a pivot table to test your measures, as shown in Figure 12-15. Remember that the measures only make sense if you filter by year. You can filter by year using a slicer. You can insert a slicer using the PivotTable Analyze tab.

Row Labels	Years Open	Was Open Prev Year
A Bicycle Association	23.00	1
A Bike Store	28.00	1
A Cycle Shop	-2.00	0
A Great Bicycle Company	27.00	1
A Typical Bike Shop	16.00	1
Acceptable Sales & Service	10.00	1
Accessories Network	14.00	1
Acclaimed Bicycle Company	19.00	1
Ace Bicycle Supply	6.00	1
Action Bicycle Specialists	6.00	1
Active Cycling	0.00	0
Active Life Toys	1.00	1
Active Systems	10.00	1
Active Transport Inc.	8.00	1
Activity Center	27.00	1
Advanced Bike Components	24.00	1

Year: 2011, 2012, 2013, 2014, 2015

Figure 12-15. *Testing the measures*

Now you can combine these measures so that you are only including resellers who have been in business for at least a year at the time of the sales. Create the following measures in the Sales table:

```
Month Sales Filtered:-Calculate([Month Sales],FILTER(Reseller,[Was Open
Prev Year]=1))
```

```
Prev Month Sales Filtered:=CALCULATE([Prev Month Sales],
    Filter(Reseller,[Was Open Prev Year]=1))
```

```
Monthly Sales Growth Filtered:=[Month Sales Filtered]-[Prev Month Sales Filtered]
```

```
Monthly Sales Growth Filtered %:=DIVIDE([Monthly Sales Growth Filtered],
    [Prev Month Sales Filtered],0)
```

Test your new measures by creating a Power Pivot table, as shown in Figure 12-16. You should see a difference between the filtered and the nonfiltered measures.

Row Labels	Month Sales	Month Sales Filtered	Monthly Sales Growth %	Monthly Sales Growth Filtered %
⊞2013	$1,702,944.54	$1,639,241.75	0.00 %	0.00 %
⊟2014	$2,185,213.21	$2,176,877.61	28.32 %	30.37 %
⊟1	$1,455,280.41	$1,406,985.25	-14.54 %	-15.74 %
January	$713,116.69	$658,066.68	-58.12 %	-60.59 %
February	$1,900,788.93	$1,897,013.80	166.55 %	188.27 %
March	$1,455,280.41	$1,406,985.25	-23.44 %	-25.83 %
⊞2	$1,001,803.77	$993,947.29	-31.16 %	-29.36 %
⊞3	$2,885,359.20	$2,872,092.51	188.02 %	188.96 %
⊞4	$2,185,213.21	$2,176,877.61	-24.27 %	-24.21 %
⊞2015	$3,510,948.73	$3,510,948.73	60.67 %	60.67 %
⊞2016			-100.00 %	-100.00 %

Figure 12-16. Testing the filtered measures

Create a Dashboard

To create a dashboard, insert a new Excel sheet. Name the sheet Reseller Sales. From the Insert tab, insert a PivotChart that shows Sales by Month and Business Type using the filtered monthly sales. Add a slicer to the report that controls the calendar year displayed by the chart. Your report should look like Figure 12-17.

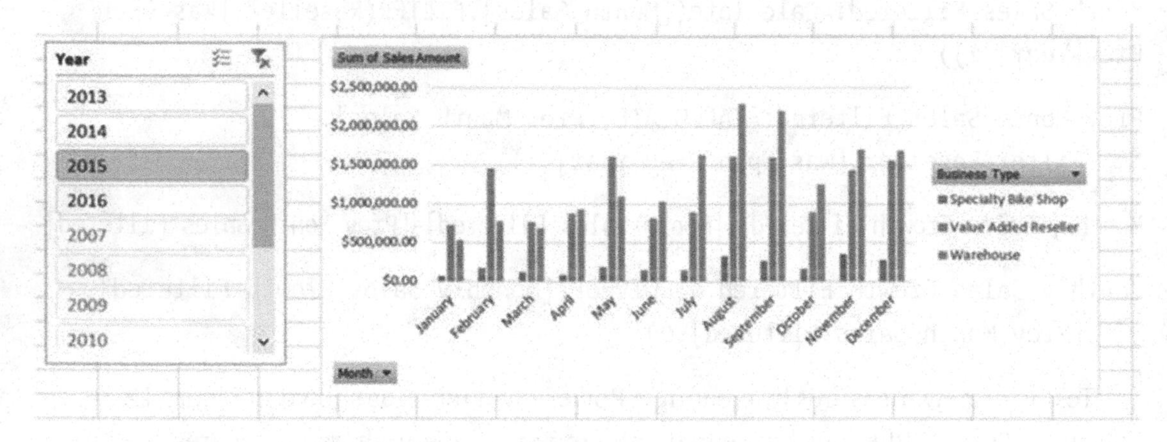

Figure 12-17. The Reseller Sales report

You now want to make another reseller analysis to find the best monthly sales for the month, quarter, and year. To do this, you need to add a new measure to the Sales table. This measure uses the VALUES and the MAXX functions to find the maximum reseller monthly sales:

```
Max Reseller Sales:=MAXX(VALUES(Reseller[ResellerKey]),[Month Sales])
```

Once you have the new measure in the model, add another column chart to the Excel sheet that shows the maximum reseller sales for each month. Make sure you update the year slicer so that it filters both charts (see Figure 12-18).

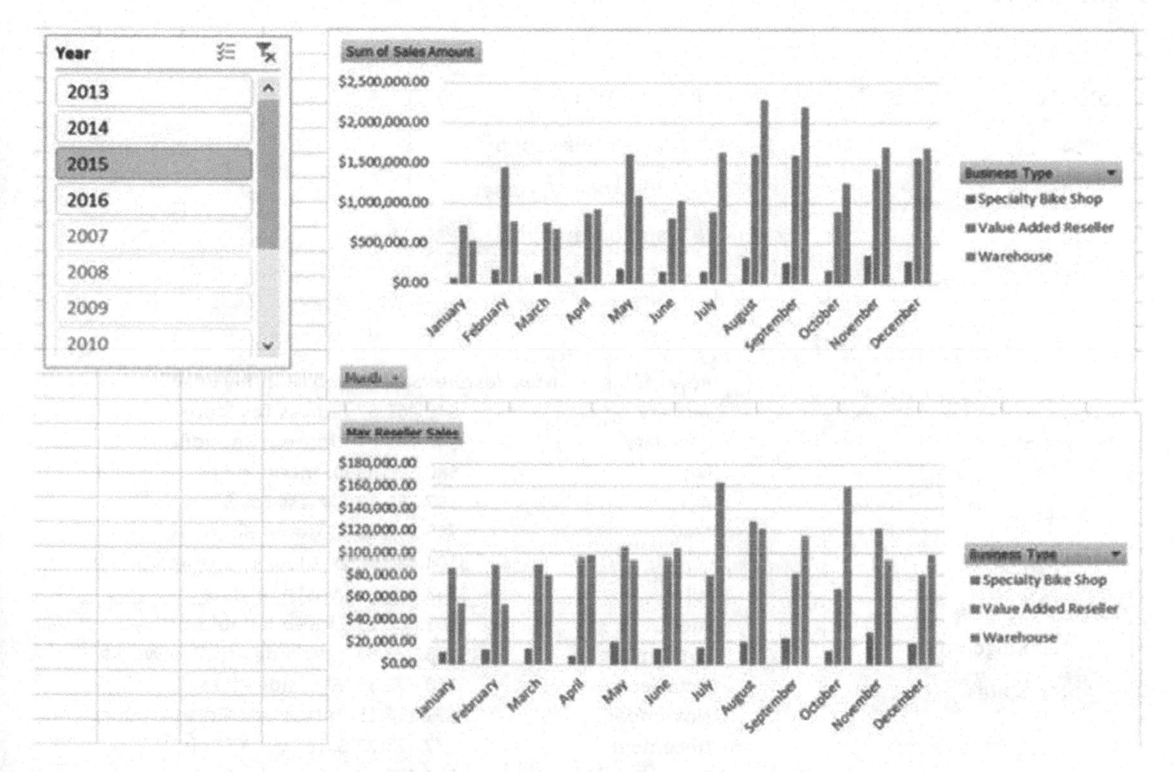

Figure 12-18. *Adding the Max Reseller Sales chart*

To take this one step further, let's say you want the name of the reseller who has the maximum monthly sales for the period. To accomplish this, you can use the TOPN function to add a measure to the Reseller table to find the top reseller for the month:

```
TOPN(1,Reseller,ResellerSales[Month Sales],DESC)
```

Because TOPN returns a table of values that could include more than one row for ties, you need to concatenate the results into a single string value. You can use the CANCATENATEX function for this. The result is as follows:

```
Top Monthly Reseller:=CONCATENATEX(TOPN(1,Reseller,Sales[Month
Sales],DESC),Reseller[Reseller Name],", ")
```

Add a new Excel worksheet named Max Resellers. Add a pivot table that shows the maximum reseller sales and reseller name for each month. Also include a slicer for the year, one for the reseller type, and one for the country. Your dashboard should look like Figure 12-19.

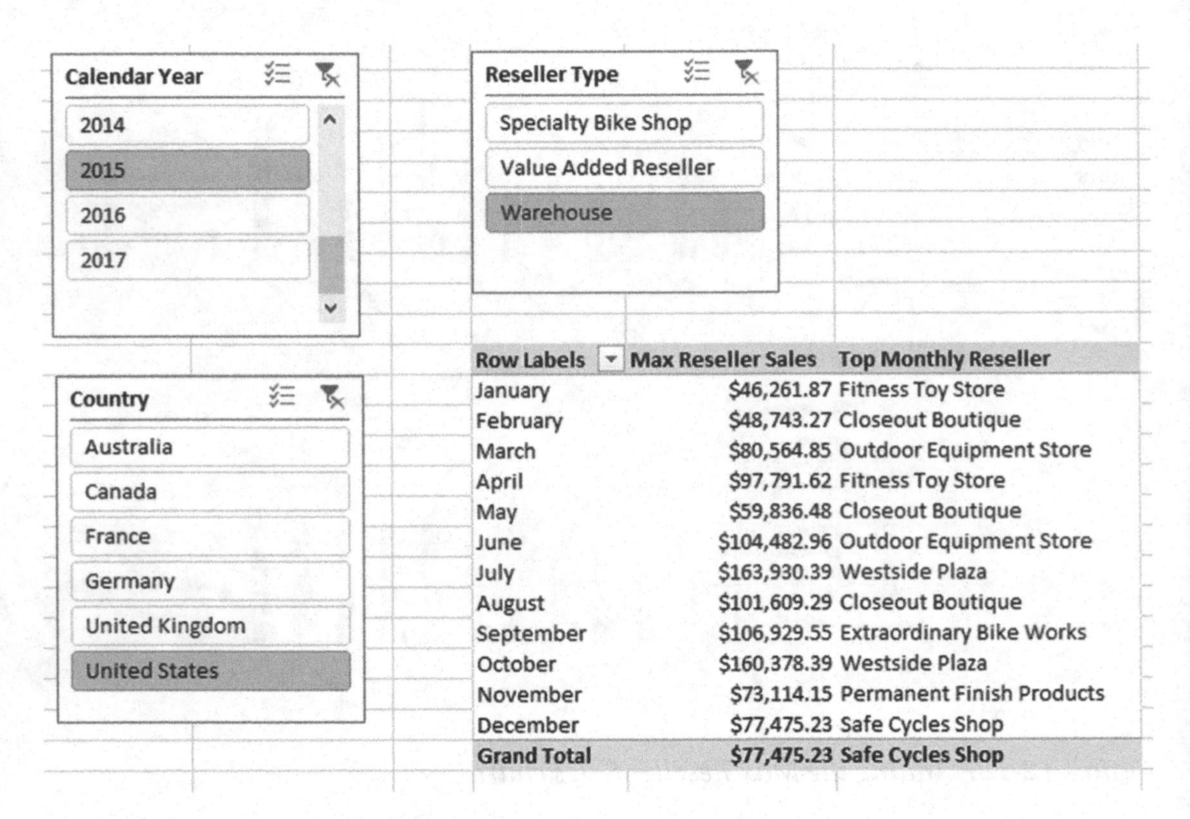

Row Labels	Max Reseller Sales	Top Monthly Reseller
January	$46,261.87	Fitness Toy Store
February	$48,743.27	Closeout Boutique
March	$80,564.85	Outdoor Equipment Store
April	$97,791.62	Fitness Toy Store
May	$59,836.48	Closeout Boutique
June	$104,482.96	Outdoor Equipment Store
July	$163,930.39	Westside Plaza
August	$101,609.29	Closeout Boutique
September	$106,929.55	Extraordinary Bike Works
October	$160,378.39	Westside Plaza
November	$73,114.15	Permanent Finish Products
December	$77,475.23	Safe Cycles Shop
Grand Total	$77,475.23	Safe Cycles Shop

Figure 12-19. *The final dashboard in an Excel worksheet*

After completing and experimenting with the dashboards, save and close Excel.

Use Case 3: Sensor Analysis

For this scenario, you work for a power company that monitors equipment using sensors. The sensors monitor various power readings, including power interruptions and voltage spikes. When the sensor senses a problem, it triggers an alarm signal that is recorded. You need to create a map that allows analysts to view and compare power interruptions and voltage spikes over time.

Load the Data

The data you will need is in several text files. Open the folder called UseCase3 in the Chapter12Labs folder. This folder has four files that contain the sensor data and the related data you need to complete the analysis. Create a new Power BI Desktop file named PowerAnalysis.pbix. Connect to the Alarms.csv file in the UseCase3 folder. You should have the following columns: PREMISE_NUMBER, METER_NUMBER, OP_CENTER, Type, and DateKey. Add a second query, `AlarmType`, which gets the alarm type data from the AlarmType.txt file.

Reopen the `Alarms` query and merge it with the `AlarmType` query using the appropriate keys (see Figure 12-20).

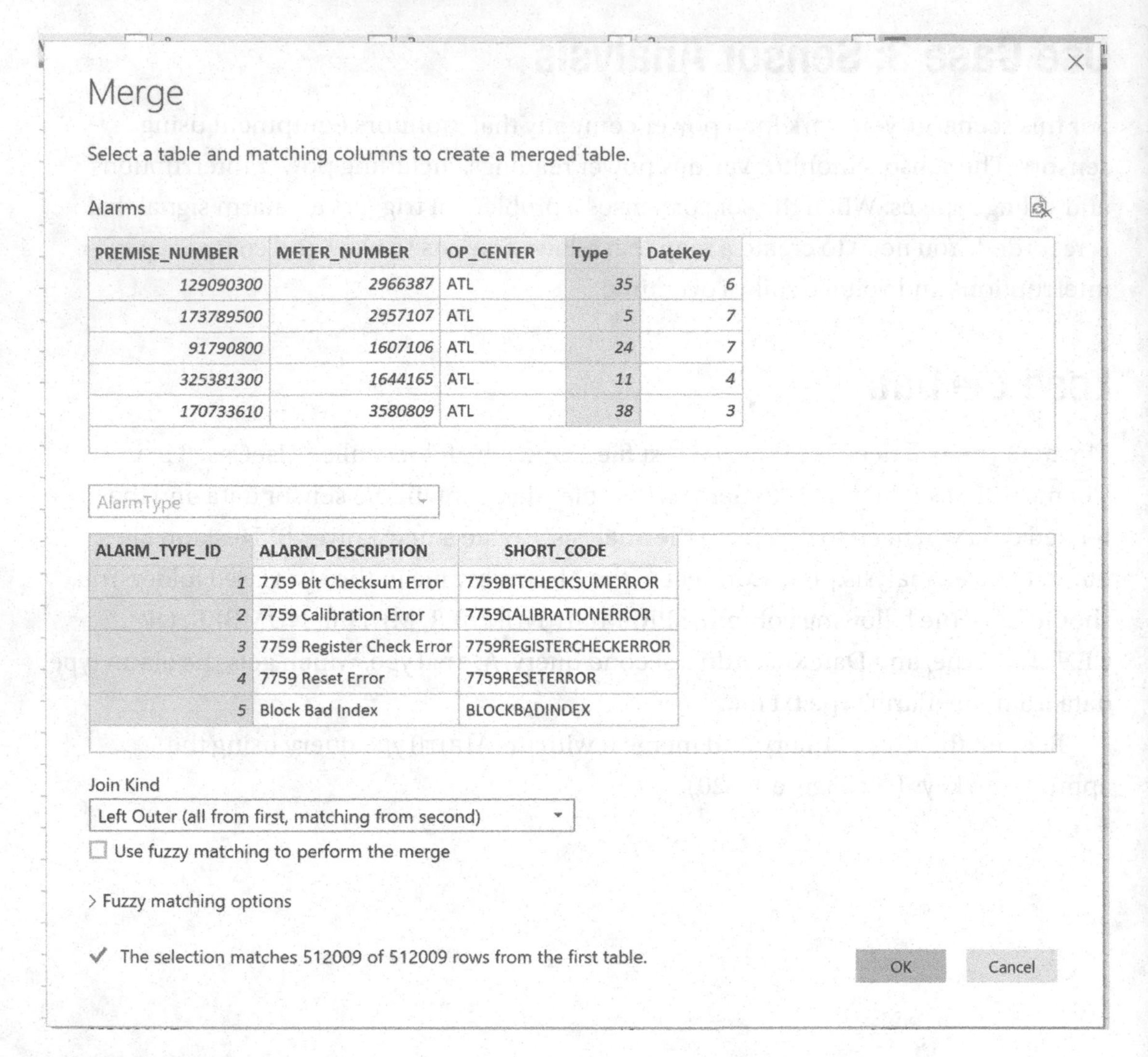

Figure 12-20. *Merging the `AlarmType` query with the `Alarms` query*

Expand the new column and select the ALARM_DESCRIPTION column. Rename this column to Alarm Type.

Repeat the previous procedures to replace the DateKey column with the dates in the Date.csv file. Use the Locations.txt file to add the longitude and latitude values to the `Alarms` query based on the OP_CENTER. You can rename the columns so that they use the same naming convention. Your alarm query data should look like Figure 12-21.

	PREMISE	METER	OP_CENTER	Alarm Type	Date	Latitude	Longitude
1	129090300	2966387	ATL	ROM Fail	7/25/2014	33.8979	-85.0988604
2	11490200	2964378	ATL	7759 Bit Checksum Error	7/25/2014	33.8979	-85.0988604
3	164889600	3043592	ATL	ROM Fail	7/22/2014	33.8979	-85.0988604
4	173789500	2957107	ATL	Block Bad Index	7/26/2014	33.8979	-85.0988604
5	275189600	4450418	ATL	7759 Calibration Error	7/26/2014	33.8979	-85.0988604
6	91790800	1607106	ATL	Meter Read Fail	7/26/2014	33.8979	-85.0988604
7	325381300	1644165	ATL	Configuration Error	7/23/2014	33.8979	-85.0988604
8	314790200	1666554	ATL	Meter Read Fail	7/24/2014	33.8979	-85.0988604
9	14574700	2111210	ATL	7759 Reset Error	7/22/2014	33.8979	-85.0988604
10	28989500	2915841	ATL	7759 Reset Error	7/25/2014	33.8979	-85.0988604
11	170733610	3580809	ATL	Soft EEPROM Error	7/22/2014	33.8979	-85.0988604
12	141668300	3113114	ATL	Power Failure	7/26/2014	33.8979	-85.0988604
13	518990500	1621877	ATL	Power Failure	7/23/2014	33.8979	-85.0988604
14	88326020	1983770	ATL	Block Buffer Size Error	7/26/2014	33.8979	-85.0988604
15	79715900	1241441	ATL	Block Buffer Size Error	7/24/2014	33.8979	-85.0988604
16	18764710	2907292	ATL	Low Battery Error	7/23/2014	33.8979	-85.0988604
17	808890300	2981854	ATL	Low Battery Error	7/26/2014	33.8979	-85.0988604
18	253990200	2893007	ATL	Block Can't Mark Bad	7/22/2014	33.8979	-85.0988604
19	50085200	2900902	ATL	Block Can't Mark Bad	7/24/2014	33.8979	-85.0988604
20	263089800	2901521	ATL	Block No Good Blocks	7/22/2014	33.8979	-85.0988604

***Figure 12-21.** The `Alarm` query with Latitude and Longitude data added*

Next, filter the data to limit it to alarm types of power failure and high AC volts. Now you can aggregate the alarm counts grouping by the Date, Op Center, Alarm Type, Longitude, and Latitude (see Figure 12-22).

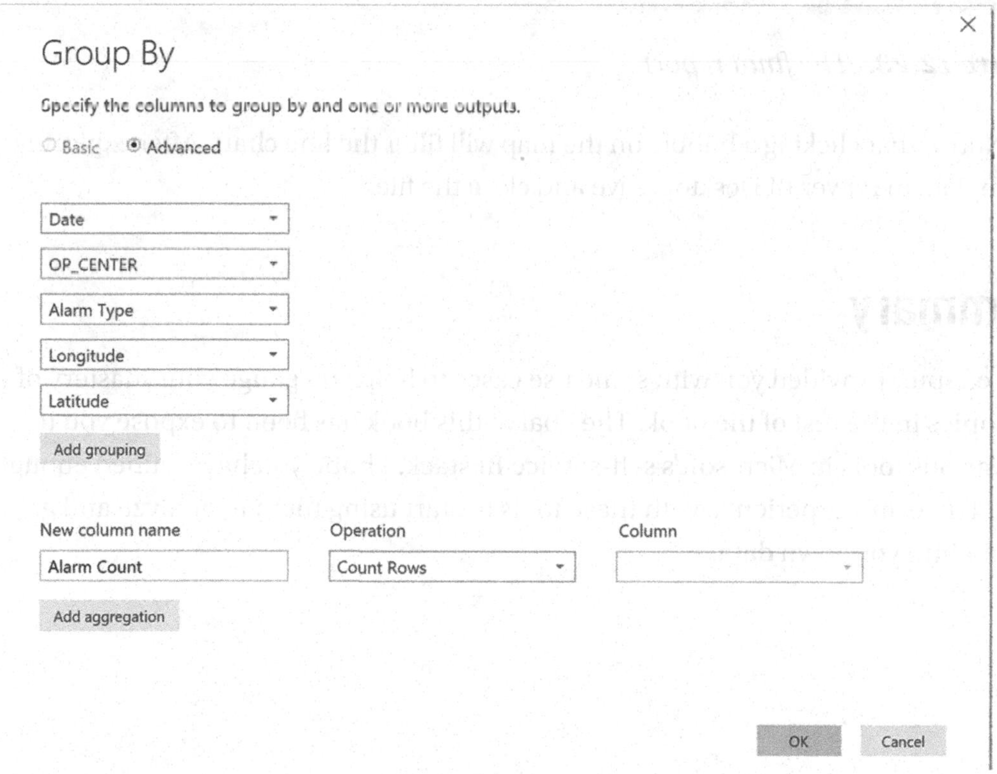

Group By

Specify the columns to group by and one or more outputs.

Basic Advanced

Date

OP_CENTER

Alarm Type

Longitude

Latitude

Add grouping

New column name | Operation | Column

Alarm Count | Count Rows |

Add aggregation

OK Cancel

***Figure 12-22.** Aggregating and grouping the alarm data*

After aggregating the data, disable the load for the `AlarmType`, `Date`, and `Locations` queries (see Figure 12-23). Select Close & Apply to load the data into the model.

On Page 1 of the report, add a map visual. Add the longitude and latitude values to the map. Add the Alarm Count to the size and the Op Center as the legend. Next, add a line chart to the report page. Use the Date as the axis, the Op Center as the legend, and the Alarm Count as the values. The report should look like Figure 12-23.

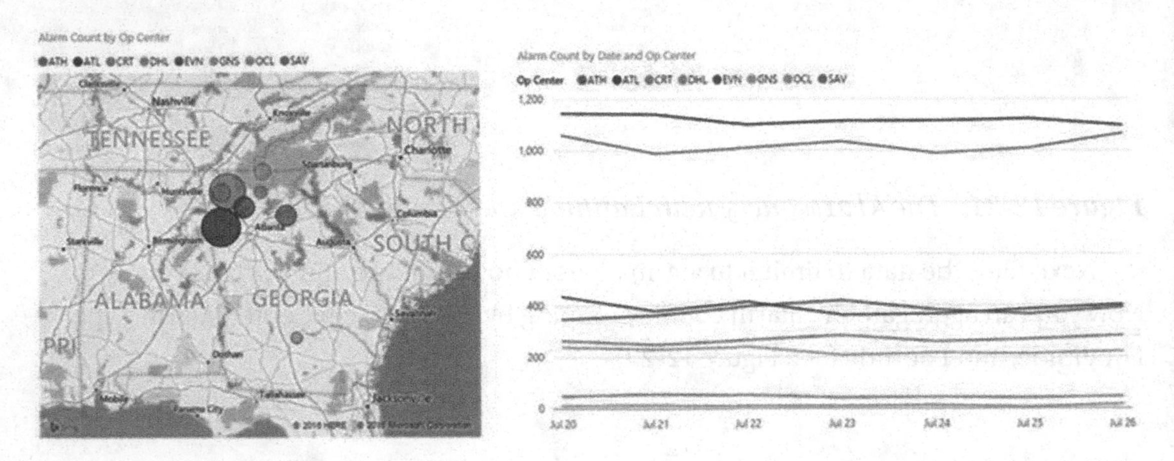

***Figure 12-23.** The final report*

Notice that clicking a bubble on the map will filter the line chart. After exploring the alarm data in Power BI Desktop, save and close the file.

Summary

This chapter provided you with some use cases to help you gauge your mastery of the topics in the rest of the book. The goal of this book has been to expose you to the various tools in Microsoft's self-service BI stack. I hope you have gained enough confidence and experience with these tools to start using them to analyze and gain insight into your own data.

Now that you have a firm understanding of how to use these tools, you should be comfortable tackling more complex topics. The next two chapters are optional advanced chapters that contain topics I think you will find useful. In addition, there are many good resources available that cover various techniques and patterns that can be used to analyze your data. Microsoft's Power BI site (`www.microsoft.com/en-us/powerbi/default.aspx`), Bill Jelen's site (`www.mrexcel.com`), and Rob Collie's Power Pivot Pro site (`www.powerpivotpro.com`) are excellent resources. For more advanced topics, check out Chris Webb's BI Blog at `https://blog.crossjoin.co.uk/` and Marco Russo's and Alberto Ferrari's SQLBI site at `www.sqlbi.com`.

CHAPTER 13

Advanced Topics in Power Query

When you build queries using the Power Query designer, the designer creates the query using the M query language. Although you can create robust queries using just the visual interface, there is a lot of useful processing that you can only complete by writing M code. This chapter goes beyond the basics and explores some of the advanced functionality in Power Query, including the M query language, parameters, and functions.

After completing this chapter, you will be able to

- Write queries with the M query language
- Create and use parameters
- Create and use functions

Writing Queries with M

When you build queries using Power Query, each step you add inserts a line of M query code. If you select the View tab in the designer and click the Advanced Editor button, you can edit the M code directly (see Figure 13-1).

D. Clark, *Beginning Microsoft Power BI*, https://doi.org/10.1007/978-1-4842-5620-6_13

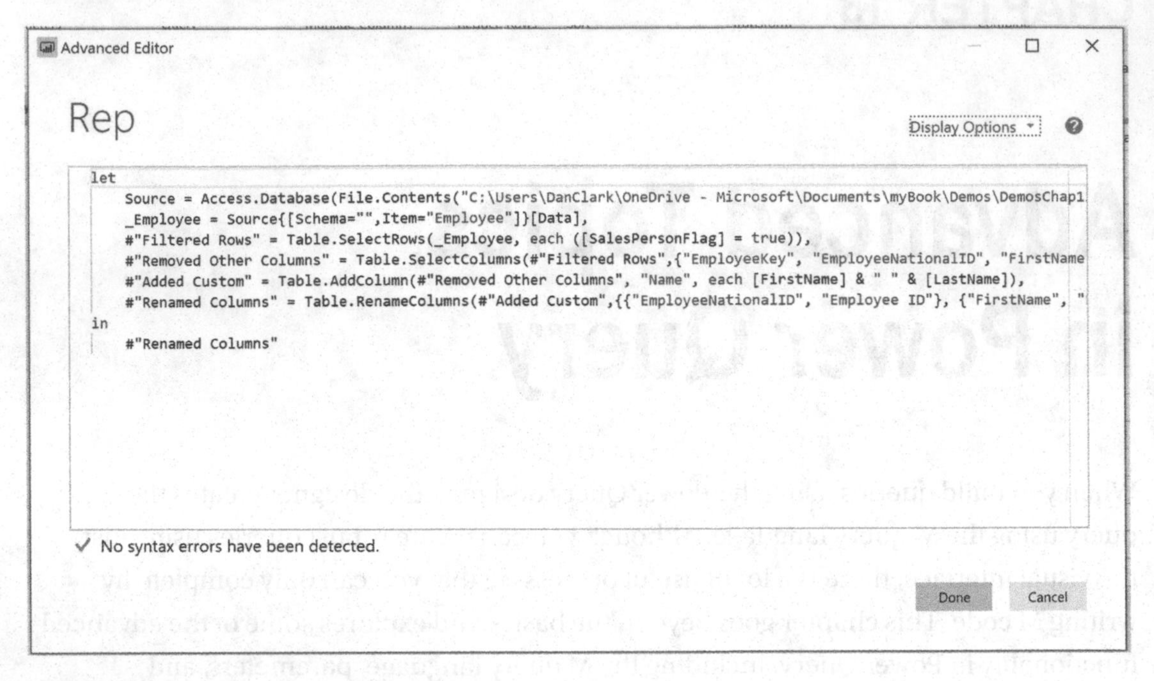
Advanced Editor
Rep
Display Options
let
Source = Access.Database(File.Contents("C:\Users\DanClark\OneDrive - Microsoft\Documents\myBook\Demos\DemosChap1
_Employee = Source{[Schema="",Item="Employee"]}[Data],
#"Filtered Rows" = Table.SelectRows(_Employee, each ([SalesPersonFlag] = true)),
#"Removed Other Columns" = Table.SelectColumns(#"Filtered Rows",{"EmployeeKey", "EmployeeNationalID", "FirstName
#"Added Custom" = Table.AddColumn(#"Removed Other Columns", "Name", each [FirstName] & " " & [LastName]),
#"Renamed Columns" = Table.RenameColumns(#"Added Custom",{{"EmployeeNationalID", "Employee ID"}, {"FirstName", "
in
#"Renamed Columns"
No syntax errors have been detected.
Done
Cancel

Figure 13-1. *Viewing the M code*

Each line of code represents the steps applied in the designer (see Figure 13-2).

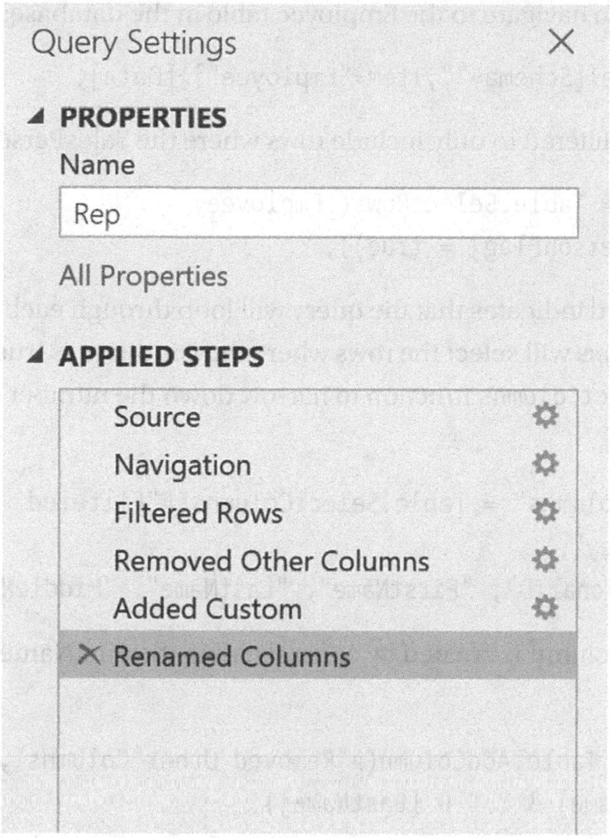

Figure 13-2. *Viewing the applied steps for a query*

The query starts with the keyword **let** followed by the steps in the query and ends with the keyword **in** followed by the name of the final table, list, or value:

```
let
      ... steps
in
      Table
```

The first step in most queries is to connect to a source. In this case, the code is connecting to an Access database file:

```
Source = Access.Database(File.Contents("C:\SalesRepAnalysis.accdb"),
    [CreateNavigationProperties=true]),
```

The next step is to navigate to the Employee table in the database:

```
_Employee = Source{[Schema="",Item="Employee"]}[Data],
```

The data is then filtered to only include rows where the SalesPersonFlag field is true:

```
#"Filtered Rows" = Table.SelectRows(_Employee,
    each ([SalesPersonFlag] = true)),
```

The each keyword indicates that the query will loop through each row, and `Table.SelectRows` will select the rows where the condition is true. The next step uses the `Table.SelectColumns` function to narrow down the number of columns selected:

```
#"Removed Other Columns" = Table.SelectColumns(#"Filtered
Rows",{"EmployeeKey",
     "EmployeeNationalID", "FirstName", "LastName", "MiddleName", "Title"}),
```

Next, a custom column is created by concatenating the FirstName and LastName columns:

```
#"Added Custom" = Table.AddColumn(#"Removed Other Columns", "Name",
     each [FirstName] & " " & [LastName]),
```

The final step is to use the `Table.RenameColumns` function to change the column names:

```
#"Renamed Columns" = Table.RenameColumns(#"Added Custom",
    {{"EmployeeNationalID", "Employee ID"}, {"FirstName", "First Name"},
     {"LastName", "Last Name"}, {"MiddleName", "Middle Name"},
     {"Name", "Sales Person"}})
```

Note that each step in the query references the table in the previous step. Also, the steps have a comma at the end of the line, except for the last step.

There are many useful functions in the M query language. For example, there are text, number, date, and time functions. The following code uses the `Text.PositionOf` function to find the colon in a text column named ProductNumber:

```
Text.PositionOf([ProductNumber],":")
```

An example of a date function is the `Date.EndOfMonth` function, which returns the date of the last day of the month of a date. For example, the following code returns the end of the month date for an OrderDate column:

```
Date.EndOfMonth([OrderDate])
```

Some other useful function categories are the splitter, combiner, and replacer functions. For example, you can combine or split text by delimiters, lengths, and positions. The following code is used to combine a list of text values using a comma:

```
Combiner.CombineTextByDelimiter(", ")
```

Note For a full list of M functions, refer to the Power Query Formula Reverence (`https://docs.microsoft.com/en-us/powerquery-m/power-query-m-function-reference`).

As you gain experience with M code, you will be able to build complex queries that go beyond what you can do with just the visual designer.

Creating and Using Parameters

The ability to use parameters in your queries is a very useful feature of Power BI Desktop. Using parameters is an easy way to update parts of your queries without having to alter the code. For example, you can create a file path using a parameter. When you share the Power BI Desktop file with a colleague, they can update the path without having to alter the query.

To create a parameter in Power BI Desktop, launch the Query Editor. On the Home tab, click the Manage Parameters drop-down and select New Parameter (see Figure 13-3).

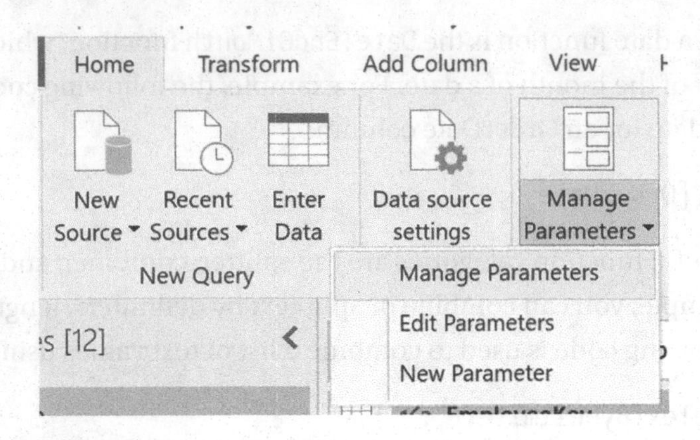

***Figure 13-3.** Adding a new parameter*

This launches a window in which you can define the parameter properties (see Figure 13-4).

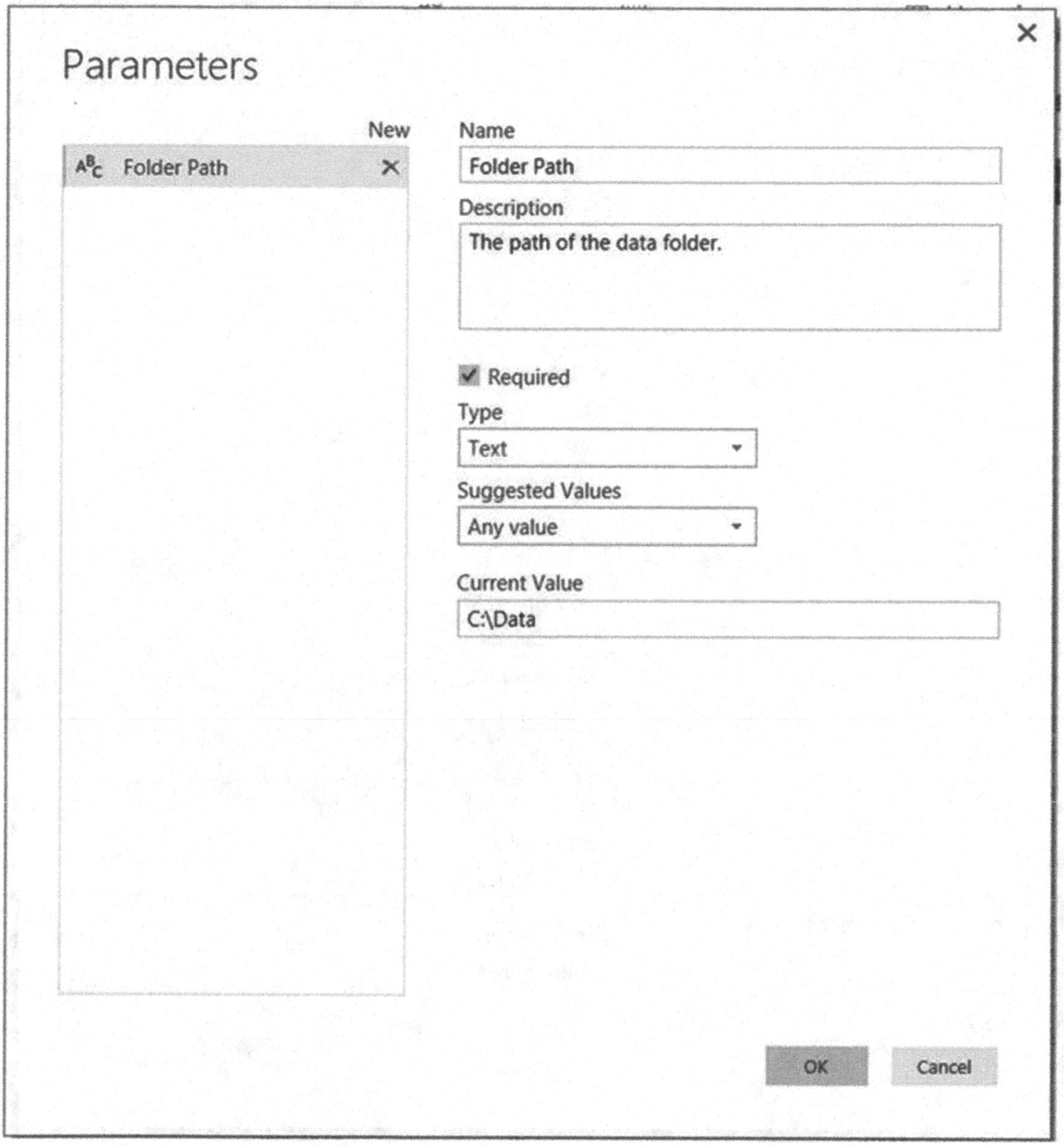

Parameters

New

Folder Path

Name

Folder Path

Description

The path of the data folder.

Required

Type

Text

Suggested Values

Any value

Current Value

C:\Data

OK Cancel

Figure 13-4. *Defining a parameter*

The type can be any type supported by Power Query, such as text, date, or decimal. The value can be any value or be restricted to a list of values. The list of values can be entered manually (see Figure 13-5) or come from a query that returns a list.

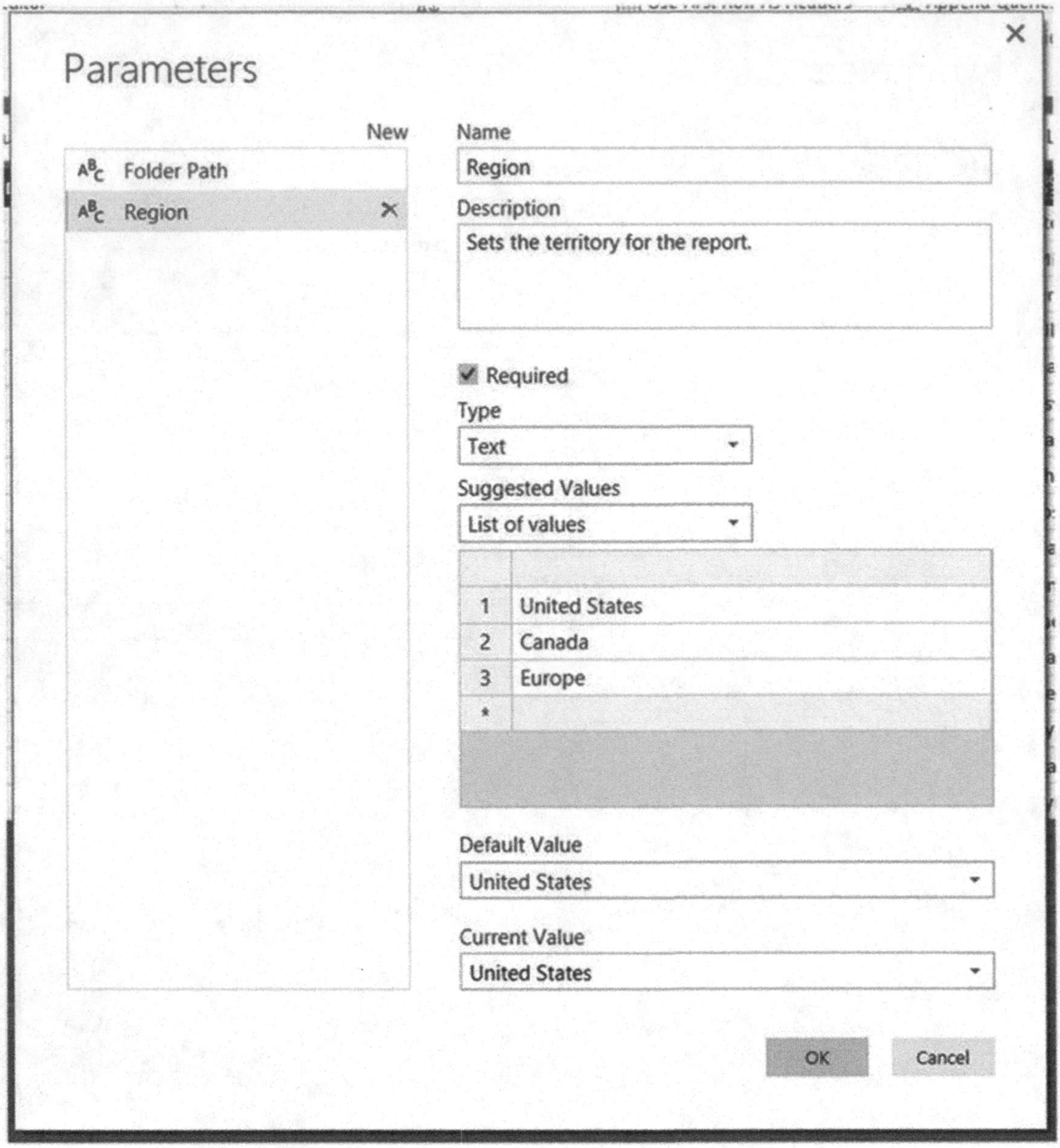

***Figure 13-5.** Defining a list of values for a parameter*

To use the parameter in your queries, you just replace the hard-coded value with the name of the parameter. For example, if you create a parameter called "Region," the following filter statement

```
Table.SelectRows(#"Removed Other Columns",
    each ([SalesTerritoryRegion] = "Canada"))
```

becomes

```
Table.SelectRows(#"Removed Other Columns",
    each ([SalesTerritoryRegion] = Region))
```

Once you create a parameter, you can edit the parameter in the Home tab of Power BI Desktop under the Edit Queries drop-down (see Figure 13-6).

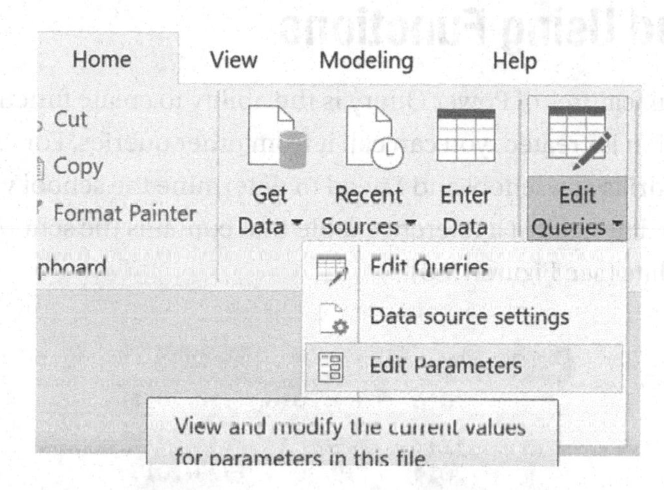

***Figure 13-6.** Editing parameters*

This launches a window where users can edit the parameter values (see Figure 13-7).

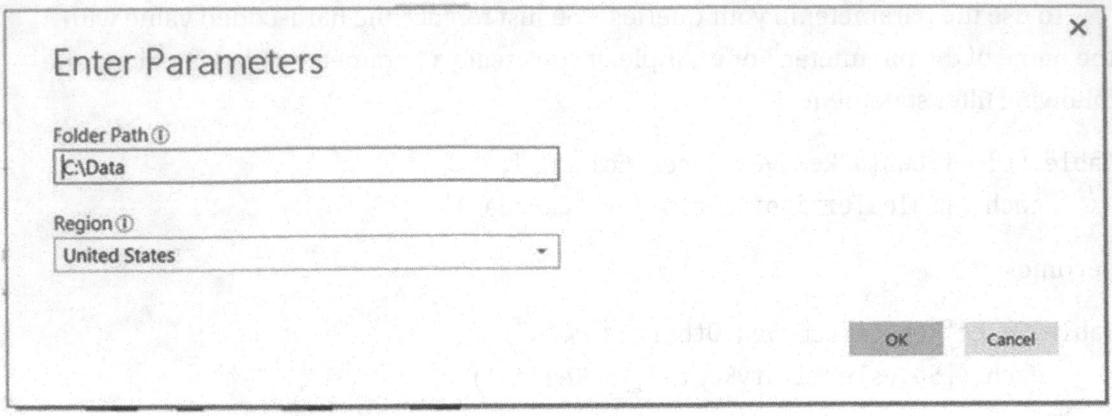

Enter Parameters

Folder Path

C:\Data

Region

United States

OK Cancel

Figure 13-7. Setting parameter values

Creating and Using Functions

One of the powerful features of Power Query is the ability to create functions using M code. Once a function is created, you can call it from other queries. For example, I have a set of dates for fire drills at a school and I need to determine the school year it belongs to. I need to check the date against a reference table that contains the school year, the start date, and the end date (see Figure 13-8).

SchoolYear	StartDate	EndDate
2016	9/7/2016	6/20/2017
2015	9/8/2015	6/22/2016
2014	9/9/2014	6/23/2015
2013	9/10/2013	6/20/2014
2012	9/6/2012	6/21/2013

Figure 13-8. School year lookup table

Say you have a date, 1/15/2015, and you need to filter this table so that the StartDate is less than or equal to the date and the EndDate is greater than or equal to the date. This will leave one row in the table, and from that row you need to select the SchoolYear value. The M code to create this query is as follows:

```
YearLookUp_Table = Source{[Item="YearLookUp",Kind="Table"]}[Data],
    #"Filtered Rows" = Table.SelectRows
        (#"YearLookUp_Table", each [StartDate] <= #date(2015, 1, 15)),
```

```
    #"Filtered Rows1" = Table.SelectRows
        (#"Filtered Rows", each [EndDate] >= #date(2015, 1, 15)),
    SchoolYear = #"Filtered Rows1"{0}[SchoolYear]
in
    SchoolYear
```

To convert this query into a function, you need to wrap the query into another `let` statement:

```
let
    LookUpSchoolYear = () =>
     [Original query]
in
    LookUpSchoolYear
```

The next step is to replace the hard-coded date in the query with a parameter you can pass in. The final query looks like the following:

```
let
    LookUpSchoolYear = (dt as date) =>
let
    Source = Excel.Workbook(File.Contents("C:\ Drills.xlsx"), null, true),
    YearLookUp_Table = Source{[Item="YearLookUp",Kind="Table"]}[Data],
    #"Filtered Rows" = Table.SelectRows
        (#"YearLookUp_Table", each [StartDate] <= dt),
    #"Filtered Rows1" = Table.SelectRows
        (#"Filtered Rows", each [EndDate] >= dt),
    SchoolYear = #"Filtered Rows1"{0}[SchoolYear]
in
    SchoolYear
in
    LookUpSchoolYear
```

Once you save the function, you can test it by clicking the Invoke button and entering a value for the parameter (see Figure 13-9).

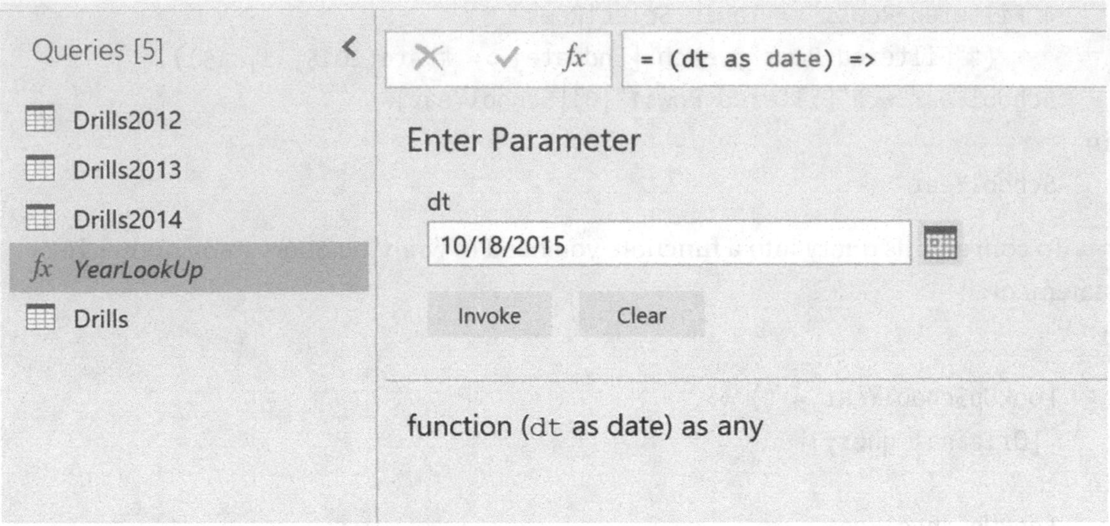

Queries [5]
Drills2012
Drills2013
Drills2014
YearLookUp
Drills

= (dt as date) =>

Enter Parameter
dt
10/18/2015
Invoke
Clear

function (dt as date) as any

***Figure 13-9.** Testing the function*

After testing the function, you can now invoke it from within another query. Figure 13-10 shows how you can use the function to create a new column called SchoolYear by passing the Date field to the function for each row.

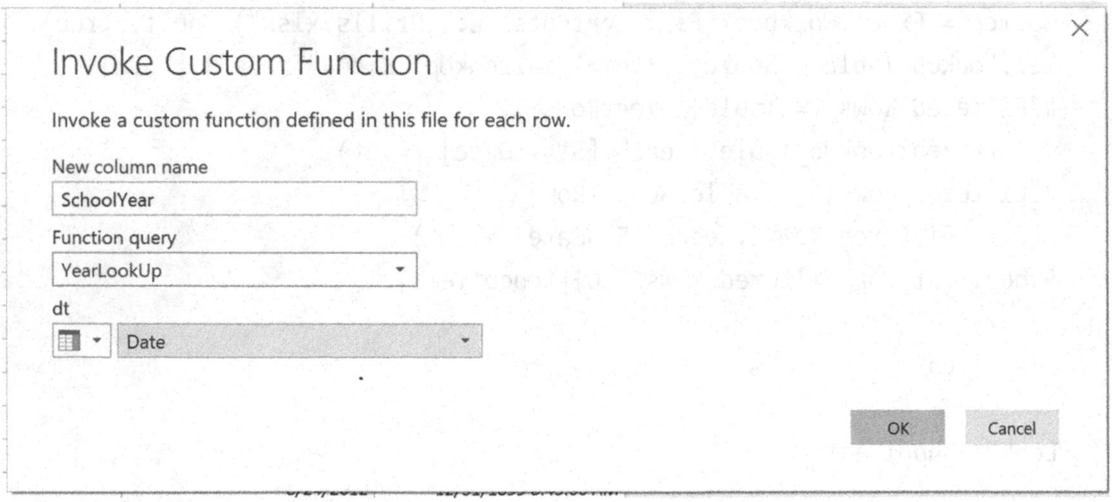

Invoke Custom Function
Invoke a custom function defined in this file for each row.
New column name
SchoolYear
Function query
YearLookUp
dt
Date
OK
Cancel

***Figure 13-10.** Using the function in a custom column*

As another example of a function, you have data in the form of the table shown in Figure 13-11.

	School	Drill Type	Date	Time	SchoolYear
1	CSH	Fire	8/27/2012	12/31/1899 8:50:00 AM	2012
2	BE	Fire	8/27/2012	12/31/1899 9:50:00 AM	2012
3	PBE	Fire	8/27/2012	12/31/1899 10:00:00 AM	2012
4	EKDE	Fire	8/28/2012	12/31/1899 8:50:00 AM	2012
5	SE	Fire	8/28/2012	12/31/1899 9:30:00 AM	2012
6	RWE	Fire	8/28/2012	12/31/1899 9:00:00 AM	2012
7	WSE	Fire	8/28/2012	12/31/1899 9:00:00 AM	2012
8	SE	Fire	8/27/2012	12/31/1899 12:10:00 PM	2012
9	SH	Fire	8/28/2012	12/31/1899 1:30:00 PM	2012
10	WE	Fire	8/28/2012	12/31/1899 9:35:00 AM	2012
11	CSH	Fire	8/28/2012	12/31/1899 2:10:00 PM	2012
12	PVE	Fire	8/29/2012	12/31/1899 10:40:00 AM	2012
13	WH	Fire	8/29/2012	12/31/1899 10:30:00 AM	2012
14	NM	Fire	8/29/2012	12/31/1899 1:05:00 PM	2012
15	HM	Fire	8/29/2012	12/31/1899 1:35:00 PM	2012
16	SAE	Fire	8/29/2012	12/31/1899 1:45:00 PM	2012

Figure 13-11. *Initial drill data*

You need to transform the data to show a list of dates for the drills grouped by school, drill type, and year, as shown in Figure 13-12.

```
= Table.Group(#"Changed Type", {"School", "Drill Type", "SchoolYear"}, {"Dates", each fCombine([Date]), type text})
```

	School	Drill Type	SchoolYear	Dates
1	AA	Duck Cover	2012	1/31/2013, 11/2/2012
2	AA	Duck Cover	2013	2/6/2014, 9/13/2013
3	AA	Duck Cover	2014	2/6/2015, 9/5/2014
4	AA	Fire	2012	5/3/2013, 4/5/2013, 3/5/2013, 2/4/2013, 1/4/2013, 12/4/2012, 11/2/2012, 10/5/2012, 9/5/2...
5	AA	Fire	2013	4/1/2014, 12/20/2013, 11/20/2013, 10/25/2013, 9/30/2013, 8/29/2013
6	AA	Fire	2014	4/13/2015, 3/9/2015, 11/24/2014, 9/5/2014
7	AA	Lockdown	2012	1/31/2013, 9/5/2012
8	AA	Lockdown	2013	2/6/2014, 9/13/2013
9	AA	Lockdown	2014	2/6/2015, 9/15/2014
10	AA	Reverse Evacuation	2012	4/5/2013, 11/2/2012
11	AA	Reverse Evacuation	2013	12/20/2013, 8/29/2013
12	AA	Reverse Evacuation	2014	3/9/2015
13	AA	Shelter In Place	2012	1/31/2013, 8/28/2012
14	AA	Shelter In Place	2013	2/6/2014, 9/13/2013
15	AA	Shelter In Place	2014	2/6/2015, 9/5/2014
16	BE	Duck Cover	2012	2/11/2013, 12/17/2012, 9/4/2012, 8/27/2012
17	BE	Duck Cover	2013	4/10/2014, 4/4/2014, 1/15/2014, 10/23/2013, 10/7/2013, 9/5/2013, 8/30/2013
18	BE	Duck Cover	2014	1/15/2015, 10/21/2014, 9/24/2014, 8/29/2014, 8/27/2014

Figure 13-12. *Summarized drill data*

To do this, rearrange the columns and sort the School, Drill Type, and Year columns ascending and the Date column descending (see Figure 13-13).

	School	Drill Type	SchoolYear	Date	Time
1	AA	Duck Cover	2012	1/31/2013	12/31/1899 2:00:00 PM
2	AA	Duck Cover	2012	11/2/2012	12/31/1899 10:35:00 AM
3	AA	Duck Cover	2013	2/6/2014	12/31/1899 1:30:00 PM
4	AA	Duck Cover	2013	9/13/2013	12/31/1899 9:25:00 AM
5	AA	Duck Cover	2014	2/6/2015	12/31/1899 9:45:00 AM
6	AA	Duck Cover	2014	9/5/2014	12/31/1899 9:40:00 AM
7	AA	Fire	2012	5/3/2013	12/31/1899 10:45:00 AM
8	AA	Fire	2012	4/5/2013	12/31/1899 1:10:00 PM
9	AA	Fire	2012	3/5/2013	12/31/1899 1:30:00 PM
10	AA	Fire	2012	2/4/2013	12/31/1899 2:00:00 PM
11	AA	Fire	2012	1/4/2013	12/31/1899 1:30:00 PM
12	AA	Fire	2012	12/4/2012	12/31/1899 12:00:00 AM
13	AA	Fire	2012	11/2/2012	12/31/1899 12:00:00 AM
14	AA	Fire	2012	10/5/2012	12/31/1899 12:15:00 PM
15	AA	Fire	2012	9/5/2012	12/31/1899 9:30:00 AM
16	AA	Fire	2013	4/1/2014	12/31/1899 1:05:00 PM
17	AA	Fire	2013	12/20/2013	12/31/1899 10:15:00 AM
18	AA	Fire	2013	11/20/2013	12/31/1899 1:00:00 PM
19	AA	Fire	2013	10/25/2013	12/31/1899 10:00:00 AM

Figure 13-13. *Preparing the data*

Next, you create a step to group the data using the following code:

```
=Table.Group(#"Changed Type", {"School", "Drill Type", "SchoolYear"},
        {"Dates", each fCombine([Date]), type text})
```

Instead of aggregating the data, you pass the list of date values for each group to the `fCombine` function. This returns a text value with the dates concatenated together. The `fCombine` function in M code is as follows:

```
let
    Source = Combiner.CombineTextByDelimiter(", ")
in
    Source
```

Now that you are familiar with writing M code, you are ready to gain some hands-on experience with the following lab.

HANDS-ON LAB: ADVANCED QUERY BUILDING WITH POWER QUERY

In the following lab, you will

- Create and use a parameter
- Alter a query using M code
- Create and use a function

1. In the LabStarterFiles\Chapter13Lab1 folder, open the Drills.xlsx file. This file contains data on drills for the various schools in a school district. Each tab in the workbook contains data for a different school year. After reviewing the data, close the file.

2. Create a new Power BI Desktop file called Chapter13Lab1.pbix. In Chapter 12, you loaded each worksheet as a separate query and appended the queries together. You can automate this using M code so that you don't have to update the query as more worksheets are added to the workbook.

3. On the Home tab, select Get Data from a Blank Query. In the Power Query Editor, change the name to Drills.

4. In the Source step, add the following code to the formula bar. Make sure you use the path to where you saved the lab files. You should see each sheet and a corresponding table in the file (see Figure 13-14).

```
= Excel.Workbook(File.Contents
  ("C:\LabStarterFiles\Chapter13Lab1\Drills.xlsx"))
```

	Name	Data	Item	Kind	Hidden
1	2012	Table	2012	Sheet	FALSE
2	2013	Table	2013	Sheet	FALSE
3	2014	Table	2014	Sheet	FALSE
4	Drills2012	Table	Drills2012	Table	FALSE
5	Drills2013	Table	Drills2013	Table	FALSE
6	Drills2014	Table	Drills2014	Table	FALSE

Figure 13-14. *Loading Excel worksheets and tables*

5. Filter the Kind column to only show sheets.

6. Keep the Name and the Data columns and remove the rest.

7. Add a custom column called TableWithHeaders and add the following code:

```
=Table.PromoteHeaders([Data])
```

8. Remove the original Data column and expand the TableWithHeaders column and don't include the original column name as a prefix. Your data should look like Figure 13-15.

	Name	School	Drill Type	Date	Time
1	2012	CE	Fire	8/23/2012	12/31/1899 1:30:00 PM
2	2012	AA	Fire	8/24/2012	12/31/1899 9:45:00 AM
3	2012	BIS	Fire	8/24/2012	12/31/1899 9:15:00 AM
4	2012	MSS	Fire	8/24/2012	12/31/1899 10:10:00 AM
5	2012	CSE	Fire	8/24/2012	12/31/1899 10:35:00 AM
6	2012	BE	Fire	8/24/2012	12/31/1899 8:05:00 AM
7	2012	SAE	Fire	8/23/2012	12/31/1899 1:50:00 PM
8	2012	PHE	Fire	8/24/2012	12/31/1899 9:00:00 AM
9	2012	BH	Fire	8/24/2012	12/31/1899 9:50:00 AM
10	2012	GE	Fire	8/24/2012	12/31/1899 1:00:00 PM
11	2012	OFE	Fire	8/23/2012	12/31/1899 2:05:00 PM
12	2012	NM	Fire	8/24/2012	12/31/1899 8:30:00 AM

Figure 13-15. *Data after expanding the TableWithHeaders column*

9. Rename the Name column to SchoolYear and change the Date and Time columns' data types to date and time.

10. Now you want to add a parameter that allows users to easily change the path to the source file. On the Home tab under the Manage Parameters drop-down, select the New Parameter option. Add a text parameter called `FilePath` that has a current value that matches the file path you used in step 4.

11. Alter the Source step in the `Drills` query to use the `FilePath` variable.

```
= Excel.Workbook(File.Contents(FilePath))
```

12. Select Close & Apply on the Home tab to close the Query Editor.

13. To test the parameter, move the Drills.xlsx file to a subfolder called Data. In the Power BI Desktop, click the Refresh button on the Home tab. You should get an error saying that it could not find the file.

14. Select the Edit Parameters option under the Edit Queries drop-down on the Home tab. Update the `FilePath`. You should now be able to refresh the data.

15. Select the Edit Queries option on the Home tab. In the Query Editor, right-click the `Drills` query and select Duplicate. Rename the new query `DrillTypeList`.

16. Remove the last two steps and edit the Expanded TableWithHeaders step so that it only selects the Drill Type column.

17. Remove the Name column and remove duplicates from the Drill Type column.

18. To convert the column to a list, right-click the header and select Drill Down (see Figure 13-16).

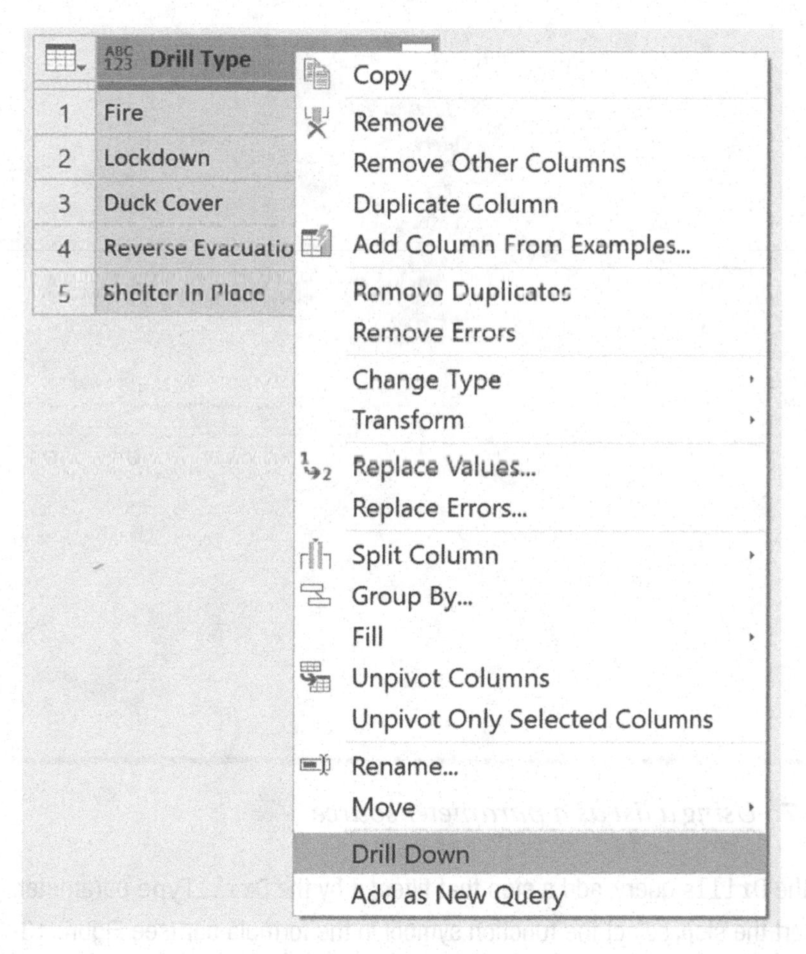

Figure 13-16. *Converting a column to a list*

19. Add a new parameter called `DrillType` that uses the list you just created as a source (see Figure 13-17).

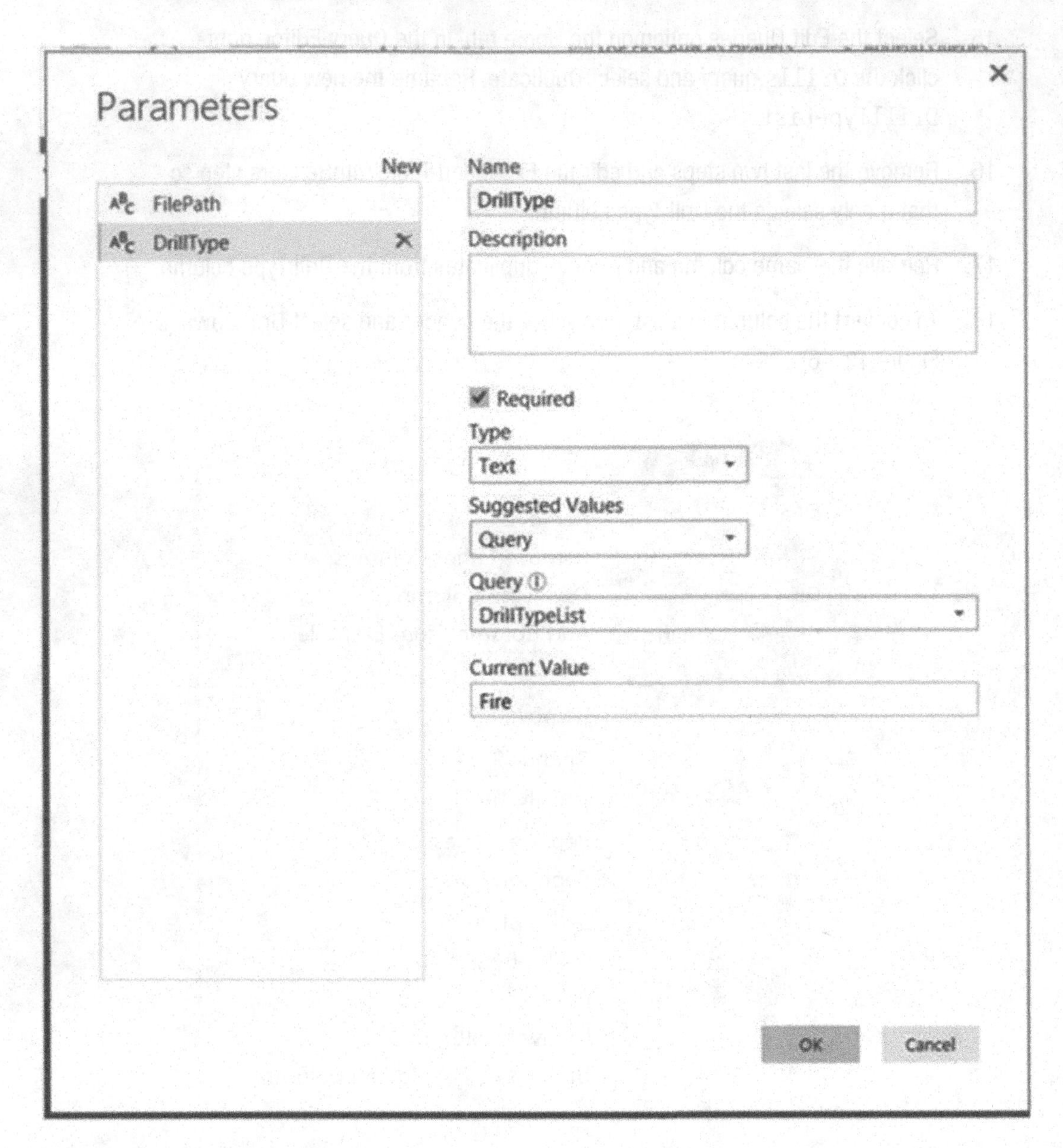

***Figure 13-17.** Using a list as a parameter source*

20. In the `Drills` query, add a step that filters it by the `DrillType` parameter. To insert the step, select the function symbol in the formula bar (see Figure 13-18).

```
= Table.SelectRows(#"Renamed Columns", each ([Drill Type] = DrillType))
```

fx = Table.SelectRows(#"Changed Type", each [Drill Type] = DrillType)

	SchoolYear	School	Drill Type	Date
1	2012	CE	Fire	
2	2012	AA	Fire	

Figure 13-18. *Inserting a step into the query*

21. Close and apply the changes in the Query Editor. Test the parameter by updating it in Power BI Desktop and refreshing the data.

22. Launch the Query Editor and create a new blank query named `CombineList`. You are going to create a function that takes a list of text values and concatenates them together.

23. Open the advanced editor and add the following code:

```
let
    Source = Combiner.CombineTextByDelimiter(", ")
in
    Source
```

24. Close the advanced editor.

25. Make a duplicate of the `Drills` query and rename it `DrillsSummary`.

26. Add a step to sort by the SchoolYear, School, Drill Type, and Date columns.

27. Remove the Time column and change the Date column to a text data type.

28. Open the advanced editor and add a grouping step using the following highlighted code:

```
#"Changed Type1" = Table.TransformColumnTypes(#"Sorted Rows",
    {{"Date", type text}}),
#"Grouped Rows" = Table.Group(#"Changed Type1", {"SchoolYear", "School",
    "Drill Type"}, {{"Dates", each CombineList([Date]), type text}})
in
    #"Grouped Rows"
```

29. Close the advanced editor. The data after grouping the rows should look like Figure 13-19.

	SchoolYear	School	Drill Type	Dates
1	2012	AA	Fire	8/24/2012, 9/5/2012, 10/5/2012, 11/2/2012, 12/4/2012, 1/4/2013, 2/4/2013, 3/5/2013, 4/5...
2	2012	BE	Fire	8/24/2012, 8/24/2012, 8/27/2012, 8/31/2012, 9/27/2012, 9/28/2012, 10/25/2012, 10/25/20...
3	2012	BH	Fire	8/24/2012, 8/30/2012, 9/20/2012, 10/23/2012, 11/15/2012, 12/12/2012, 1/31/2013, 2/20/2...
4	2012	BIS	Fire	8/24/2012, 9/4/2012, 9/7/2012, 11/2/2012, 11/14/2012, 12/11/2012, 1/7/2013, 2/28/2013, ...
5	2012	BM	Fire	8/24/2012, 8/29/2012, 9/25/2012, 10/22/2012, 11/20/2012, 12/19/2012, 1/14/2013, 2/12/2...
6	2012	CE	Fire	8/23/2012, 8/27/2012, 8/30/2012, 8/30/2012, 9/14/2012, 9/28/2012, 10/3/2012, 10/25/201...
7	2012	CSE	Fire	8/24/2012, 8/30/2012, 9/27/2012, 10/25/2012, 11/21/2012, 12/14/2012, 1/10/2013, 2/7/20...
8	2012	CSH	Fire	8/27/2012, 8/28/2012, 9/26/2012, 11/1/2012, 11/29/2012, 12/20/2012, 1/24/2013, 2/12/20...

Figure 13-19. *The final grouped data*

30. When you're done, save the file and close Power BI Desktop.

Summary

Power Query is a very powerful tool used to transform data before it is loaded into the Power Pivot data model. Although a lot of functionality is exposed by the visual designer, even more functionality is exposed by the M query language. This chapter went beyond the basics and explored some of the advanced functionality in Power Query, including the M query language, parameters, and functions. This chapter only scratched the surface of advanced querying with M. If you want to learn more, I strongly recommend *Power Query for Power BI and Excel* by Chris Webb (Apress, 2014).

The next chapter covers some advanced topics in Power BI that I think you may find useful, including using custom visuals, advanced mapping, row-based security, and templates.

CHAPTER 14

Advanced Topics in Power BI Desktop

This chapter covers some advanced topics in Power BI that I think you may find useful. It includes using custom visuals, advanced mapping, row-based security, templates, and content packs.

After completing this chapter, you will be able to

- Use custom visuals
- Implement geospatial analysis
- Implement row-based security
- Create templates and content packs

Using Custom Visuals

Although Power BI includes an impressive set of visuals out of the box, there are times when they might not fit your needs. The great thing is that Microsoft has provided the source code for the visuals it ships and have provided an open source framework so that developers can create their own visuals. Although you probably won't create your own visuals, there is an active developer community creating and sharing what they have done. You can download and import these into Power BI Desktop for use in your own reports. To view the available visuals, go to the Visuals Gallery at `http://app.powerbi.com/visuals` (see Figure 14-1).

D. Clark, *Beginning Microsoft Power BI*, https://doi.org/10.1007/978-1-4842-5620-6_14

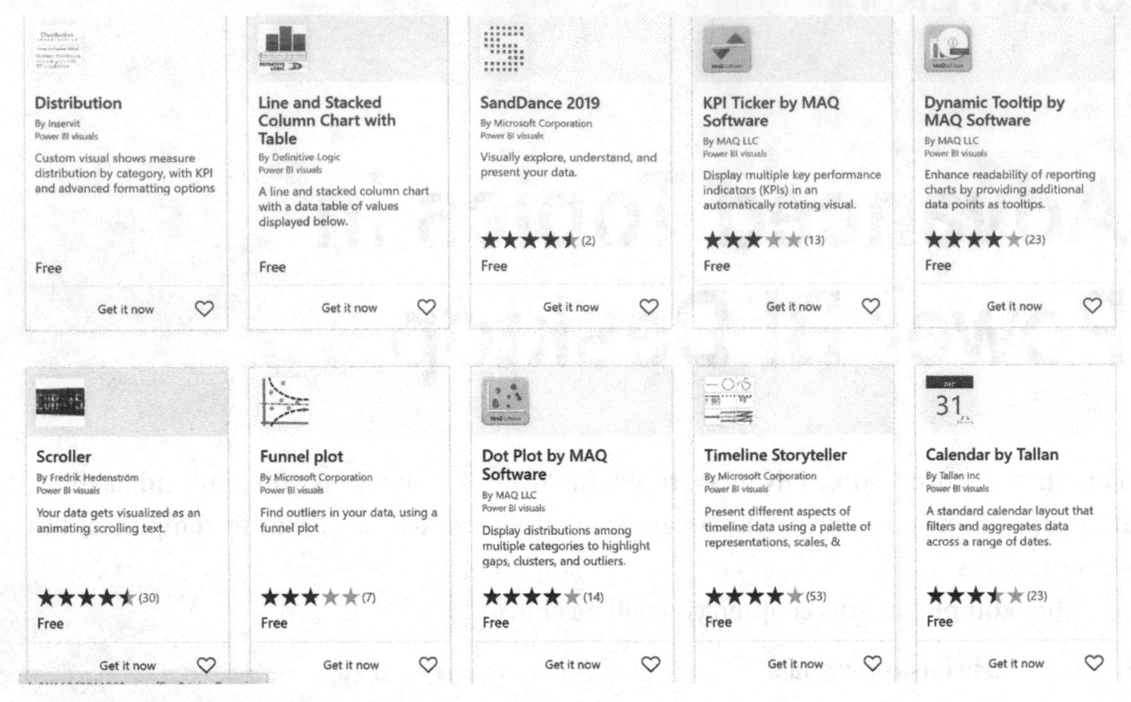

Figure 14-1. Some of the custom visuals available

If you click a visual in the gallery, you can download the visual and a sample Power BI Desktop file that shows how it is used (see Figure 14-2).

Figure 14-2. Downloading a custom visual

The custom visual is contained in a file with a pbiviz extension. After downloading the file, open an instance of Power BI Desktop. At the lower right corner of the Visualizations toolbox, click the ellipses (see Figure 14-3). This will give you the option of adding or removing a custom visual.

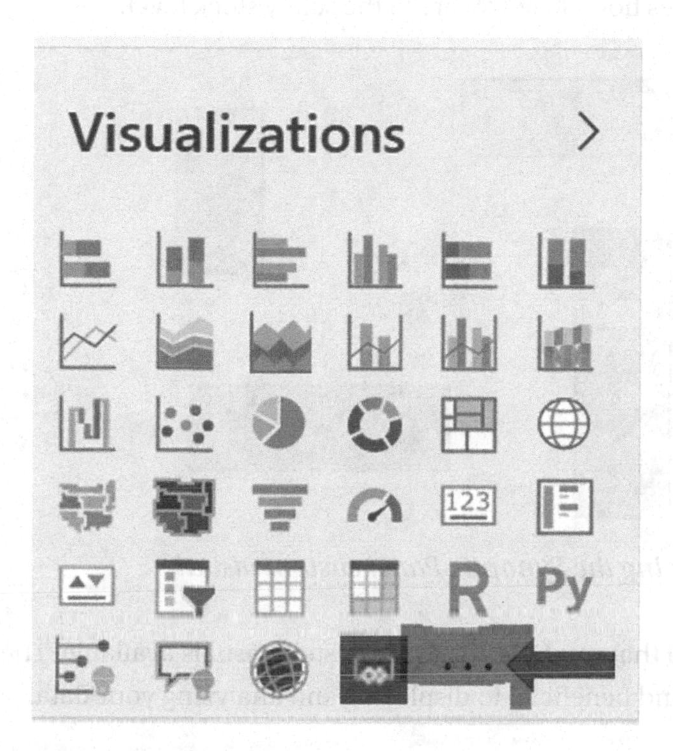

***Figure 14-3.** Adding a custom visual*

To add the visual, you just import the pbiviz file. After importing the visual, you will see a new visual icon in the toolbox.

It's a good idea to download and experiment with the demo file. They usually give you hints on using the visual and explain the various properties. Figure 14-4 shows the Bullet Chart in a demo report.

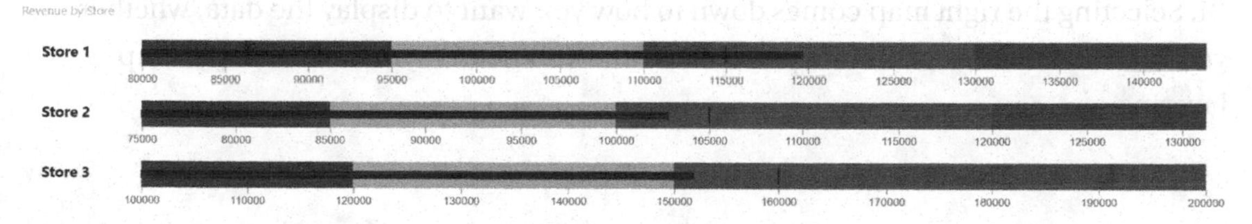

***Figure 14-4.** The Bullet Chart in a demo report*

Another place to get some interesting visuals is at the OKViz web site (`http://okviz.com`). They have a Synoptic Panel visual that allows you to create areas on images and assign colors and display information driven by your data. For example, you can create a map of a warehouse and display the number of parts in stock (see Figure 14-5). The color indicates how close you are to the safety stock level.

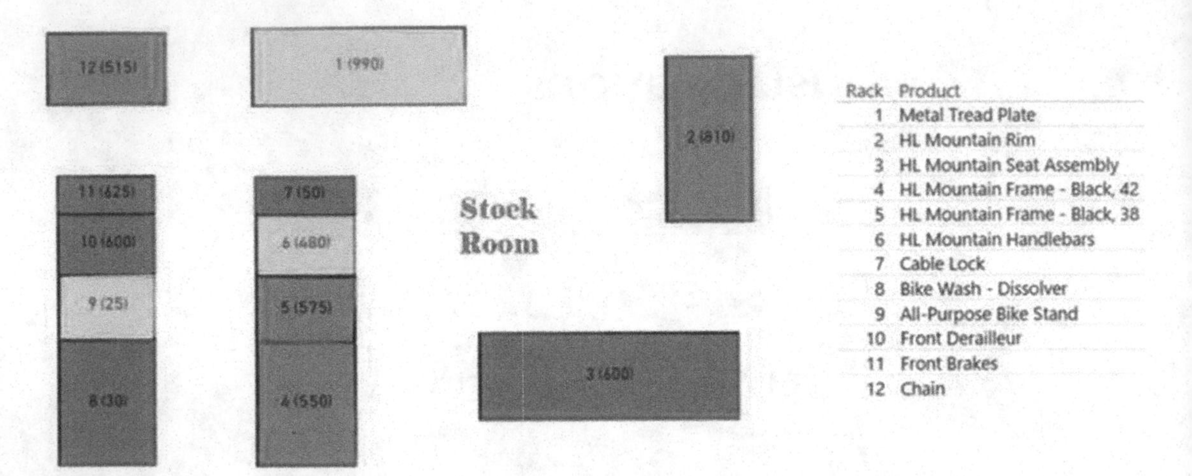

Figure 14-5. *Using the Synoptic Panel custom visual*

I recommend that you investigate the custom visuals available. There are many visuals you will find beneficial to displaying and analyzing your data.

Implementing Geospatial Analysis

As you saw in Chapter 8, if the data you are analyzing has a location component, displaying the data on a map is a powerful way to gain insight and analyze the data. There are many different options available when mapping data in Power BI. You can use the built-in map, filled map, or shape map. You can also use custom visuals such as the Synoptic Panel or the Globe Map. There is also an ESRI ArcGIS visual for Power BI. Selecting the right map comes down to how you want to display the data, whether you need to include custom maps, and whether you need to include multiple map layering.

If you need advance mapping capabilities such as clustering, heat maps, filtering, and multiple layers, there are several good third-party controls you can use. Two that are very good are the ArcGIS Map visual for Power BI by ESRI (`www.esri.com/software/arcgis/arcgis-maps-for-power-bi`) and the Mapbox visual by Mapbox (`https://www.mapbox.com/`). Figure 14-6 shows an ArcGIS Map depicting pothole locations imposed over a map layer showing population density.

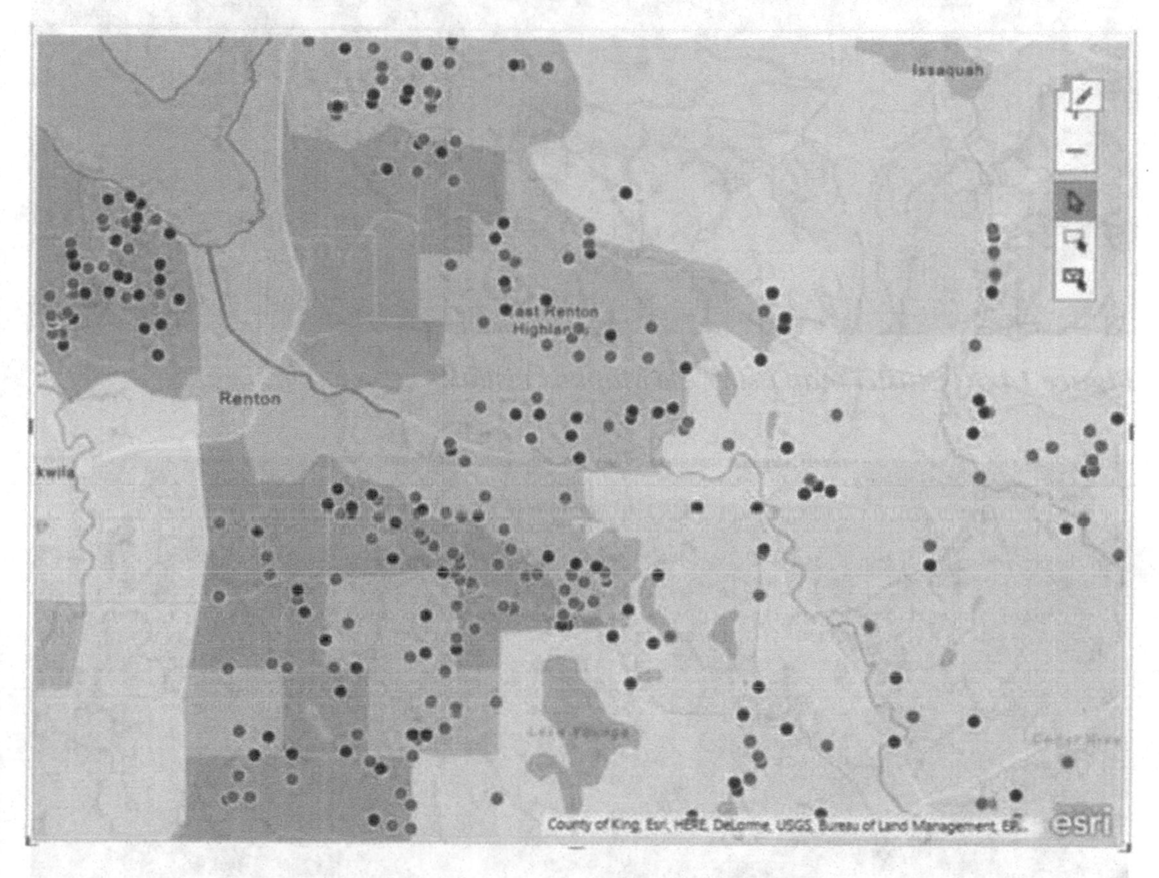

Figure 14-6. *A multilayer ArcGIS Map*

Figure 14-7 shows a clustered map using the Mapbox visual; as you drill into the map, you can see the individual data points that make up the clusters (see Figure 14-8).

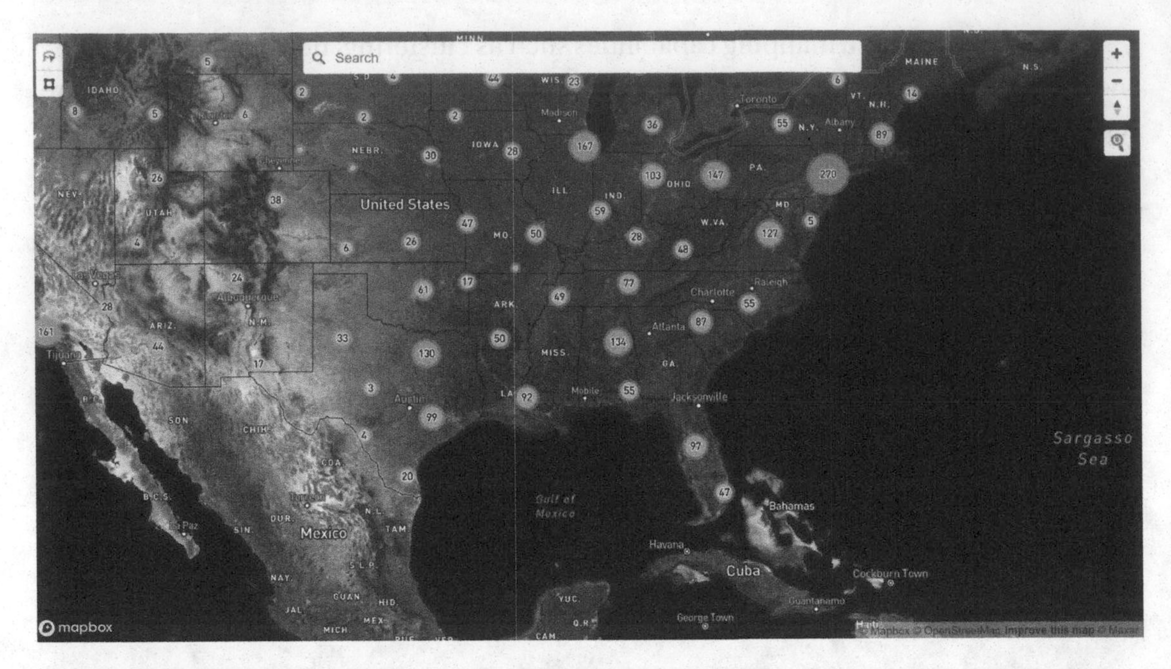

Figure 14-7. *Cluster map using the Mapbox visual*

Figure 14-8. *Drilling down to individual points*

You have many options when considering geospatial analysis using Power BI. If you just need basic mapping, the native map visuals are very capable. If you need more powerful features such as multilayering, clustering, and map-based filtering, there are third-party tools that are up to the task.

Implementing Row-Based Security

Security is an important aspect of any reporting system. It is critical that users only see data that they are authorized to see. To implement security in Power BI, you set up roles and use DAX to enforce data access rules. For example, you can create a role for USA Sales and authorize them to see sales for stores in the United States (see Figure 14-9).

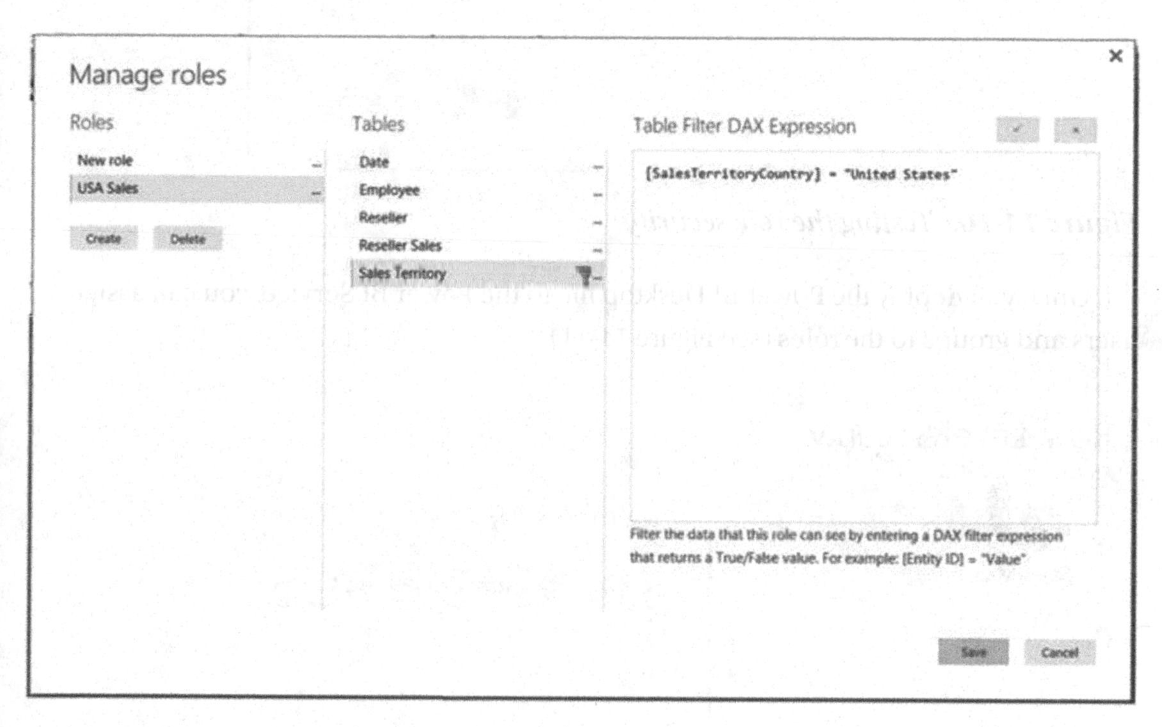

Figure 14-9. *Creating roles and DAX filters*

Once you set up the roles and rules, you can test the security by viewing the reports as a member of the role would (see Figure 14-10).

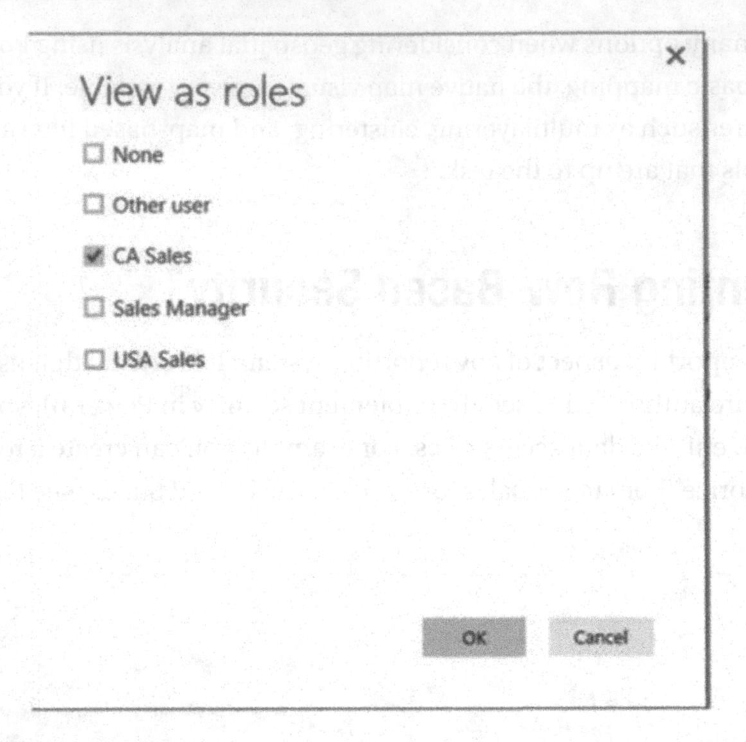

***Figure 14-10.** Testing the role security*

Once you deploy the Power BI Desktop file to the Power BI Service, you can assign users and groups to the roles (see Figure 14-11).

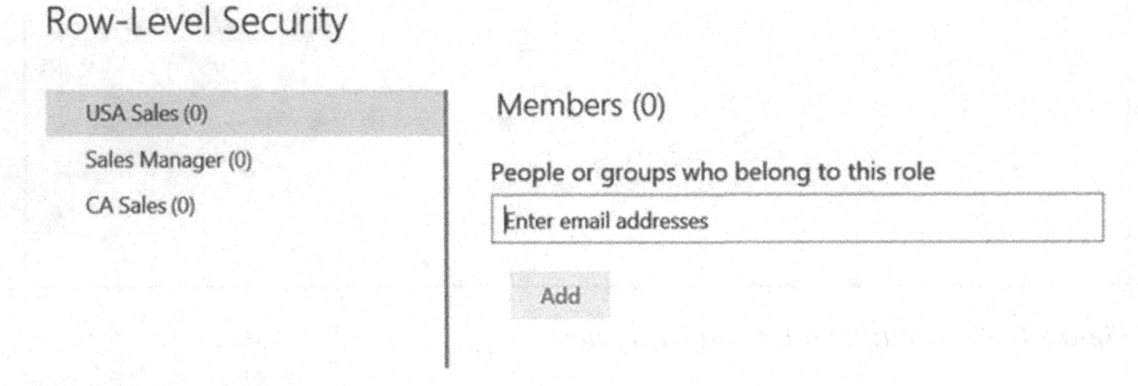

***Figure 14-11.** Assigning users and groups to a role*

Being able to secure your data is crucial to any successful enterprise reporting/analytics solution. Using row-level security is a great way to ensure that data is only exposed to users authorized to see it.

Note In order to implement row-level security, you need to have a Power BI Pro license.

Creating Templates and Content Packs

When sharing your Power BI Desktop files, you can make a copy of the pbix file, but this also contains the data. A better way is to create a template file based on the Power BI Desktop and share that. The template doesn't contain the data but does contain all definitions of the data model, reports, queries, and parameters. When you import the template, you are prompted for any parameters defined in the file. This is very useful when the file is moved to a new environment. For example, when you share a file with a colleague, they can easily specify the file path to a local data source.

To create a template file, click File ➤ Export in Power BI Desktop and choose the Power BI Template (see Figure 14-12).

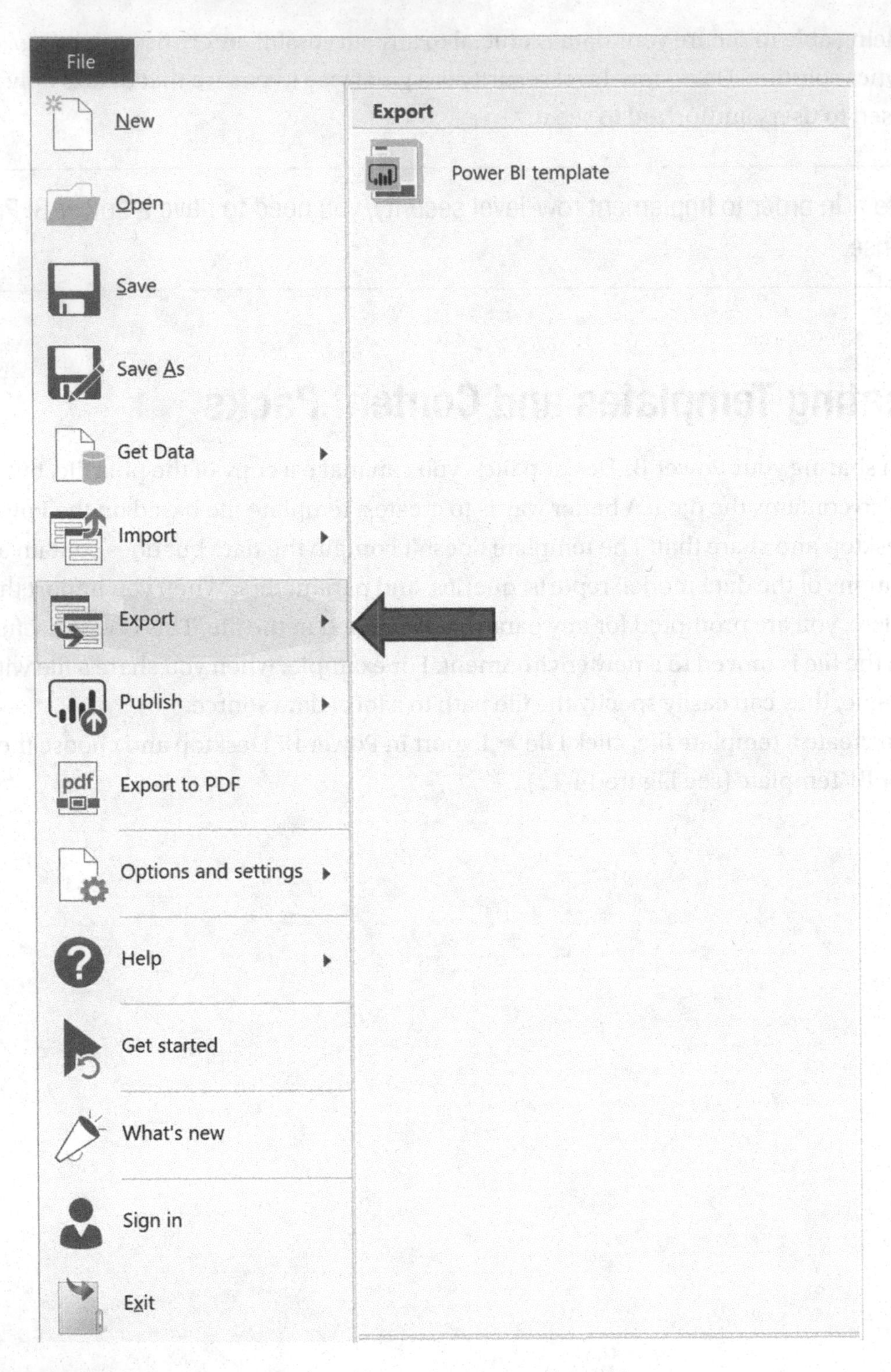

***Figure 14-12.** Creating a Power BI Desktop template file*

Once you save the file, it will have a pbit extension. When a user opens the template file, they are asked to provide any parameter values (see Figure 14-13).

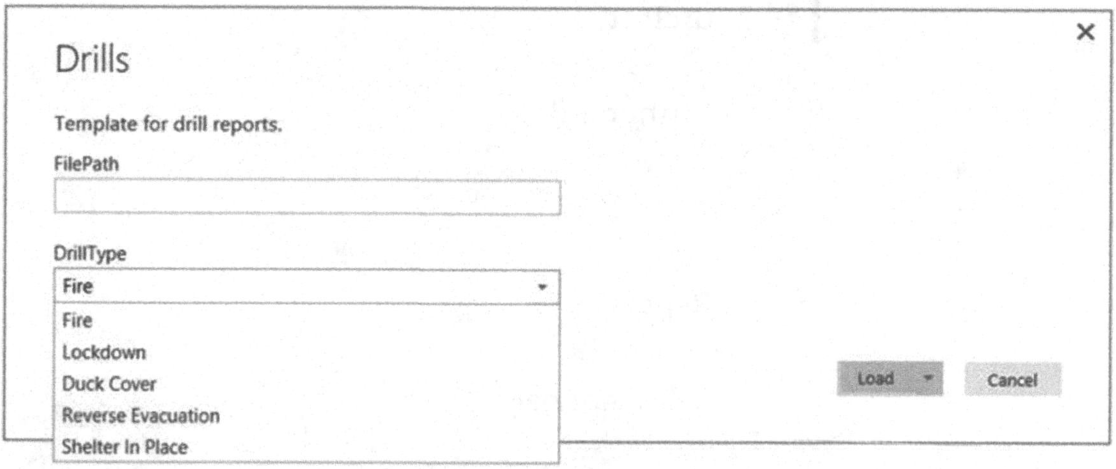

Figure 14-13. *Providing parameter values*

When the user saves the file, it becomes a Power BI Desktop file with the pbix extension.

Although using templates is a great way to share Power BI Desktop files, if you want to package up a complete solution for deployment to the Power BI Service, content Power BI Apps are the way to go. With Power BI Apps, you can package up your dashboards and reports for your colleagues to use. Once you create the app, you publish it to the Power BI Service, where you can expose it to the entire organization or members of a security group. If you restrict the app to a specific group, only members of the group will be able to see it in the app library. Members of the group have read-only access to the reports and dashboards. The app owner can modify and republish the app. Users of the app will automatically see the updates.

To create an app, deploy the Power BI Desktop file to a Power BI Service workspace. Create any dashboards you want to include in the app (see Figure 14-14).

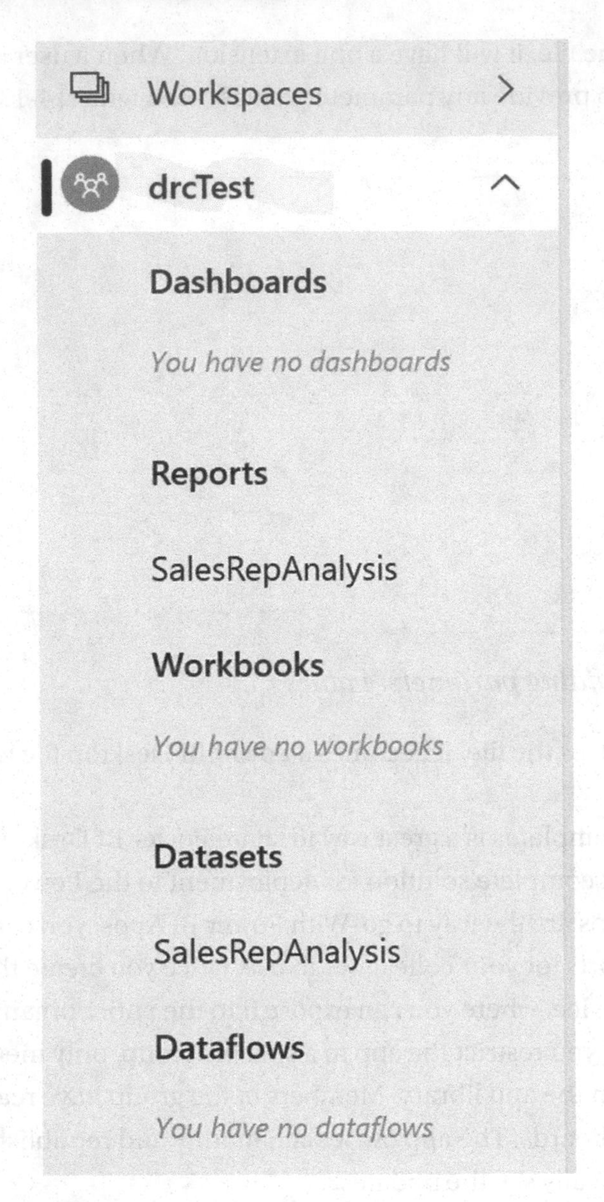

***Figure 14-14.** Selecting the workspace node*

Choose the reports and dashboards you want to include in the app (see Figure 14-15).

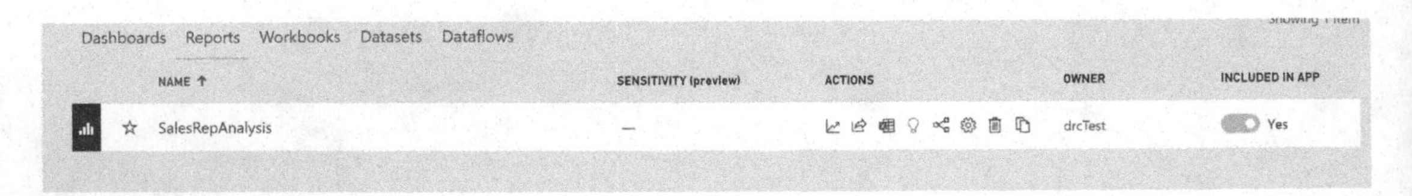

***Figure 14-15.** Choosing the reports and dashboards*

Once you select the content for the app, click the Publish app link in the upper right-hand location of the portal (see Figure 14-16).

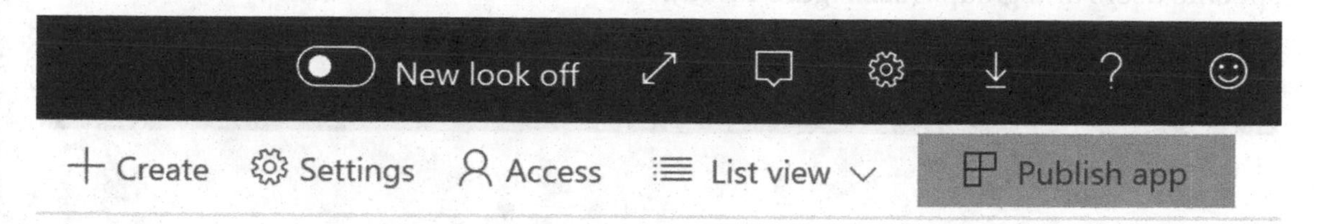

Figure 14-16. *Publishing the app*

Next, you set the app name, description, and optional support site. You can also upload a logo and pick a color theme (see Figure 14-17).

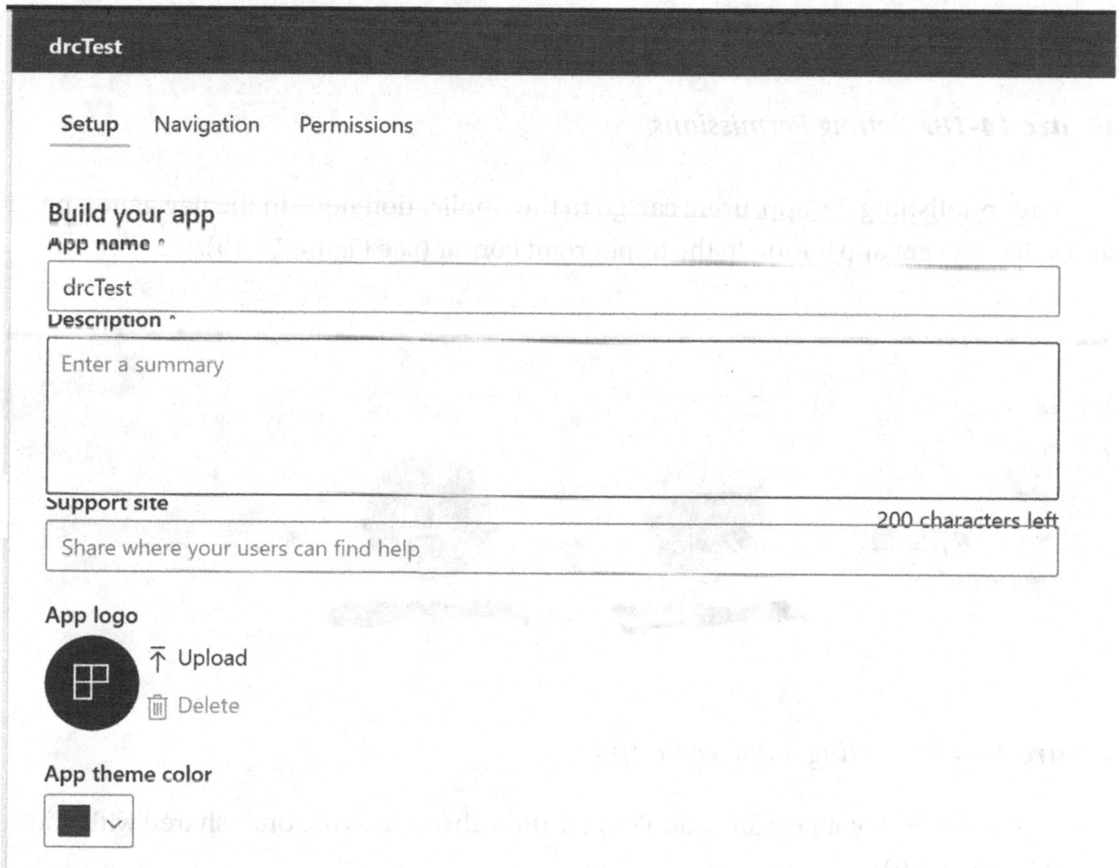

drcTest
Setup Navigation Permissions
Build your app
App name *
drcTest
Description *
Enter a summary
Support site
200 characters left
Share where your users can find help
App logo
Upload
Delete
App theme color

Figure 14-17. *Setting app properties*

On the navigation tab, you can set up a custom navigation pane for the app. Using the permissions pane, you can choose to expose the app to the entire organization or specific users and groups (see Figure 14-18).

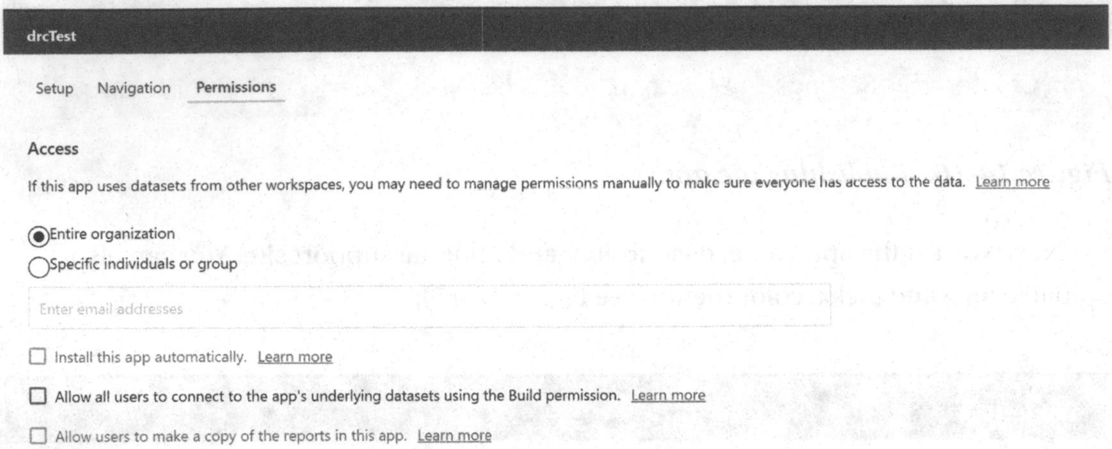

drcTest

Setup Navigation Permissions

Access

If this app uses datasets from other workspaces, you may need to manage permissions manually to make sure everyone has access to the data. Learn more

Entire organization

Specific individuals or group

Enter email addresses

Install this app automatically. Learn more

Allow all users to connect to the app's underlying datasets using the Build permission. Learn more

Allow users to make a copy of the reports in this app. Learn more

Figure 14-18. *Setting Permissions*

After publishing the app, users can go to the Application node in the navigation pane and select the get app button in the upper right corner (see Figure 14-19).

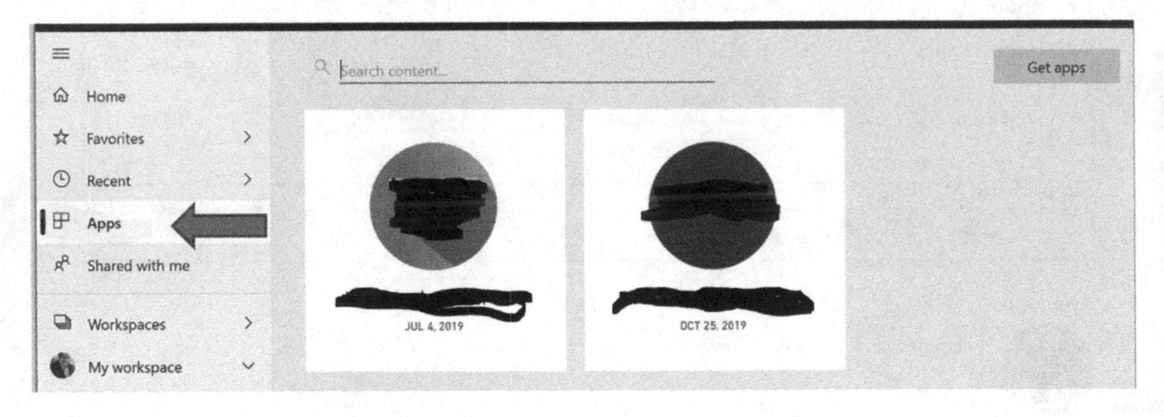

Figure 14-19. *Getting a Power BI App*

You can search your organization's apps and subscribe to the ones shared with you (see Figure 14-20).

***Figure 14-20.** Selecting apps shared with you*

In addition to your organizational apps, many apps from Microsoft and other vendors are available (see Figure 14-21). For example, there are Power BI Apps for Salesforce, Dynamics CRM, Office 365, and Azure Audit Logs. These allow you to get up and running quickly and deploy dashboards related to the service.

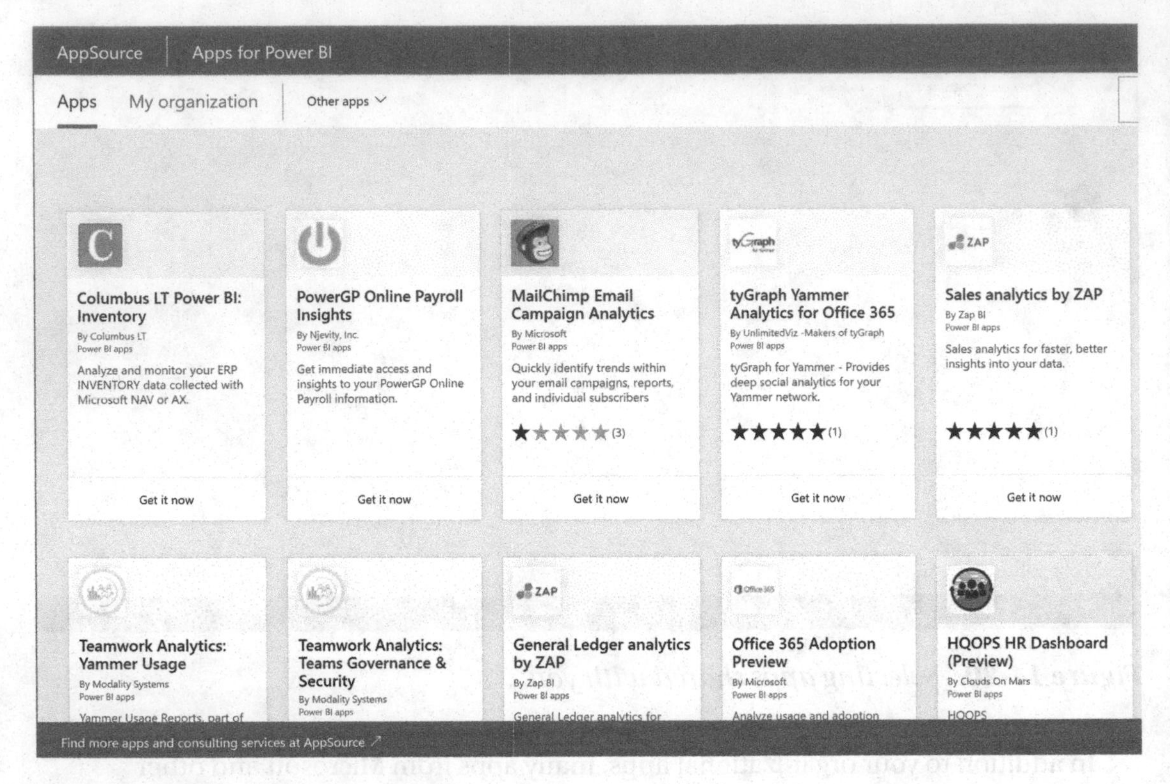

Figure 14-21. *Selecting a Power BI App from vendors*

Note To create Power BI Apps, you need to have a Power BI Pro license.

Now that you are familiar with some advanced features of Power BI Desktop, you are ready to gain some hands-on experience with these features in the following lab.

HANDS-ON LAB: ADVANCED TOPICS IN POWER BI

In the following lab, you will

- Use some custom visuals
- Implement row-level security
- Create a Power BI template

1. In the LabStarterFiles\Chapter14Lab1 folder, open the CustomVisuals.pbix file. This file contains data on sales and quotas for sales reps.

2. In the LabStarterFiles\Chapter14Lab1 folder, there is a Visuals folder that contains several custom visual pbviz files. Add these visuals to the Visualizations toolbox in Power BI Desktop. After importing the visuals, you should see the new icons in the toolbox (see Figure 14-22).

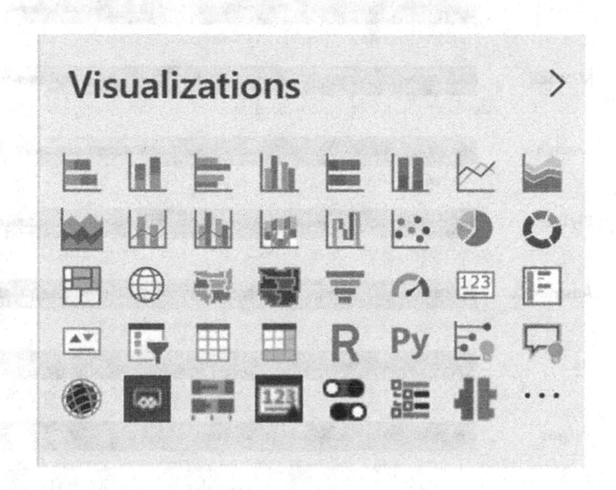

***Figure 14-22.** Adding custom visuals to the toolbox*

3. Add the following visuals and fields to the wells, as indicated in Table 14-1.

***Table 14-1.** Adding Visuals and Fields to the Wells*

Visual	Well	Fields
HierarchySlicer	Fields	Period.Year Period.Quarter
BulletChart	Category	Rep.LastName
	Value	Sales.TotalSales
	TargetValue	Quota.Quota
	Minimum	Quota.SeventyPercentQuota
	Satisfactory	Quota.NinetyPercentQuota
	VeryGood	Quota.Quota

4. The final report page should be similar to Figure 14-23. Investigate the different formatting available in the visuals.

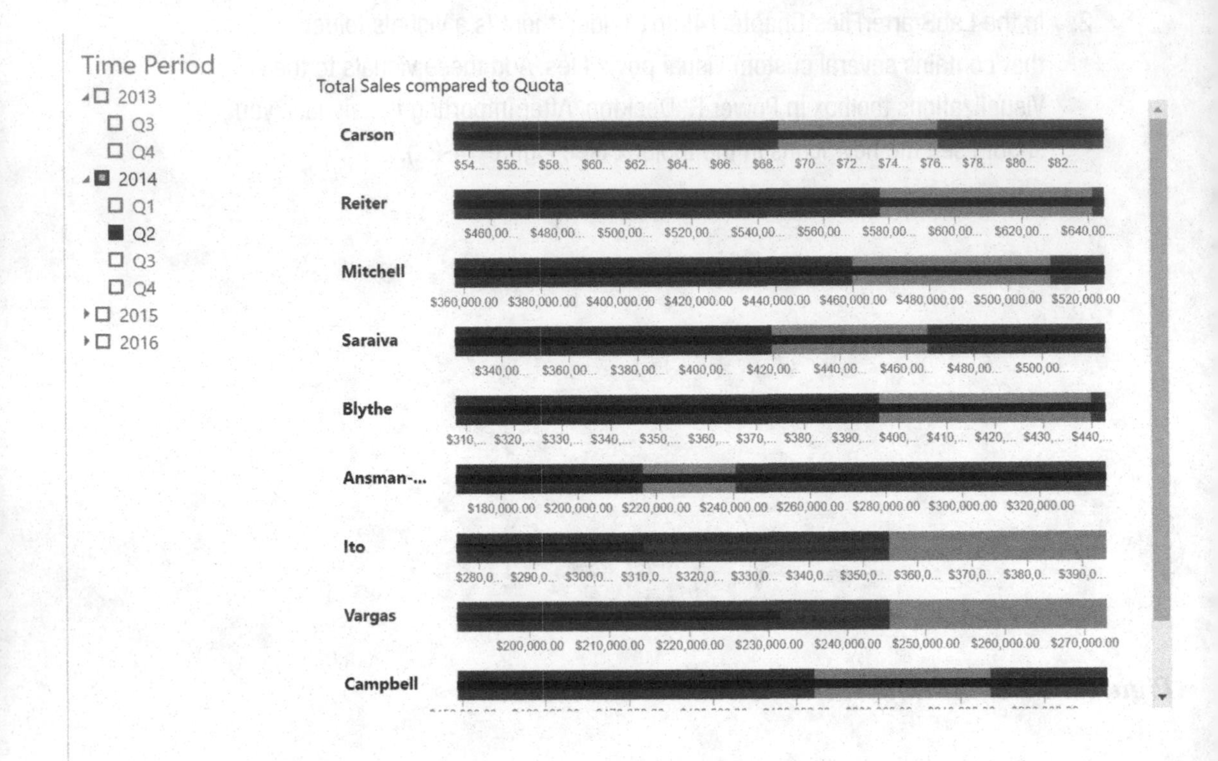

Figure 14-23. Report, Page 1, with custom visuals

5. Add a second page to the report and add the visuals and fields to the wells as shown in Table 14-2.

Table 14-2. Adding Page 2 Visuals and Fields to the Wells

Visual	Well	Fields
Tornado	Group	Territory.Region
	Values	Sales.Sales2015Q1
		Sales.Sales2016Q1
WordCloud	Category	Rep.LastName
	Values	Sales.TotalSales

6. The final report page should be similar to Figure 14-24. Take some time to investigate the different formatting available in the visuals. Close and save the file when finished.

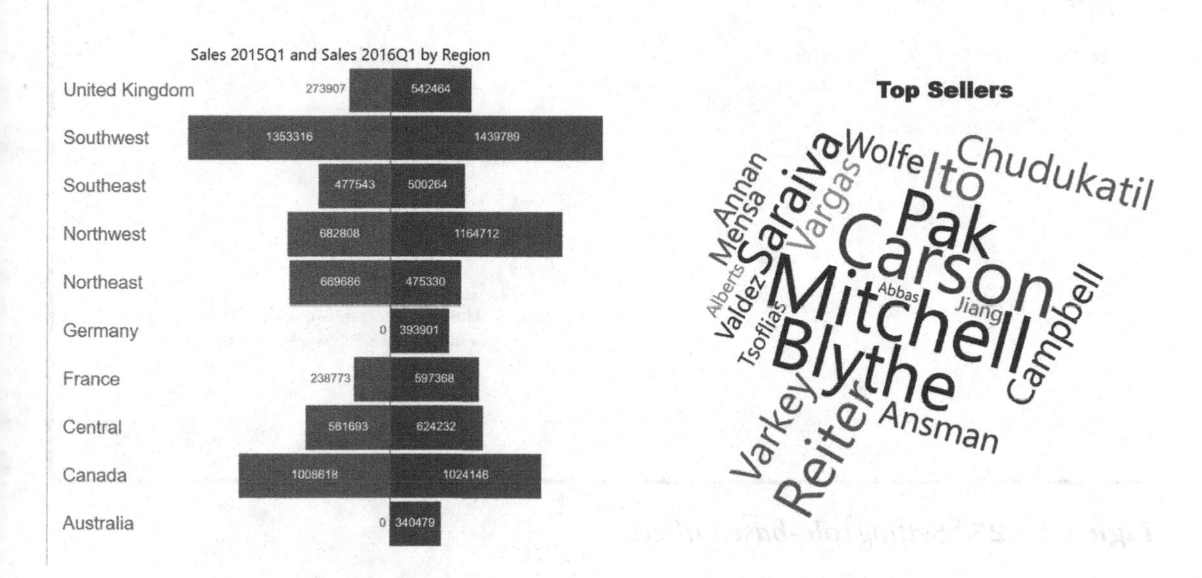

Figure 14-24. Report, Page 2, with custom visuals

7. In the LabStarterFiles\Chapter14Lab1 folder, open the RowLevelSecurity.pbix file.
8. On the Modeling tab, select Manage Roles. Add the NorthAmericanSales, PacificSales, and EuropeanSales roles.
9. Using a DAX filter, restrict the territories they can see. Figure 14-25 shows the DAX filter for the PacificSales role.

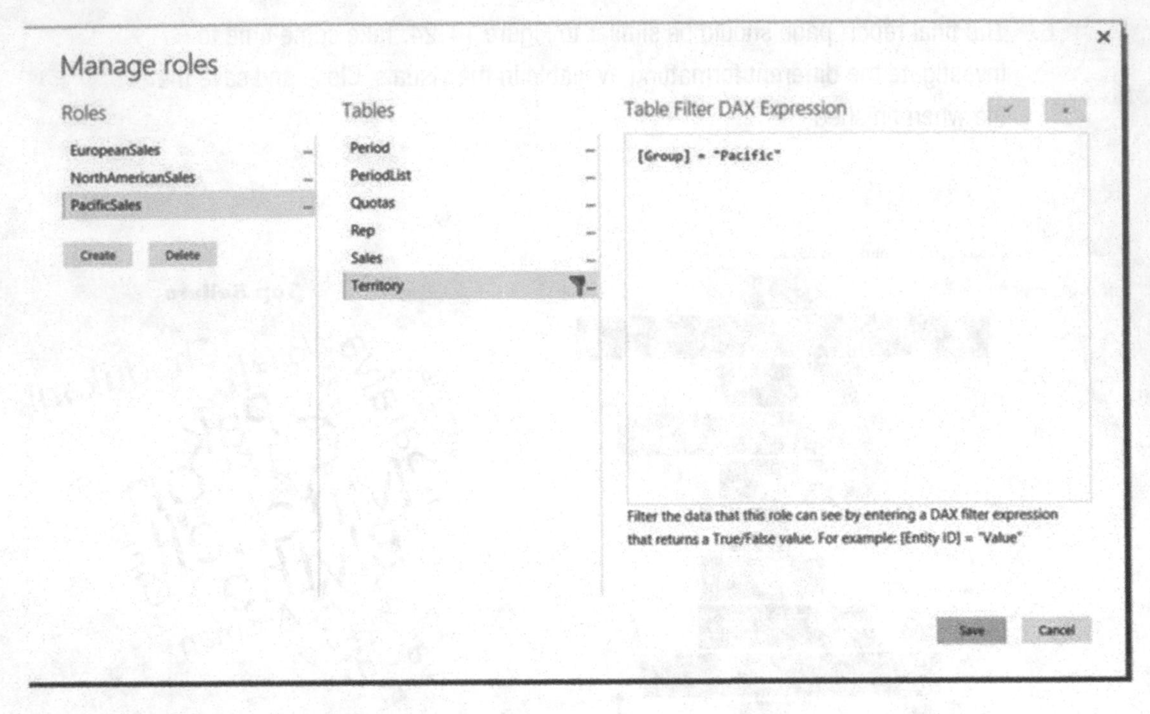

***Figure 14-25.** Setting role-based filters*

10. Test the filter by selecting the View as Roles. If you view Page 1 of the report as the EuropeanSales role, you should only see France, Germany, and United Kingdom data (see Figure 14-26).

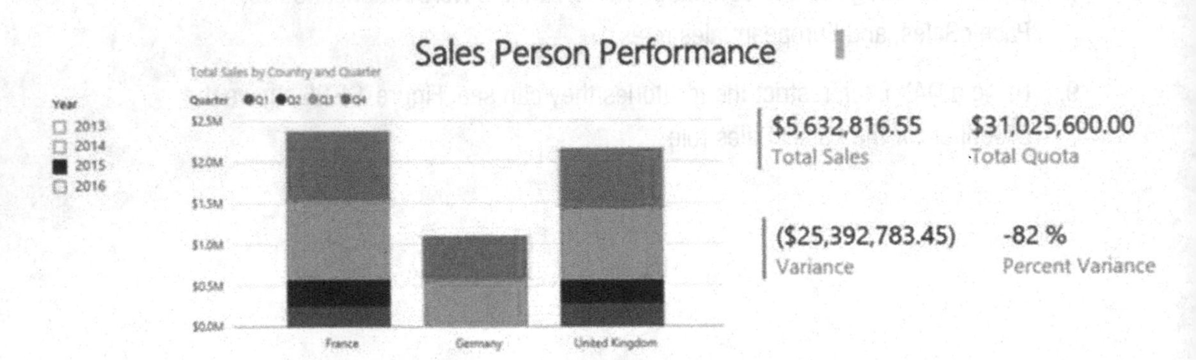

***Figure 14-26.** Testing the EuropeanSales role*

11. Once done testing the roles, save the file. On the File tab, click Export and export the file as a Power BI Template file. Add a template description and save the file as TemplateDemo.pbit.
12. Close Power BI Desktop and double-click the template file. This should launch Power BI Desktop and ask you for the folder path. Provide the path to the data folder which holds the SalesRepAnalysis.accdb file and click Load.
13. After the data loads, save the file as TemplateDemo.pbix.
14. Close Power BI Desktop and compare the size of the template file with the original pbix file. The template file doesn't contain data, so it should be much smaller.

Summary

This chapter covered some advanced topics in Power BI, including using custom visuals, advanced mapping, row-based security, templates, and Power BI Apps.

The next chapter is a new chapter I have added to this edition of the book. It covers some advanced topics in Power BI data modeling that I think you will find useful. It includes direct queries, composite models, and dataflows. These topics are very important as you move to larger data models and toward a centralized data model for the enterprise.

CHAPTER 15

Advanced Topics in Power BI Data Modeling

This chapter covers some advanced topics in Power BI data modeling that I think you may find useful. These are important for analyzing data over very large data sets. It includes direct queries, aggregation tables, and dataflow.

After completing this chapter, you will be able to

- Determine when to use DirectQuery
- Use aggregation tables
- Use dataflows to populate a Common Data Model

Direct Queries

Throughout this book we have been importing the data into the Power BI tabular model. For most cases this is the preferred method for connecting to data sources and working with the data. If the data in the source changes, we need to reconnect to it and import the data. There is no permanent connection maintained to the source data. After importing the data, it is highly compressed and the size of the data in Power BI is much smaller than the size of the data in the source system. By importing the data, you get the full functionality of Power BI and Power Query. You can import data from multiple sources and combine them into the model. You can schedule refreshes up to eight times per day if needed (more if you use Power BI Premium). Because the data is contained in the model, this gives you the maximum performance and response time.

Although in most cases importing the data is your best option, there are some use cases when you need to choose the direct query option for connecting to the data. Power BI has a limit of 1 GB per model (unless you are using Power BI Premium). So, if you

D. Clark, *Beginning Microsoft Power BI*, https://doi.org/10.1007/978-1-4842-5620-6_15

need to deal with very large data sets, you must use the DirectQuery option. Another use case for using the Direct Query option is the volatility of your data. If the data in your data source is continuously updating and you need to reflect the changes in your reports, then you need to use the direct query option.

Although you need to use DirectQuery in these limited use cases, it comes with many downsides. The biggest problem with DirectQuery is performance. Every time you interact with your report, several queries are issued back to the data source and, depending on the source and your network connection, this can be very slow and even time out. Another big disadvantage with DirectQuery is that it only supports a subset of the DAX functions.

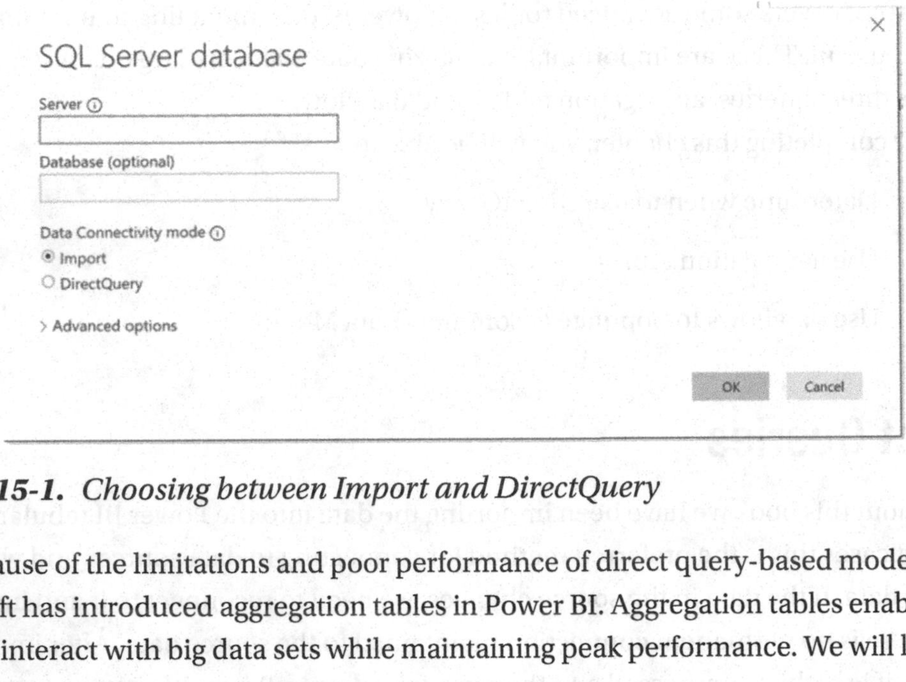

Figure 15-1. *Choosing between Import and DirectQuery*

Because of the limitations and poor performance of direct query-based models, Microsoft has introduced aggregation tables in Power BI. Aggregation tables enable users to interact with big data sets while maintaining peak performance. We will look at these next.

Using Aggregation Tables

Analyzing big data sets is a real challenge when it comes to performance. If the data set is too big to import into the data model, you will need to use DirectQuery which has some real performance issues. One way to get around this is to analyze your queries and preaggregate the data before it is imported into the model. While this works well, what happens when you need to drill down to the individual record level? This is where aggregate tables built into and managed by the Power BI model shine.

Aggregate tables boost query performance by caching data at the aggregate level and issuing direct queries at the detail level. So, if you wanted to look at sales data rolled up by month and year, it would come from the aggregate table. At the same time, you could drill down to see the sales for a single day. This would issue a direct query that would perform quickly because you are only returning a small subset of the data.

As an example, suppose we have the data model shown in Figure 15-2. It contains a sales fact table and several dimension tables. The sales table contains billions of rows, so we are initially using a direct query to the data source.

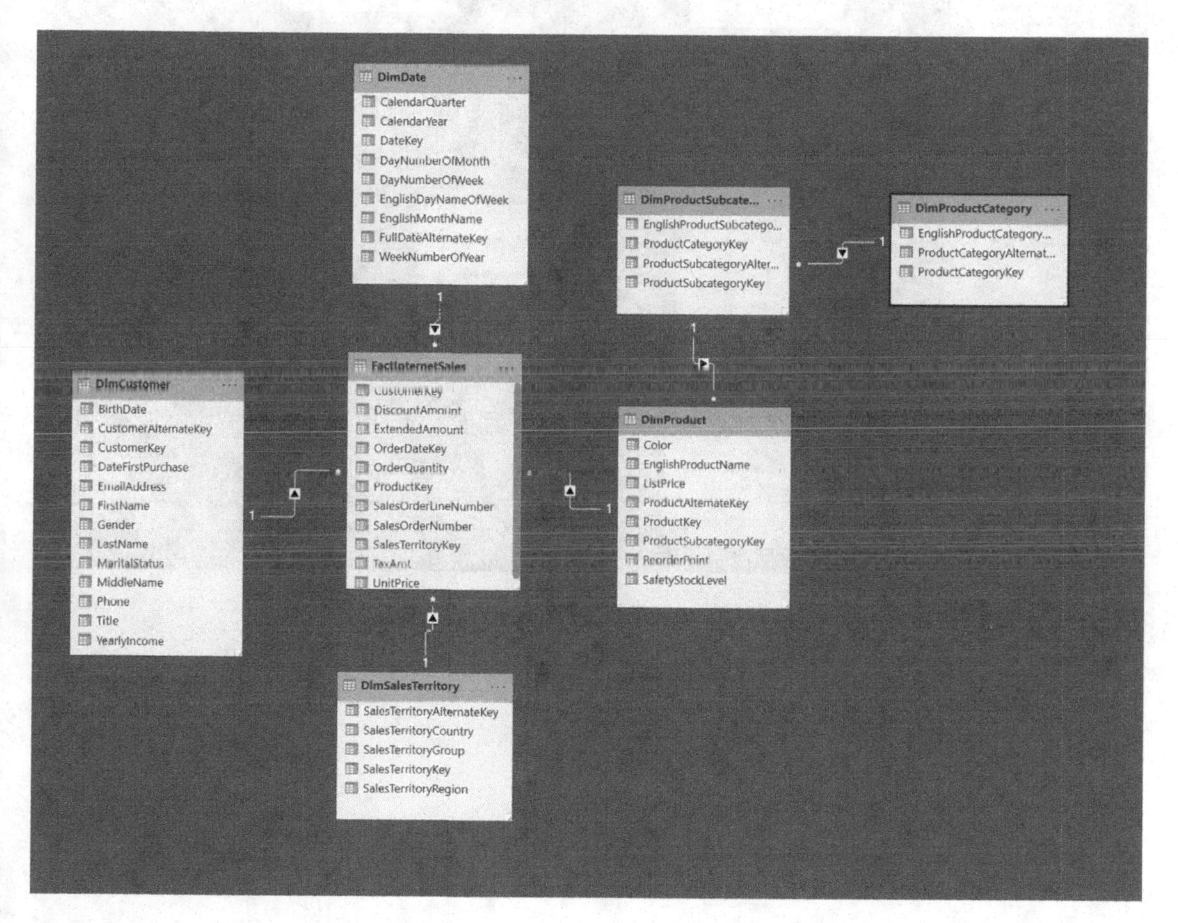

Figure 15-2. *A big data model*

Since most of the time the sales are analyzed by grouping at the customer, date, and/or the product subcategory levels, we can create a sales fact table rolled up to the CustomerKey, DateKey, and ProductSubcategoryKey levels. The aggregate table can be created in the data source or in Power Query. After creating the sales aggregate table, it is added to the model as shown in Figure 15-3.

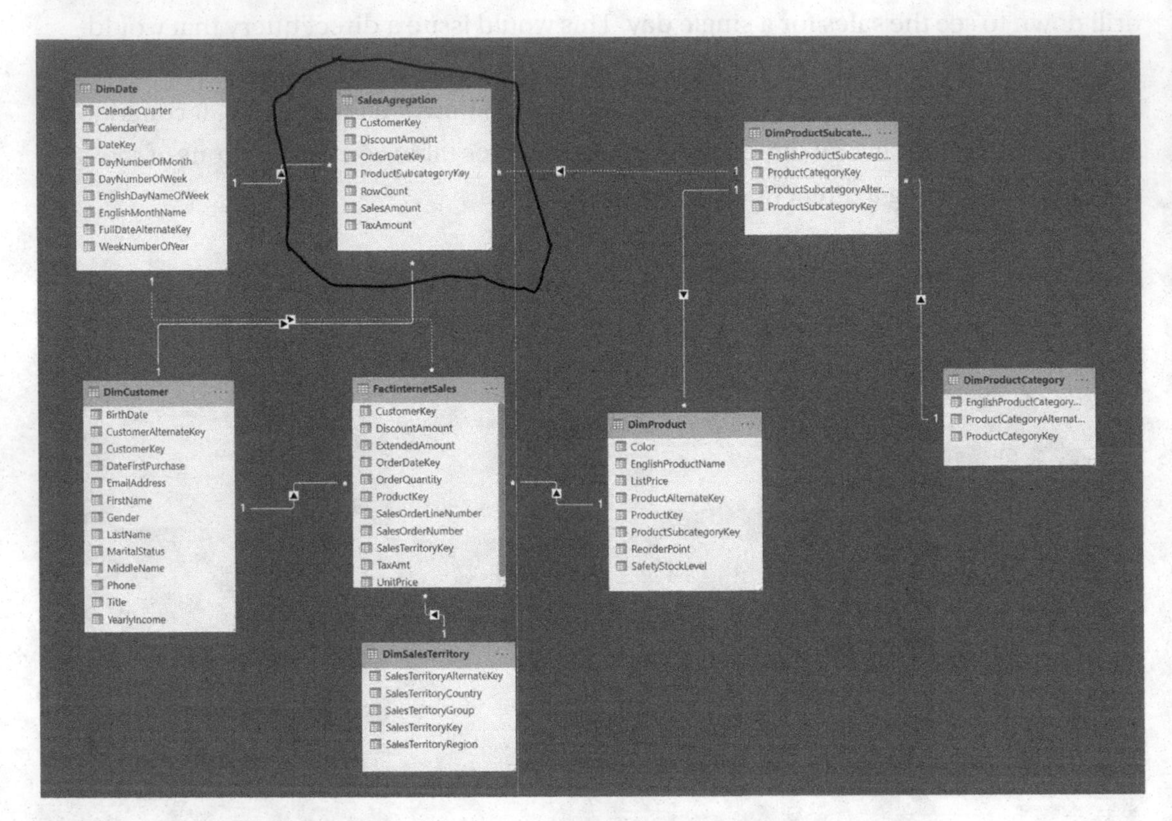

Figure 15-3. *Adding an aggregate table to the model*

Next, we set the storage mode of the sales aggregation to import (see Figure 15-4).

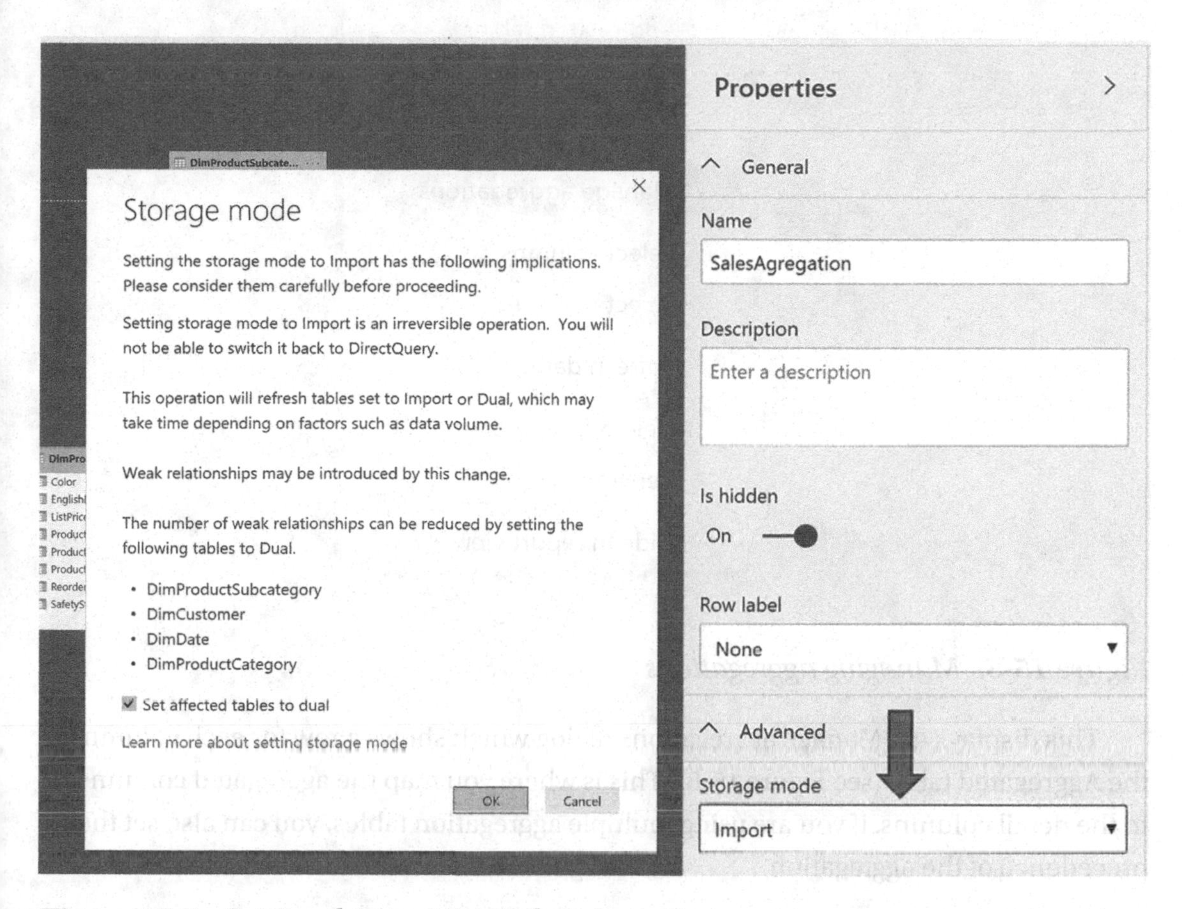

Figure 15-4. Setting the storage mode to import

Once you set the storage mode, you will see a dialog asking if we want to set the related tables to the dual storage mode. This will allow the related tables to either get data from imported data or from a direct query depending on whether data is retrieved from the sales aggregate table or the sales table. To set this up, you need to right-click the table in the field list pane and select the manage aggregation option (see Figure 15-5).

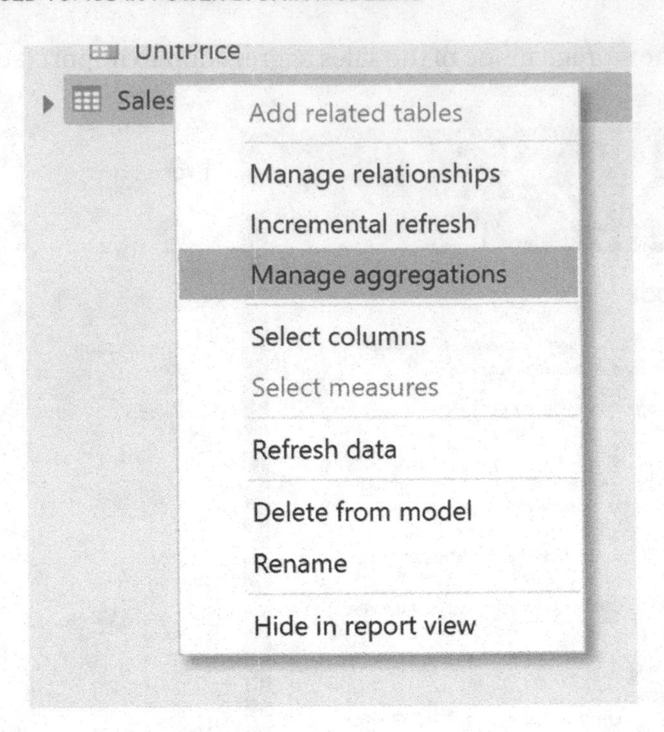

Figure 15-5. *Managing aggregations*

This displays the Manage aggregations dialog which shows a row for each column in the Aggregated table (see Figure 15-6). This is where you map the aggregated columns to the detail columns. If you are using multiple aggregation tables, you can also set the precedence of the aggregation.

Manage aggregations

Aggregations accelerate query performance to unlock big-data sets. Learn more

Aggregation table: SalesAgregation

Precedence: 0

AGGREGATION COLUMN	SUMMARIZATION	DETAIL TABLE	DETAIL COLUMN
CustomerKey	GroupBy	FactInternetSales	CustomerKey
OrderCount_Sum	Sum	FactInternetSales	OrderQuantity
OrderDateKey	GroupBy	FactInternetSales	OrderDateKey
ProductSubcategoryKey	GroupBy	DimProduct	ProductSubcategory...
SalesAmount_Sum	Sum	FactInternetSales	SalesAmount

This table will be hidden if aggregations are set because aggregation tables must be hidden.

Apply all | Cancel

***Figure 15-6.** Mapping the aggregations*

Looking at the visuals in Figure 15-7, the visual on the left uses the aggregation table while the one on the right needs to issue a direct query back to the data source.

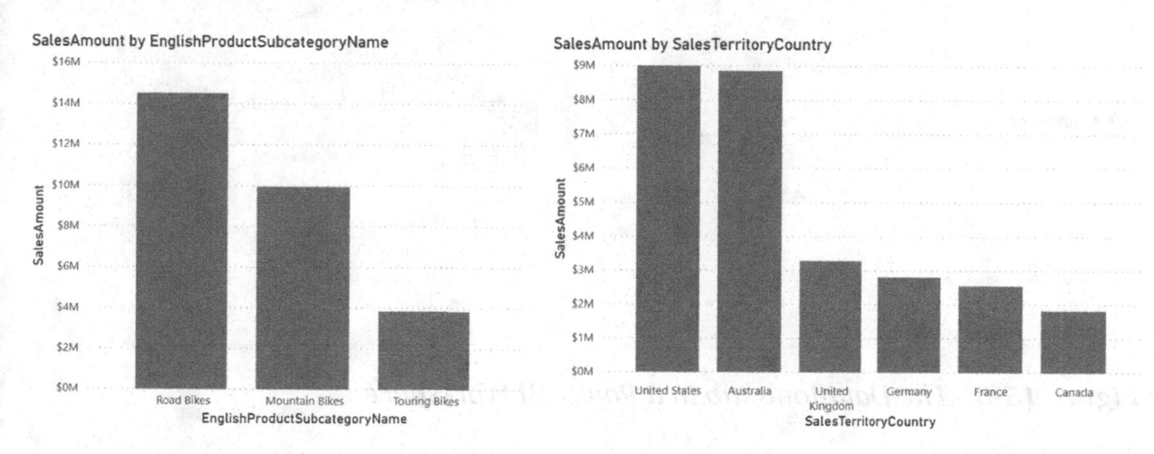

***Figure 15-7.** Comparing visual queries*

Aggregation tables are a great way to increase performance when working with large data sets. Another challenge you run into when working with large data sets is prepping the data. This is where dataflows come into play as you will see next.

Implementing Dataflows

Dataflows are used to prepare big data for use in reporting. Instead of working with data in Power BI Desktop, you work with the data in the Power BI Service. The cool thing about dataflows is that it uses Power Query for data preparation and provides a very similar experience to working with Power Query in the Power BI Desktop. Dataflows are designed to use the Common Data Model. Data in the Common Data Model is stored in Azure Data Lake, which is designed to work efficiently with big data. The Common Data Model is also a central repository where you can share data across the enterprise as well-defined entities that have consistent schemas and data linage.

To create a dataflow, you need to go to the Power BI Service and create/select a workspace other than the default My workspace (see Figure 15-8).

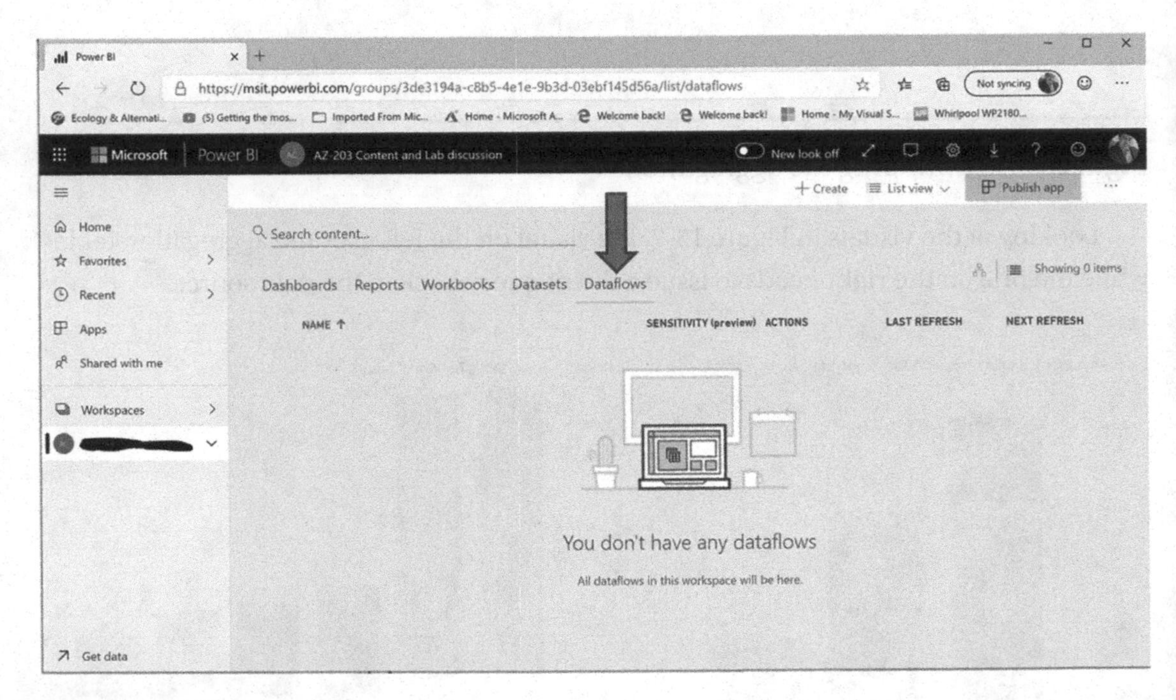

***Figure 15-8.** The Dataflows tab in a Power BI Workspace*

Use the Create button in the top right of the page and select Dataflow in the drop-down list (see Figure 15-9).

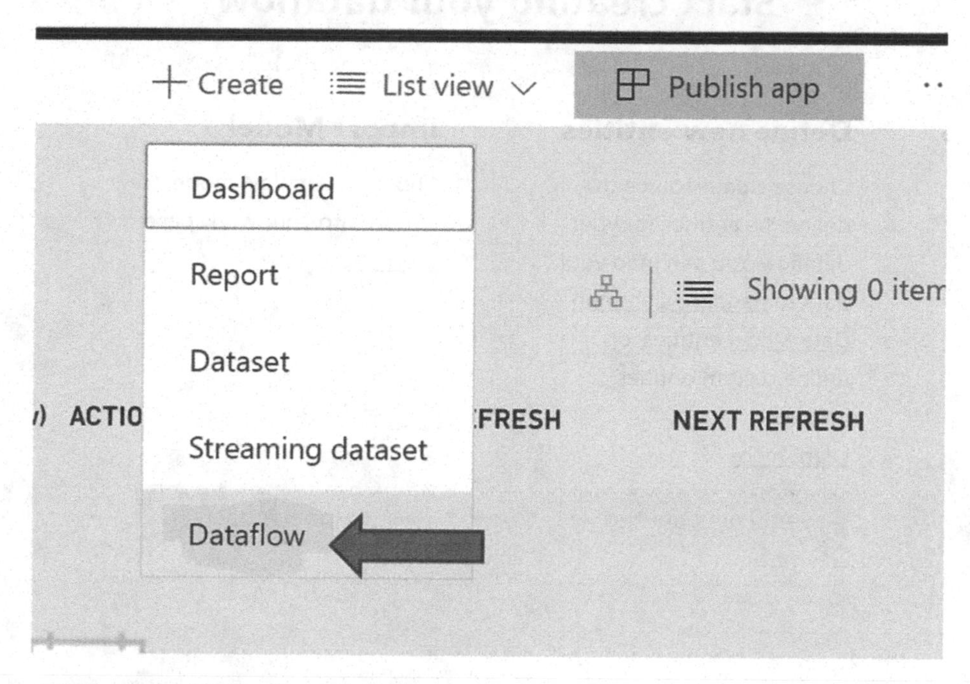

***Figure 15-9.** Creating a new dataflow*

You will then be given the option to create or import an existing entity (see Figure 15-10).

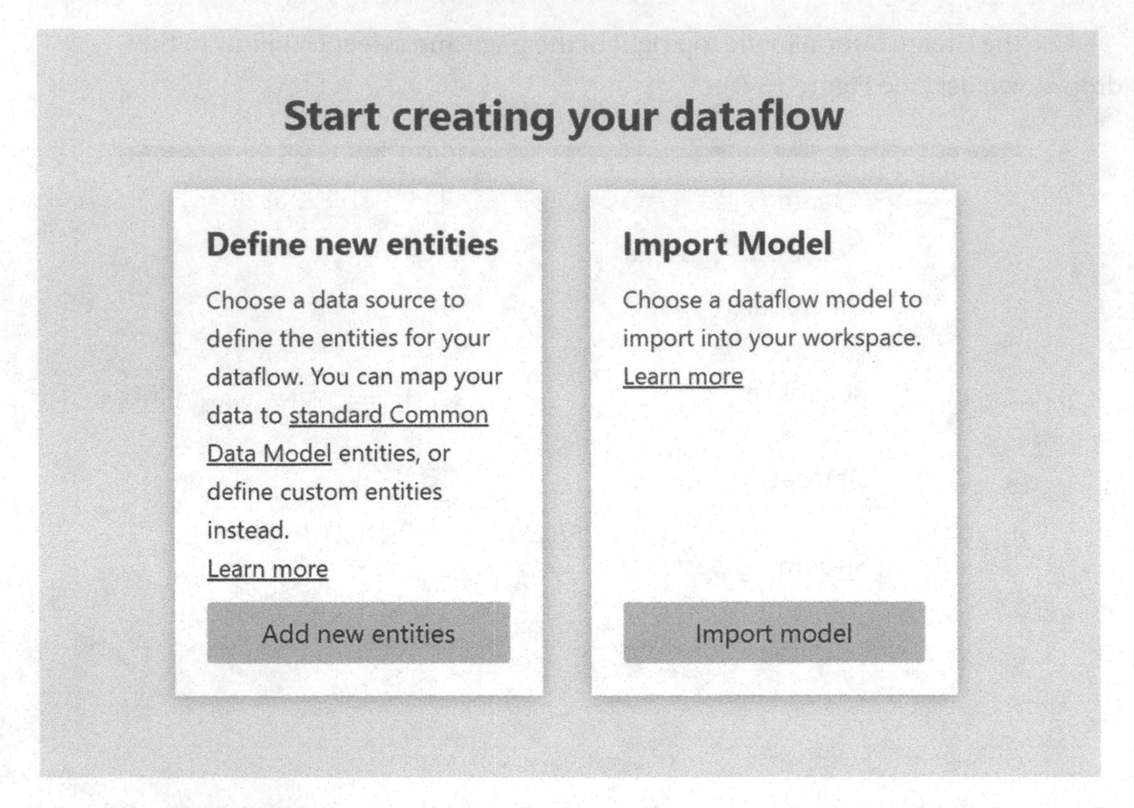

Figure 15-10. *Defining a new entity*

Selecting Add new entities will present you with a screen where a data source can be selected (see Figure 15-11).

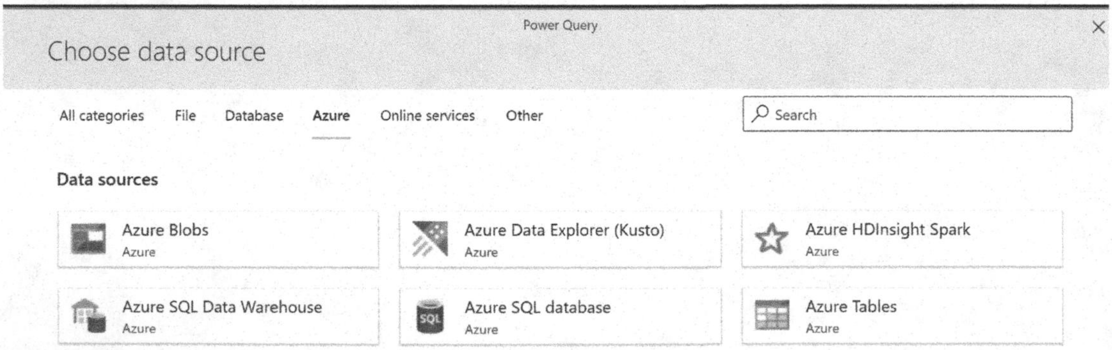

Power Query
Choose data source

All categories File Database Azure Online services Other

Search

Data sources

Azure Blobs
Azure

Azure Data Explorer (Kusto)
Azure

Azure HDInsight Spark
Azure

Azure SQL Data Warehouse
Azure

Azure SQL database
Azure

Azure Tables
Azure

Figure 15-11. *Selecting a data source*

After selecting a data source, you provide connection information and connect to the data source (see Figure 15-12).

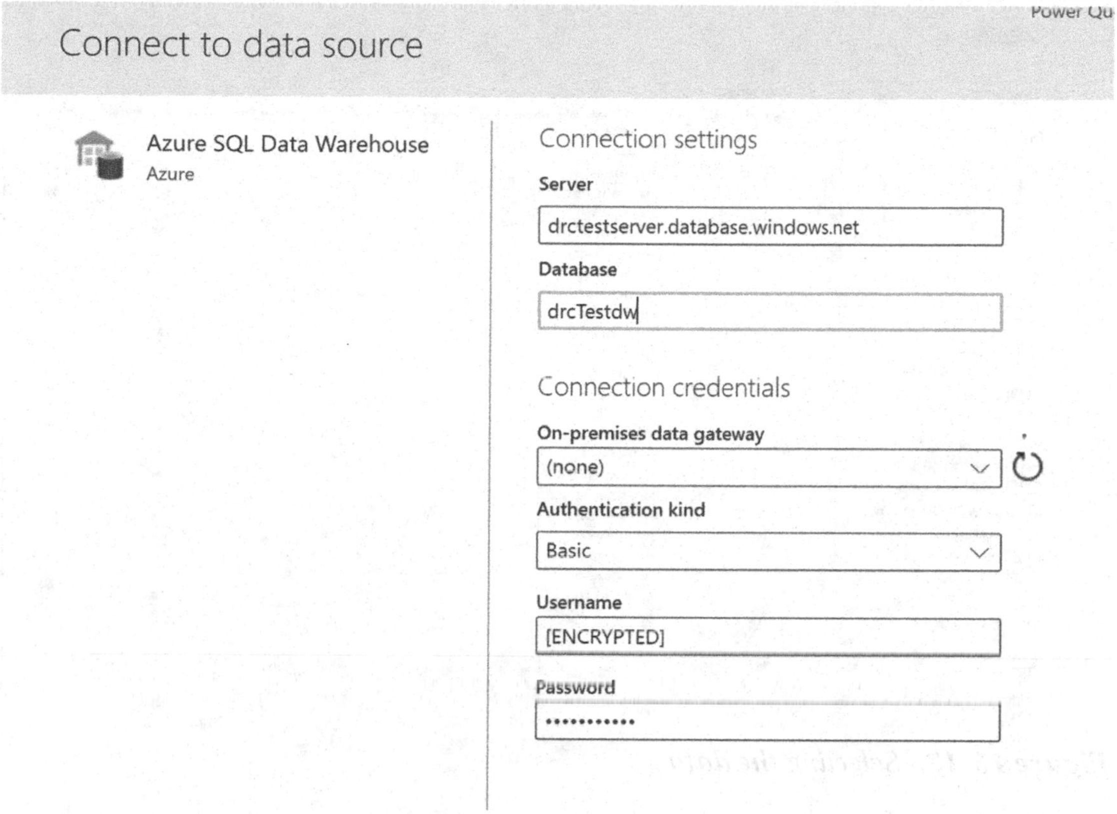

Figure 15-12. Connecting to a data source

Once you connect to the data source and select the data, you can transform the data (see Figure 15-13).

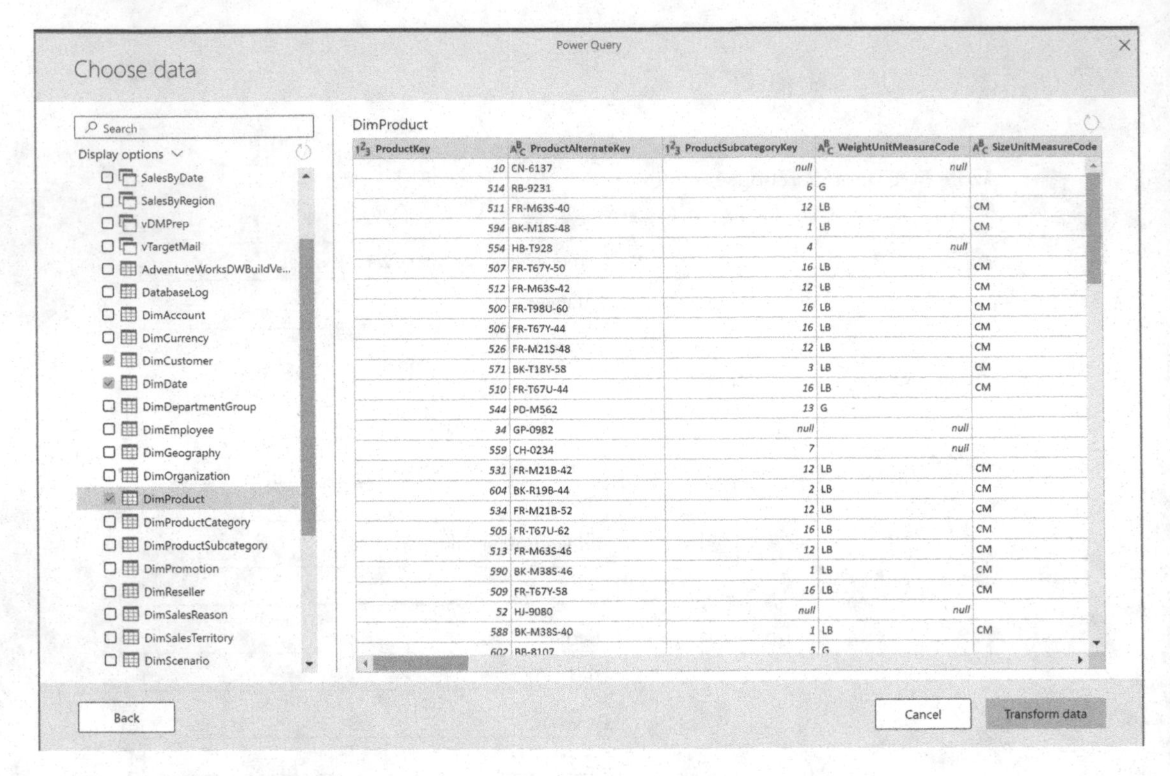

Figure 15-13. *Selecting the data*

Selecting the Transform data button launches the Power Query interface (see Figure 15-14).

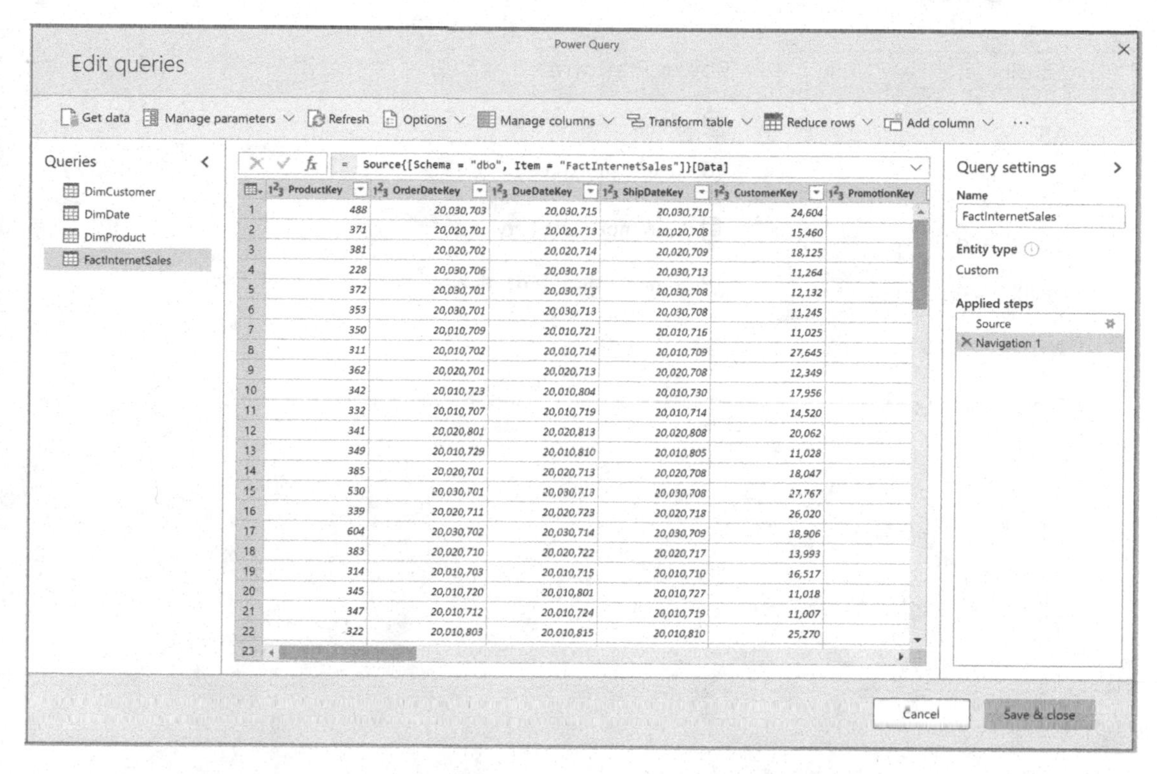

Figure 15-14. *Power Query online*

Although not exactly the same as Power Query in the Power BI Desktop, it is very similar, and you should be quite comfortable with transforming data using Power Query online. Once you save and close the queries, they become entities stored in Azure Data Lake and can be integrated into the Common Data Model.

The Common Data Model is used to streamline data management and app development by combining data into a known form and employing structural and semantic reliability across numerous apps and deployments. It provides the link between self-service BI and enterprise governance over the vast amount of data available.

You can connect to a dataflow or the Common Data Service (which hosts the Common Data Model) just as you would any other data source in Power BI (see Figure 15-15).

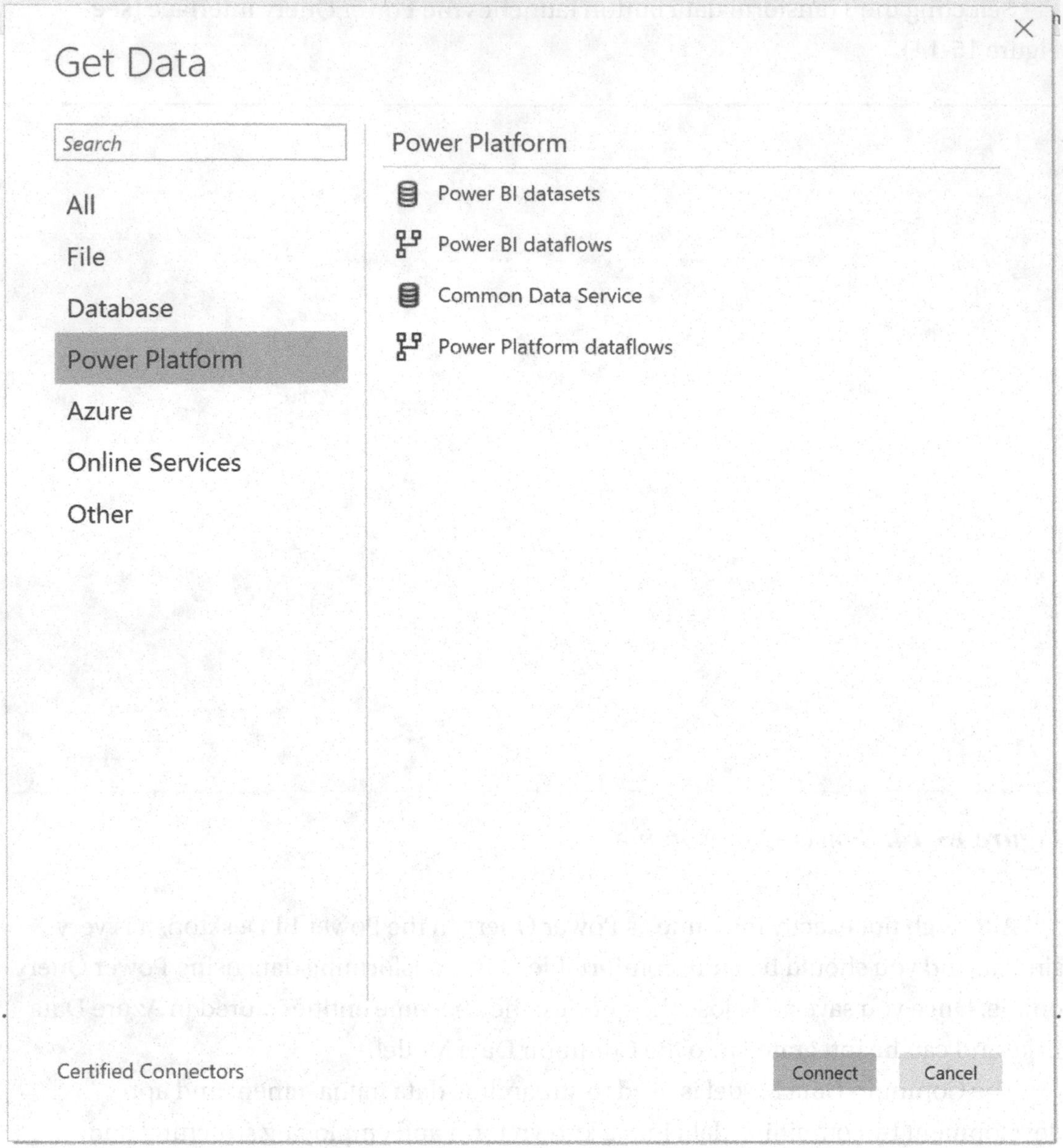

Figure 15-15. *Connecting to dataflows*

Now that you have some understanding of aggregations and dataflows, let's get some hands-on experience with these topics.

Note To complete the following lab, you need to have access to an AdventureworksDW2017 demo database. You can find installation instructions for installing the SQL Server 2017 Developer edition at `http://www.sqlservertutorial.net/install-sql-server/`. You can find instructions to install the AdventureworksDW2017 demo database at `https://docs.microsoft.com/en-us/sql/samples/adventureworks-install-configure?view=sql-server-ver15`

HANDS-ON LAB: ADVANCED TOPICS IN POWER BI

In the following lab, you will

- Create and use an aggregate table
- Create a dataflow

1. In the LabStarterFiles\Chapter15Lab1 folder, open the Chapter15Lab1.pbix file. This file uses direct query to get data from the AdventureworksDW2017 database.
2. You will need to point your data source settings to the location of your instance of the data warehouse. To do this, select the Edit Queries drop-down on the Home tab and choose Data source settings. In the pop-up window, select Change source and enter your server information (see Figure 15-16).

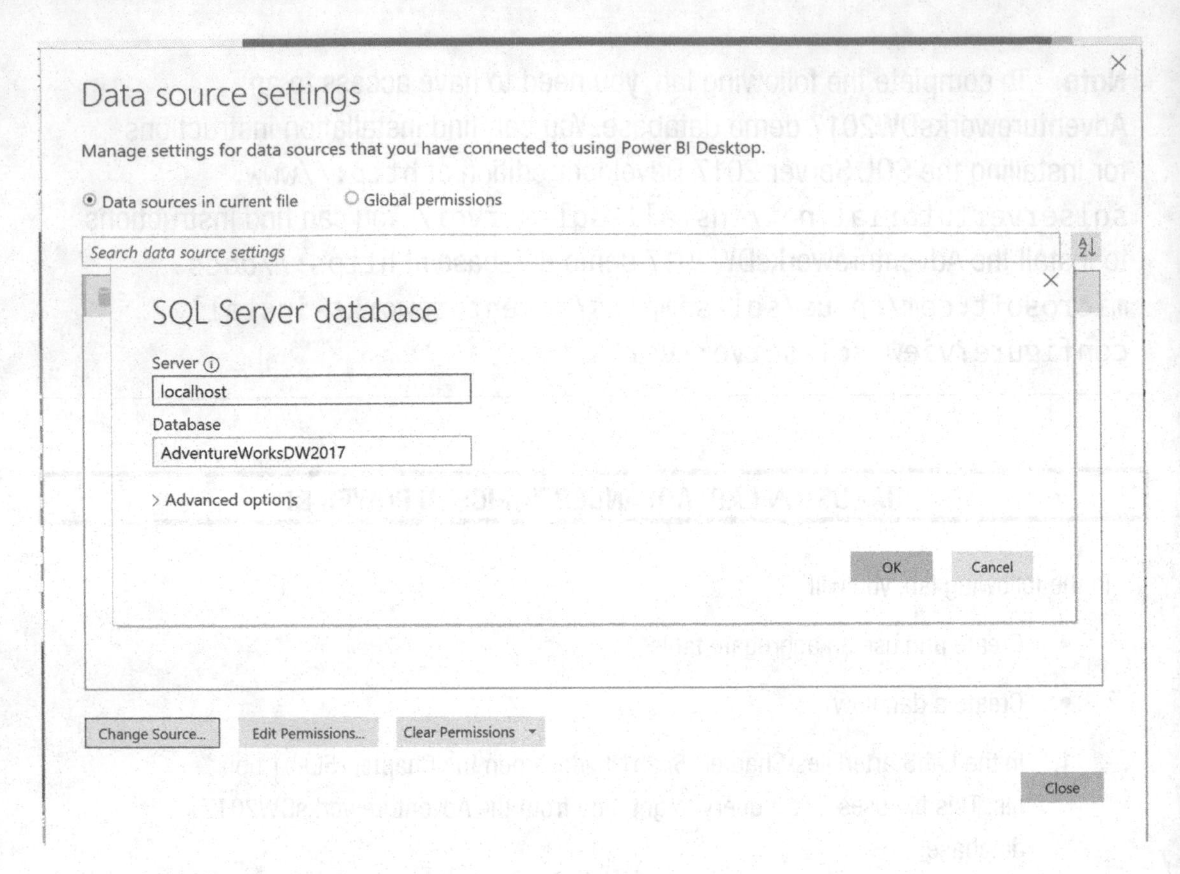

Figure 15-16. Updating the data source setting

3. If you successfully connect to the database, you should see data listed in the Power Query program window. Choose Close & Apply. (If you get a warning about not being able to make an encrypted connection, select OK.)

4. Select the data modeling tab and observe the data model (see Figure 15-17).

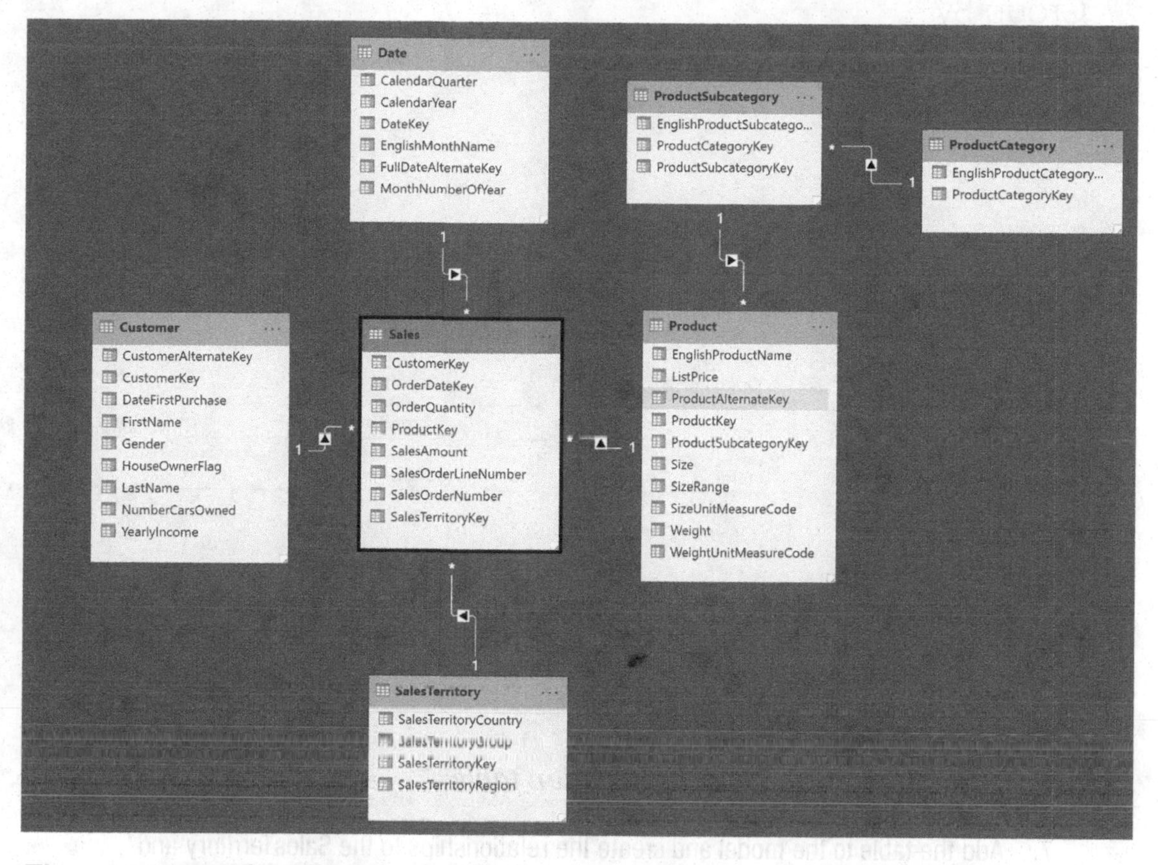

***Figure 15-17.** The Adventureworks data model*

5. Using Power Query, right-click the Sales table query and select Duplicate.

6. Using the duplicate table, create an aggregated sales table that sums up the sales amount and order count by SalesTerritoryKey and OrderDateKey using the Import mode (see Figure 15-18).

Group By

Specify the columns to group by and one or more outputs.

Basic Advanced

OrderDateKey

SalesTerritoryKey

Add grouping

New column name	Operation	Column
Quant_agg	Sum	OrderQuantity
SalesAmount_agg	Sum	SalesAmount

Add aggregation

OK Cancel

***Figure 15-18.** Creating the sales aggregation table*

7. Add the table to the model and create the relationships to the SalesTerritory and Date tables (see Figure 15-19). Make sure the data types in the aggregation table match those in the details table.

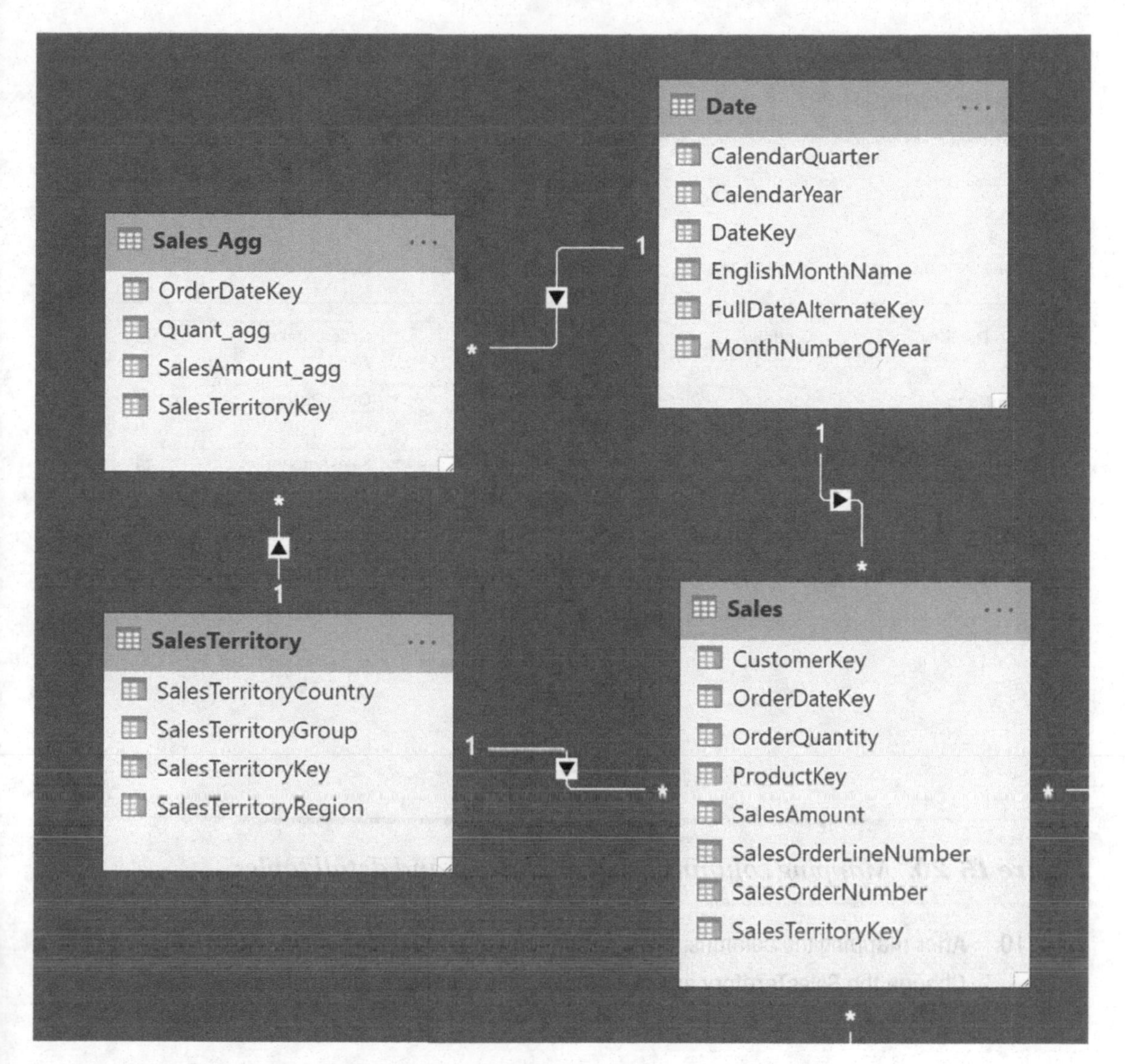

Figure 15-19. *Adding the aggregate table*

8. Right-click the Sales_Agg table and select Manage aggregations.
9. Map the aggregation columns to the detail columns. If you can't select the detail column, double-check the data types for a mismatch.

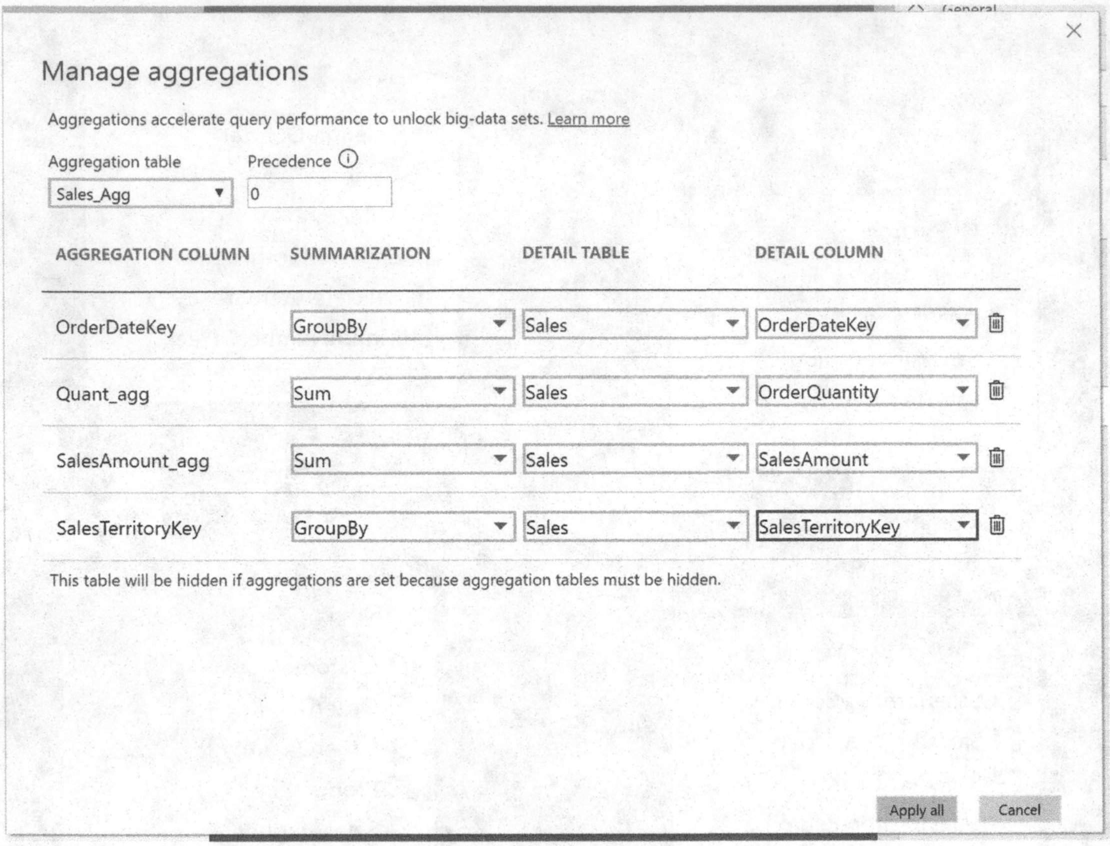

Manage aggregations

Aggregations accelerate query performance to unlock big-data sets. Learn more

Aggregation table: Sales_Agg

Precedence: 0

AGGREGATION COLUMN	SUMMARIZATION	DETAIL TABLE	DETAIL COLUMN
OrderDateKey	GroupBy	Sales	OrderDateKey
Quant_agg	Sum	Sales	OrderQuantity
SalesAmount_agg	Sum	Sales	SalesAmount
SalesTerritoryKey	GroupBy	Sales	SalesTerritoryKey

This table will be hidden if aggregations are set because aggregation tables must be hidden.

Apply all Cancel

Figure 15-20. *Mapping columns between the agg and detail tables*

10. After mapping the columns, the aggregation table will be hidden in Report view. Change the SalesTerritory and Date tables so they have a storage mode of dual (see Figure 15-21).

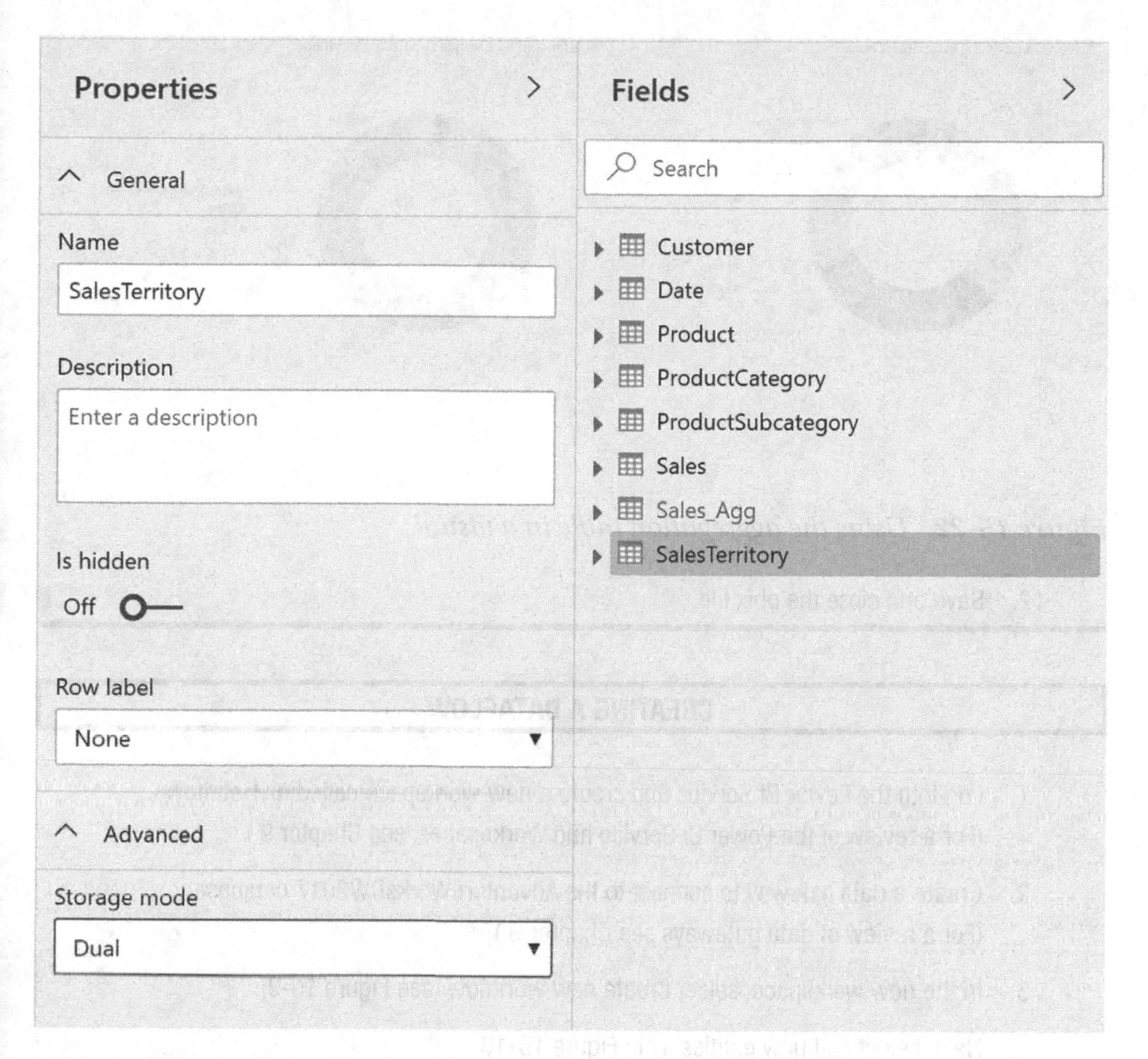

Figure 15-21. *Setting Dual storage mode*

11. Creating a visual that sums up sales by country uses the Sales_Agg table, while one that sums up sales by product category uses the Sales table and issues a direct query back to the database (see Figure 15-22).

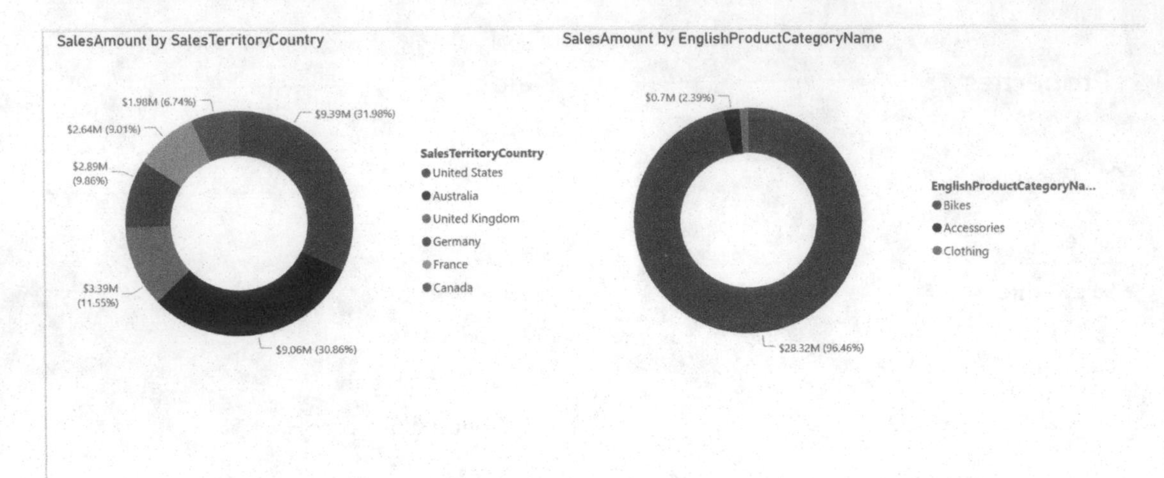

Figure 15-22. *Using the aggregation table in a visual*

12. Save and close the pbix file.

CREATING A DATAFLOW

1. Log into the Power BI Service and create a new workspace called myDataflows. (For a review of the Power BI Service and Workspaces, see Chapter 9.)
2. Create a data gateway to connect to the AdventureWorksDW2017 database. (For a review of data gateways see Chapter 9.)
3. In the new workspace, select Create new workflow (see Figure 15-9).
4. Next, select Add new entities as in Figure 15-10.
5. In the Choose data source screen on the Database tab, select the SQL Server database (see Figure 15-23).

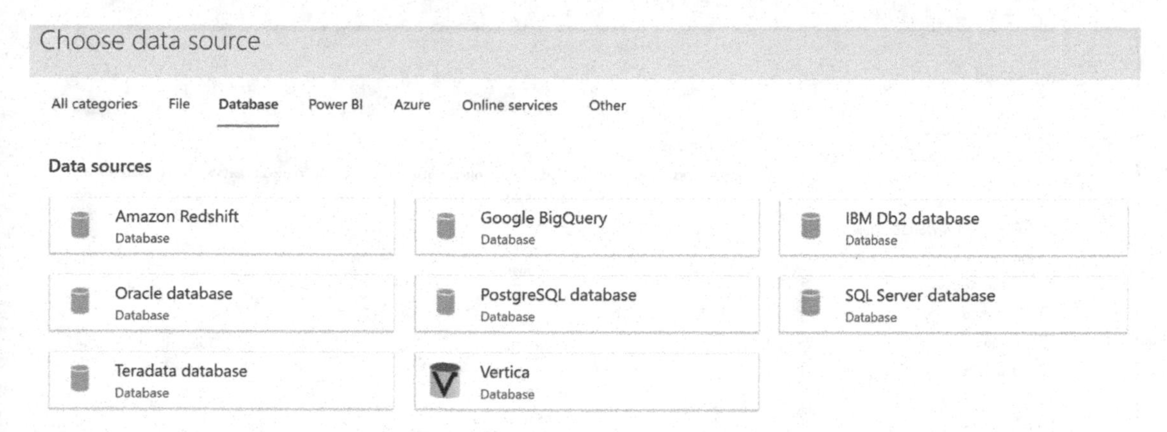

***Figure 15-23.** Selecting the data source*

6. Enter the connection information to the SQL Database (see Figure 15-24).

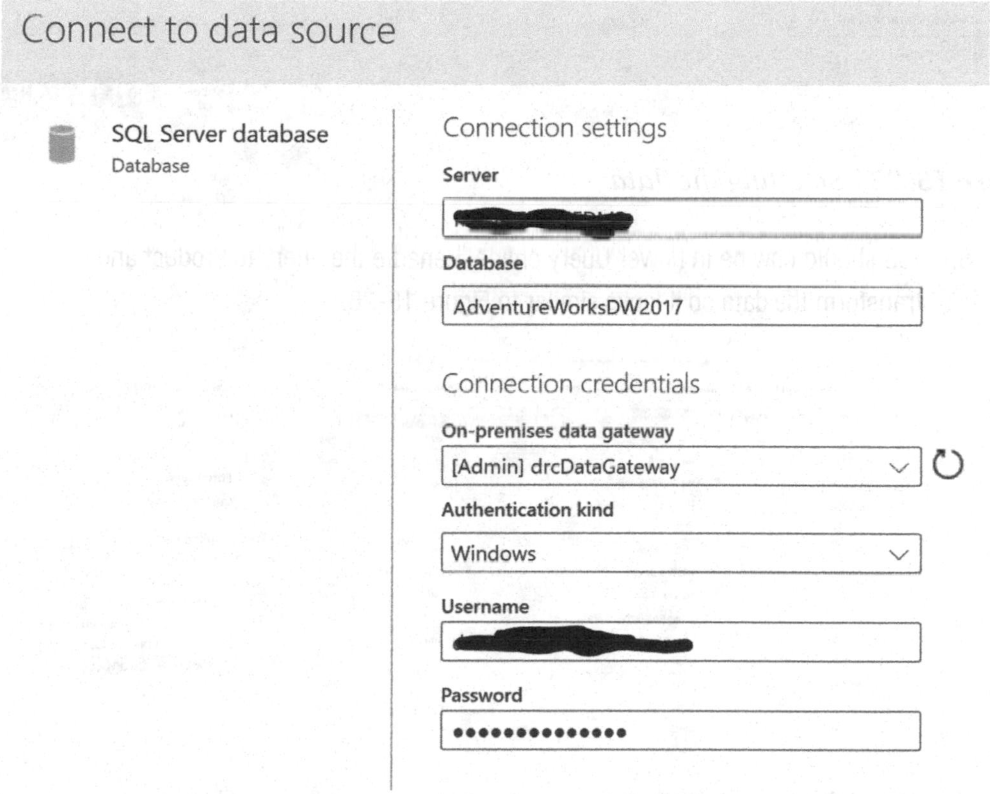

***Figure 15-24.** Connecting to the database*

7. Once connected to the database, select the DimProducts table and click the Transform data button (see Figure 15-25).

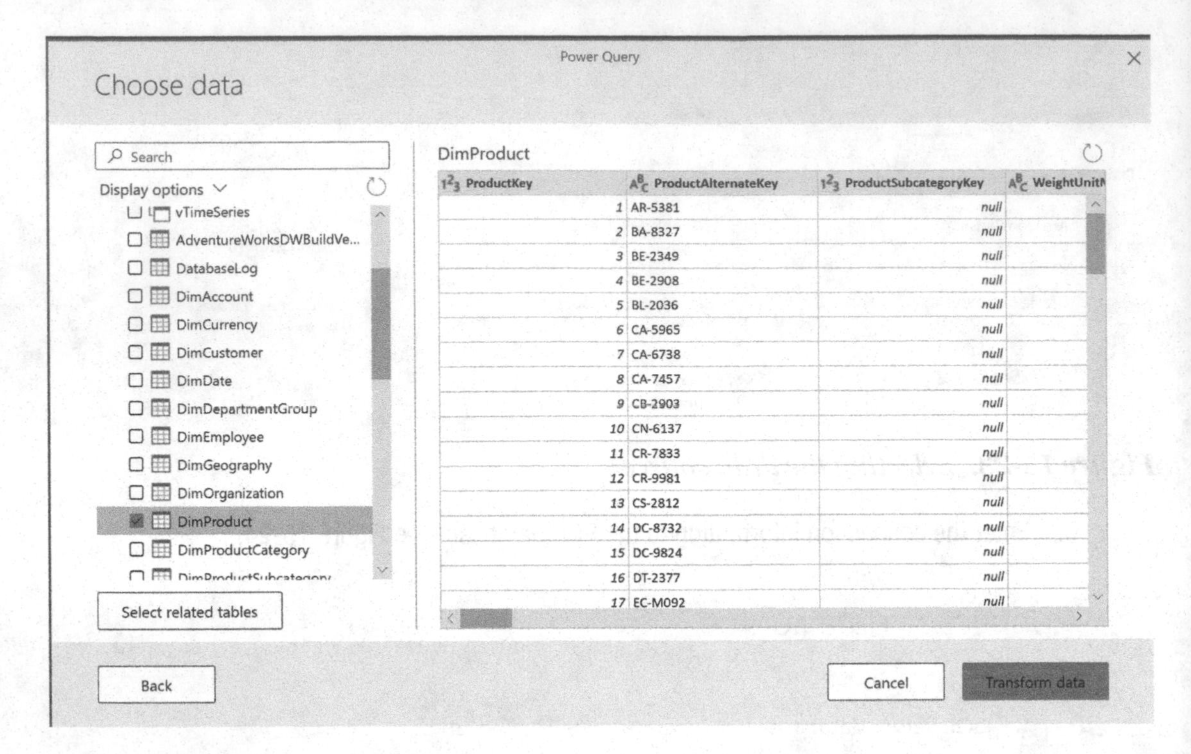

Figure 15-25. *Selecting the data*

8. You should now be in Power Query online. Rename the query to Product and transform the data so it looks similar to Figure 15-26.

```
= Table.RenameColumns(#"Expanded DimProductCategory", {{"EnglishProductName",
```

	ProductKey	ProductAlternateKey	Product	Subcategory	Category
1	1	AR-5381	Adjustable Race	null	null
2	2	BA-8327	Bearing Ball	null	null
3	3	BE-2349	BB Ball Bearing	null	null
4	4	BE-2908	Headset Ball Bearings	null	null
5	5	BL-2036	Blade	null	null
6	6	CA-5965	LL Crankarm	null	null
7	7	CA-6738	ML Crankarm	null	null
8	8	CA-7457	HL Crankarm	null	null
9	9	CB-2903	Chainring Bolts	null	null
10	10	CN-6137	Chainring Nut	null	null
11	11	CR-7833	Chainring	null	null
12	12	CR-9981	Crown Race	null	null
13	13	CS-2812	Chain Stays	null	null
14	14	DC-8732	Decal 1	null	null
15	15	DC-9824	Decal 2	null	null
16	16	DT-2377	Down Tube	null	null
17	17	EC-M092	Mountain End Caps	null	null
18	18	EC-R098	Road End Caps	null	null

Query settings
Name: Product
Entity type: Custom
Applied steps: Source, Navigation, Choose columns, Expanded DimProductSubcategory, Expanded DimProductCategory, Renamed columns

Figure 15-26. *Transforming the database*

9. Save and close Power Query. Name the dataflow Adventureworks. You now have a dataflow Adventureworks with an entity Product.

10. Open Power BI Desktop and select Get Data; then select Power BI dataflows (see Figure 15-27).

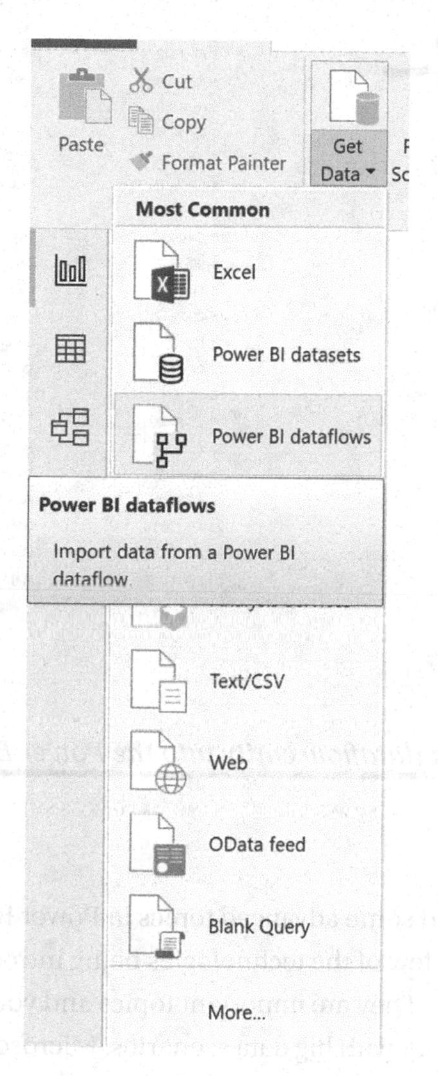

Figure 15-27. *Selecting Power BI dataflow as a data source*

11. You should then be able to select the Product entity in the Adventureworks dataflow (see Figure 15-28). Once imported into the model, it acts just like any data source in Power BI desktop.

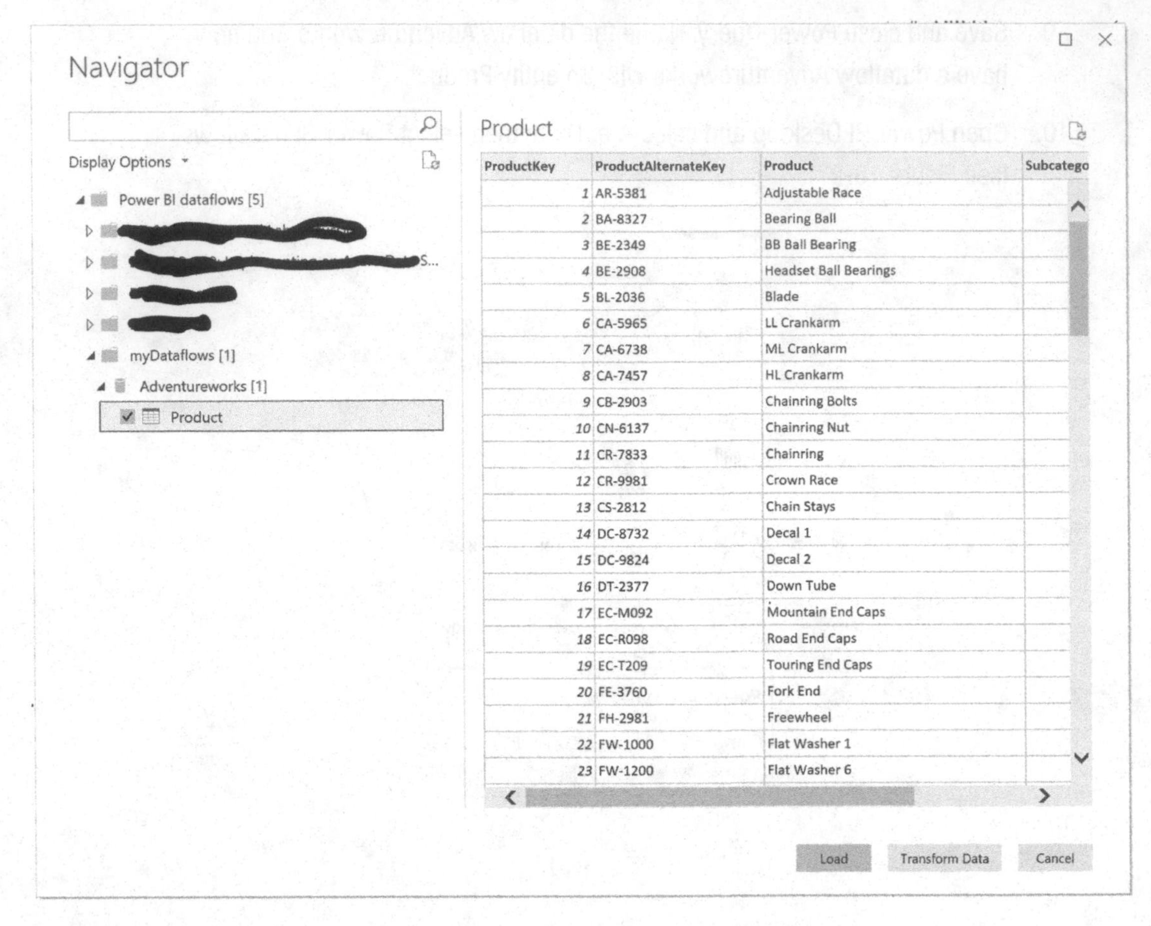

***Figure 15-28.** Loading a dataflow entity into the Power BI data model*

Summary

This chapter briefly covered some advanced topics in Power BI, including aggregations and dataflows. These are a few of the technologies being introduced to facilitate big data and enterprise governance. They are important topics and you should delve deeper into them if you need to work with big data scenarios. Microsoft is also committed to providing tools and technologies like dataflows and the Common Data Model to simplify reusability, security, and tracking data lineage throughout the organization.

Microsoft is committed to making Power BI the best reporting/analytics tool available for both self-service and enterprise level. New features and improvements are being released monthly. Remember, this book is the first step to learning the Power BI stack, not the last. Hopefully, I have given you a firm foundation on which to build. I wish you well on your journey and don't forget you have the Power (BI) to succeed!

Index

D. Clark, *Beginning Microsoft Power BI*, https://doi.org/10.1007/978-1-4842-5620-6

D

E

F

G, H

I, J, K

L

X, Y, Z

Printed in the United States
By Bookmasters